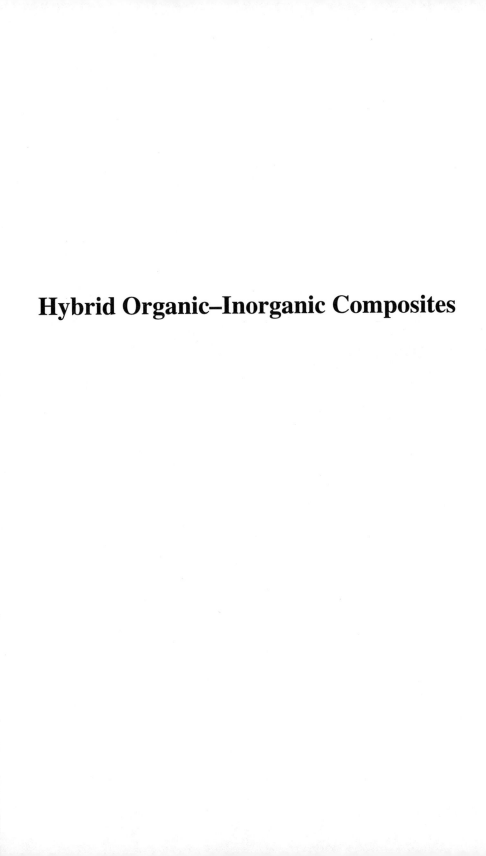

Hybrid Organic–Inorganic Composites

ACS SYMPOSIUM SERIES **585**

Hybrid Organic–Inorganic Composites

J. E. Mark, EDITOR
University of Cincinnati

C. Y-C Lee, EDITOR
U.S. Air Force

P. A. Bianconi, EDITOR
Pennsylvania State University

Developed from a symposium sponsored
by the Division of Polymeric Materials:
Science and Engineering, Inc.,
at the 207th National Meeting
of the American Chemical Society,
San Diego, California,
March 13–17, 1994

American Chemical Society, Washington, DC 1995

Library of Congress Cataloging-in-Publication Data

Hybrid organic–inorganic composites / J. E. Mark, editor, C. Y-C Lee, editor, P. A. Bianconi, editor.

p.   cm.—(ACS symposium series, ISSN 0097–6156; 585)

"Developed from a symposium sponsored by the Division of Polymeric Materials, Science and Engineering, Inc., at the 207th National Meeting of the American Chemical Society, San Diego, California, March 13–17, 1994."

Includes bibliographical references and index.

ISBN 0–8412–3148–6

1. Composite materials. 2. Organic compounds. 3. Inorganic compounds. 4. Polymeric composites. 5. Nanostructure materials. I. Mark, James E., 1934–   . II. Lee, C. Y-C, 1947–   . III. Bianconi, P. A., 1957–   . IV. American Chemical Society. V. American Chemical Society. Meeting (207th: 1994: San Diego, Calif.) VI. Series.

TA418.9.C6H96   1995
620.1′18—dc20
                                                    95–7108
                                                    CIP

This book is printed on acid-free, recycled paper.

Foreword

THE ACS SYMPOSIUM SERIES was first published in 1974 to provide a mechanism for publishing symposia quickly in book form. The purpose of this series is to publish comprehensive books developed from symposia, which are usually "snapshots in time" of the current research being done on a topic, plus some review material on the topic. For this reason, it is necessary that the papers be published as quickly as possible.

Before a symposium-based book is put under contract, the proposed table of contents is reviewed for appropriateness to the topic and for comprehensiveness of the collection. Some papers are excluded at this point, and others are added to round out the scope of the volume. In addition, a draft of each paper is peer-reviewed prior to final acceptance or rejection. This anonymous review process is supervised by the organizer(s) of the symposium, who become the editor(s) of the book. The authors then revise their papers according to the recommendations of both the reviewers and the editors, prepare camera-ready copy, and submit the final papers to the editors, who check that all necessary revisions have been made.

As a rule, only original research papers and original review papers are included in the volumes. Verbatim reproductions of previously published papers are not accepted.

M. Joan Comstock
Series Editor

Contents

Preface xi

1. Some General Trends in the Area of Organic–Inorganic
   Composites.. 1
       J. E. Mark

BIOMIMICRY AND TEMPLATE-BASED SYSTEMS

2. Organic–Inorganic Hybrids with a Crystalline Polymer
   Matrix.. 6
       Jeremy Burdon, Jeffrey Szmania, and Paul Calvert

3. Inorganic–Protein Interactions in the Synthesis
   of a Ferrimagnetic Nanocomposite 19
       T. Douglas, J. W. M. Bulte, D. P. E. Dickson, R. B. Frankel,
       Q. A. Pankhurst, B. M. Moskowitz, and S. Mann

4. Ion-Exchange Intercalation into the MPS_3 Layered
   Compounds: Design of Nanocomposites with Unusual
   Magnetic, Electrical, and Nonlinear Optical Properties................. 29
       R. Clément

5. Preparation and Characterization of Nanocomposites
   of Poly(ethylene oxide) with Layered Solids 43
       John P. Lemmon, Jinghe Wu, and Michael M. Lerner

6. Polymer–Clay Hybrids... 55
       Akane Okada, Arimitsu Usuki, Toshio Kurauchi,
       and Osami Kamigaito

7. (Perfluorosulfonate Ionomer)–(Inorganic Oxide)
   Nanocomposites: Organic Modification of Surfaces
   of Silicon Oxide Nanoparticles Grown In Situ 66
       Q. Deng, K. A. Mauritz, and R. B. Moore

NEW TECHNIQUES AND APPROACHES

8. **Nanostructured Organic–Inorganic Hybrid Materials Synthesized Through Simultaneous Processes** 86
   Bruce M. Novak, Mark W. Ellsworth, and Celine Verrier

9. **Multiple Size Scale Structures in Silica–Siloxane Composites Studied by Small-Angle Scattering** 97
   G. Beaucage, T. A. Ulibarri, E. P. Black, and D. W. Schaefer

10. **Composite Polymer Colloid Nucleated by Functionalized Silica** 112
    E. Bourgeat-Lami, P. Espiard, A. Guyot, S. Briat,
    C. Gauthier, G. Vigier, and J. Perez

11. **Vinyl-Polymer-Modified Hybrid Materials and Photoacid-Catalyzed Sol–Gel Reactions** 125
    Yen Wei, Wei Wang, Jui-Ming Yeh, Bin Wang,
    Dachuan Yang, James K. Murray, Jr., Danliang Jin,
    and Gu Wei

12. **Hybrid Organic–Inorganic Interpenetrating Networks** 142
    A. B. Brennan, T. M. Miller, and R. B. Vinocur

13. **A New Route to Polymer-Filled Glass: Hybrid Interpenetrating Networks with Appreciable Toughness** 163
    Kenneth G. Sharp

14. **Solidification of Colloidal Crystals of Silica** 181
    Hari Babu Sunkara, Jagdish M. Jethmalani,
    and Warren T. Ford

15. **Thermo-irreversible Gelation and Percolation-Based Mechanical Response via Metal–Olefin Coordination in Diene Polymers** 192
    Francis Bossé, Pronab Das, and Laurence A. Belfiore

BONDING AND OTHER INTERACTIONS

16. **Hybrid Organic–Inorganic Silica Materials: Chemical Evidence for Organization in the Solid** 210
    G. Cerveau, C. Chorro, R. Corriu, C. Lepeytre,
    J. P. Lère-Porte, J. Moreau, P. Thepot,
    and M. Wong Chi Man

17. Hybrid Organic–Inorganic Materials: The Sol–Gel Approach...... 226
    J. D. Mackenzie

18. Sol–Gel-Derived Silica–Siloxane Composite Materials:
    Effect of Reaction Conditions in Polymer-Rich Systems 237
    E. P. Black, T. A. Ulibarri, G. Beaucage, D. W. Schaefer,
    Roger A. Assink, D. F. Bergstrom, P. A. Giwa-Agbomeirele,
    and G. T. Burns

LADDER AND BRIDGED STRUCTURES

19. Hypervalent Spiro Polysiliconate and Polygermylate Ionomers:
    Novel Ladder and Network Materials 248
    James H. Small, Kenneth J. Shea, Douglas A. Loy,
    and Gregory M. Jamison

20. Hexylene- and Phenylene-Bridged Polysiloxane Network
    Materials 264
    Douglas A. Loy, Gregory M. Jamison, Roger A. Assink,
    Sharon Myers, and Kenneth J. Shea

HIGH-TEMPERATURE POLYMERS

21. Structural Design of High-Performance Polymers for Sol–Gel
    Processing 280
    T. D. Dang, J. P. Chen, and F. E. Arnold

22. Preparation and Properties of High-Clarity Polyamide–Silica
    Hybrid Materials 291
    Z. Ahmad, Shuhong Wang, and J. E. Mark

23. Preparation and Mechanical Properties of Polybenzoxazole–
    Silica Hybrid Materials 297
    J. P. Chen, Z. Ahmad, Shuhong Wang, J. E. Mark,
    and F. E. Arnold

ELECTRICAL, OPTICAL, AND INTERFACIAL PROPERTIES

24. Morphological Studies of Conductive Polymers Deposited
    onto High-T_c Superconductors 308
    Steven G. Haupt, Rung-Kuang Lo, Jianai Zhao,
    and John T. McDevitt

25. **Novel Organic–Inorganic Composite Materials for Photonics**...... 317
    Paras N. Prasad, Frank V. Bright, Upvan Narang, Run Wang,
    Richard A. Dunbar, Jeffrey D. Jordan, and Raz Gvishi

26. **Inorganic–Organic Hybrid Coatings for Metal and Glass
    Surfaces**.. 331
    H. Schmidt, R. Kasemann, T. Burkhart, G. Wagner,
    E. Arpac, and E. Geiter

27. **Surface Modification of Carbon Fibers for Advanced
    Composite Materials**... 348
    Yuechuan Wang and Roderic P. Quirk

Author Index... 363

Affiliation Index... 364

Subject Index.. 364

Preface

THE UNIQUE PROPERTIES OF ORGANIC AND INORGANIC materials can
be synergistically combined in a number of ways. The purpose of the
March 1994 symposium on hybrid organic–inorganic composites and this
resulting book was to illustrate some of these ways. Most of the chapters
in this book are based on contributions to the symposium, but two of the
chapters (6 and 13) were commissioned separately to improve the balance
in the subject matter.

The chapters in this book cover a wide range of topics, as can be seen
from the overview chapter, which previews some of the contents of the
book. The variety of topics is certainly desirable and stems in part from
the various backgrounds of the people working in this area. Contributors
include inorganic chemists, organic chemists, physical chemists, polymer
scientists and engineers, chemical engineers, and materials scientists.

We hope that this book will be of use to these various constituencies
and to others in related disciplines. Although our primary goal was to
report on the status of the rapidly expanding area of organic–inorganic
composites, there is also much discussion of unsolved problems that
should inspire readers to enter and contribute to this area as well.

J. E. MARK
Department of Chemistry and Polymer Research Center
University of Cincinnati
Cincinnati, OH 45221–0172

C. Y-C LEE
Office of Scientific Research
U.S. Air Force
Bolling Air Force Base
Washington, DC 20332–6448

P. A. BIANCONI
Department of Chemistry
Pennsylvania State University
University Park, PA 16802

October 11, 1994

Chapter 1

Some General Trends in the Area of Organic–Inorganic Composites

J. E. Mark

**Department of Chemistry and Polymer Research Center,
University of Cincinnati, Cincinnati, OH 45221–0172**

This overview chapter first reviews the sol-gel process as it is being used by ceramists, and then describes how it is being modified by the incorporation of organic materials. Six general topics involving the organic-inorganic composites resulting from these new approaches are then illustrated using examples chosen from the chapters of the present book.

The Sol-Gel Process in General

Most organic-inorganic hybrid composites are prepared by introducing polymeric components into the sol-gel technology which is now much used to prepare ceramic materials. It is therefore appropriate to first describe how this sol-gel approach has been used by ceramists to prepare ceramics having unusually attractive properties[1-9]. In this technology, organo-silicates, -titanates, -aluminates, etc. are typically hydrolyzed to multi-hydroxy compounds which then condense into gel-like structures that can be dried and fired into ceramic coatings, foams, or monolithic objects. The classic example is the reaction of tetraethoxysilane (TEOS) [$Si(OC_2H_5)_4$] to yield silica (SiO_2), and ethanol as a volatile and easily-removed byproduct. A variety of acids and bases, and even some salts, greatly catalyze the process.

There are a number of advantages to this new technique for preparing ceramic materials. First, the high purity of the chemical reactants insures higher ceramic purities than can generally be obtained in the usual melting of minerals such as common sand. The temperatures involved are very much lower than those required to melt typical ceramics, and this encourages the incorporation of organic phases. In some cases, the porous ceramic precursors formed as intermediates can be pervaded by polymers (such as a polysilane) that can be converted into reinforcing phases (such as β silicon carbide) during a subsequent firing process. Also, it is much easier to imagine placing a thin coating of silica onto a surface to be used as part of a wave-guide by this technique, than by a dipping process involving molten silica! Finally, co-hydrolysis of different organometallics can be used to obtain ceramic "alloys" that are essentially unobtainable by the usual ceramic technologies.

0097–6156/95/0585–0001$12.00/0

The Sol-Gel Process in the Preparation of Organic-Inorganic Composites

One indirect advantage of including polymers in this technology is facilitation of the processing techniques involved. Of greater interest, however, are attempts to obtain synergistic effects, specifically to produce materials that have an optimized combination of the best properties of polymers with the best properties of ceramics[10-29]. A general example would be to maintain some of the best properties of a ceramic material but to improve its mechanical strength. This could be done by having a tough polymer present to provide an additional mode for absorbing impact energy, thereby reducing brittleness.

Some General Trends

Biomimicry and Template-Based Systems. The goal in biomimicry is to understand the structure of biomaterials to the extent that some of the same guiding principles can be used to prepare better synthetic (non-biological) materials. One approach is illustrated in this volume by the precipitation of an irorganic phase between the crystalline regions of a partially-crystalline organic polymer by Calvert et al., in a manner reminiscent of the way bone structures are produced. Another example, by Mann and coworkers, involves the generation of a ferrimagnetic phase within the constraining environment of the protein ferritin.

Since template-directed syntheses are commonplace in biosytems, there is a strong connection between these two topics. One approach involving intercalation of organic phases between layered structures is illustrated by the contributions of Clement, by Lerner et al., and by Okada et al. Some related experiments in which silica-like particles are grown within the pores of a membrane are described by Mauritz and coworkers.

New Techniques and Approaches. The complexity of many organic-inorganic composites, and the very diverse backgrounds of many of the people working in this area has encouraged a wide range of new techniques and approaches. This has occurred with regard to both synthesis and characterization. An example of a new synthetic technique is the solution to the shrinkage problem generally occurring in the preparation of some composites, by Novak et al. In their approach, alkoxide groups that would otherwise have to be removed from the reacting system are instead polymerized in-situ to give the desired organic phase. An example of a new characterization technique in this area is the extensive use of small-angle scattering measurements by Beaucage, Ulibarri, Black, and Schaefer to investigate composite structures.

Two novel ways for the generation of organic phases are then described. In the first, by Guyot and coworkers, functionalized silica is used as nucleation sites in the emulsion polymerization of a typical vinyl monomer, ethyl acrylate. In the second, Wei et al. use a photochemical polymerization to synthesize the polymeric phases in a variety of composites. The use of a different type of radiation, γ photons, to polymerize a second monomer absorbed into a composite is outlined by Brennan, Miller, and Vinocur.

Techniques for preparing organic-inorganic composites from non-aqueous sytems is the subject of one contribution, by Sharp, and the use of polymeric phases to lock regular arrays of silica particles in place is the subject of another, by Ford et al. A final example in this category, by Belfiore and coworkers, describes the thermo-irreversible gelation of diene polymers in the presence of an inorganic salt.

Bonding and Other Interactions. The two phases in these composites are very disparate and, since good miscibility is generally desirable, much attention is being paid to the nature of the interactions occurring in these materials. The chemical bonding occurring in some of these systems is discussed by Corriu and colleagaues, and a broader discussion of the structures and interactions occuring in three types of organic-inorganic composites is given by Mackenzie. A study showing how the maximum size of silica particles growing in an elastomeric matrix is controlled by the constraining effects of the stretched-out polymer chains is given jointly by the Schaefer group at Sandia and the Burns group at Dow Corning.

Ladder and Bridged Structures. A number of ionic polysiliconate and polygermylate compounds form unusual penta- and hexacovalent compounds, and these have been used to form novel ladder, bridged, and network materials. Some of this work is described in two articles by Shea, Small, and Meyers at UC Irvine collaborating with Loy, Jamison, and Assink at Sandia.

High-Temperature Polymers. Using high-performance, high-temperaure polymers as the organic phase in these composites presents several special problems. For one thing, such polymers are generally almost intractable even under the best of conditions. Second, polymers of this type are designed to be unreactive, since this is required in most of their applications. This unreactivity becomes a problem, however, in getting good bonding between the polymer phase and the inorganic, ceramic phase. Finally, there is what is likely to become part of the central dogma in the area of organic-inorganic composises: the better the properties of the polymer to begin with, the more difficult it is to achieve further improvements. Some descriptions of recent work in this area, emphasizing polymers of interest to the US Air Force, are given in three chapters by Arnold et al., and by Mark et al.

Electrical, Optical, and Interfacial Properties. The characterization of these properties is extremely active in polymer science in general, so it will come as no surprise that organic-inorganic composites are playing increasingly-important roles in this area. Examples of electrical and optical properties are the chapters on conductive polymer/superconductive bilayer systems by McDevitt and coworkers, and on composite materials in photonics by Prasad and coworkers.

The use of this technology for interfacial properties is illustrated using coatings for metals and glass by Schmidt et al., and the chemical modification of carbon fibers for composites by Wang and Quirk.

References

1. *Ultrastructure Processing of Ceramics, Glasses, and Composites*; Hench, L. L.; Ulrich, D. R., Eds.; Wiley & Sons: New York, 1984.
2. *Better Ceramics Through Chemistry*; Brinker, C. J.; Clark, D. E.; Ulrich, D. R., Eds.; North Holland: New York, 1984.
3. *Science of Ceramic Chemical Processing*; Hench, L. L.; Ulrich, D. R., Eds.; Wiley & Sons: New York, 1986.
4. Ulrich, D. R. *CHEMTECH* **1988**, *18*, 242.
5. Ulrich, D. R. *J. Non-Cryst. Solids* **1988**, *100*, 174.
6. *Better Ceramics Through Chemistry IV*; Zelinski, B. J. J.; Brinker, C. J.; Clark, D. E.; Ulrich, D. R., Eds.; Materials Research Society: Pittsburgh, 1990.
7. Ulrich, D. R. *J. Non-Cryst. Solids* **1990**, *121*, 465.
8. Brinker, C. J.; Scherer, G. W. *Sol-Gel Science*; Academic Press: New York, 1990.
9. *Ultrastructure Processing of Advanced Materials*; Uhlmann, D. R.; Ulrich, D. R., Eds.; Wiley & Sons: New York, 1992.
10. Mark, J. E.; Erman, B. *Rubberlike Elasticity. A Molecular Primer*; Wiley-Interscience: New York, 1988.

11.   McGrath, J. E.; Pullockaren, J. P.; Riffle, J. S.; Kilic, S.; Elsbernd, C. S. In *Ultrastructure Processing of Advanced Ceramics*; Mackenzie, J. D.; Ulrich, D. R., Eds.; Wiley-Interscience: New York, 1988; p 55.
12.   Mark, J. E. *Chemtech* **1989**, *19*, 230.
13.   Chung, Y. J.; Ting, S.-J.; Mackenzie, J. D. In *Better Ceramics Through Chemistry IV*; Zelinski, B. J. J.; Brinker, C. J.; Clark, D. E.; Ulrich, D. R., Eds.; Materials Research Society: Pittsburgh, 1990; Vol. 180; p 981.
14.   Saegusa, T.; Chujo, Y. *J. Macromol. Sci. - Chem.* **1990**, *A27*, 1603.
15.   Mann, S. *Nature* **1991**, *349*, 285.
16.   Wung, C. J.; Pang, Y.; Prasad, P. N.; Karasz, F. E. *Polymer* **1991**, *32*, 605.
17.   Fitzgerald, J. J.; Landry, C. J. T.; Pochan, J. M. *Macromolecules* **1992**, *25*, 3715.
18.   Sarikaya, M.; Aksay, I. A. In *Chemical Processing of Advanced Materials*; Hench, L. L.; West, J. K., Eds.; Wiley & Sons: New York, 1992; p 543.
19.   Abramoff, B.; Klein, L. C. In *Ultrastructure Processing of Advanced Materials*; Uhlmann, D. R.; Ulrich, D. R., Eds.; Wiley & Sons: New York, 1992; p 401.
20.   *Chemical Processing of Advanced Materials*; Hench, L. L.; West, J. K., Eds.; Wiley & Sons: New York, 1992.
21.   Schmidt, H. In *Ultrastructure Processing of Advanced Materials*; Uhlmann, D. R.; Ulrich, D. R., Eds.; Wiley & Sons: New York, 1992; p 409.
22.   *Submicron Multiphase Materials*; Baney, R. H.; Gilliom, L. R.; Hirano, S.-I.; Schmidt, H. K., Eds.; Materials Research Society: Pittsburgh, PA, 1992; Vol. 274, p 85.
23.   Schaefer, D. W.; Jian, L.; Sun, C.-C.; McCarthy, D.; Jiang, C.-Y.; Ning, Y.-P.; Mark, J. E.; Spooner, S. In *Ultrastructure Processing of Advanced Materials*; Uhlmann, D. R.; Ulrich, D. R., Eds.; Wiley & Sons: New York, 1992.
24.   Heuer, A. H.; Fink, D. J.; Laraia, V. J.; Arias, J. L.; Calvert, P. D.; Kendall, K.; Messing, G. L.; Blackwell, J.; Rieke, P. C.; Thompson, D. H.; Wheeler, A. P.; Weiss, A.; Caplan, A. I. *Science* **1992**, *255*, 1098.
25.   Novak, B. M. *Adv. Mats.* **1993**, *5*, 442.
26.   Landry, C. J. T.; Coltrain, B. K.; Landry, M. R.; Fitzgerald, J. J.; Long, V. K. *Macromolecules* **1993**, *26*, 3702.
27.   Wang, B.; Wilkes, G. L. *J. Macromol. Sci., Pure Appl. Chem.* **1994**, *A31*, 249.
28.   Mark, J. E.; Calvert, P. D. *J. Mats. Sci., Part C* **1994**, *1*, 159.
29.   Kojima, Y.; Usuki, A.; Kawasumi, M.; Okada, A.; Kurauchi, T.; Kamigaito, O.; Kaji, K. *J. Polym. Sci., Polym. Phys. Ed.* **1994**, *32*, 625.

RECEIVED October 24, 1994

BIOMIMICRY AND
TEMPLATE-BASED SYSTEMS

Chapter 2

Organic–Inorganic Hybrids with a Crystalline Polymer Matrix

Jeremy Burdon, Jeffrey Szmania, and Paul Calvert

Department of Materials Science and Engineering, University of Arizona, Tucson, AZ 85721

Many nanoscale hybrids of organic and inorganic materials have now been prepared. In general the materials become stiffer but significantly more brittle as increasing quantities of the inorganic phase are introduced into the organic polymer. This behavior is similar to that found in conventional particle-reinforced polymers. The properties of bone, as a polymer reinforced with nanometer-sized ribbon-shaped crystals of mineral, are good when compared with the properties of these synthetic polymer composites. In an effort to induce more structured, and possibly elongated, precipitation of the inorganic phase in hybrids, a crystalline polymer matrix is used. It was expected that the inorganic phase would be constrained to deposit between the crystalline polymer lamellae. Titania-polypropylene composites have been studied. They show an increase in stiffness but there is a loss of strength, apparently due to segregation of the titania to the spherulite boundaries. This study also shows that melt processing of hybrids is feasible as opposed to the usual casting of films.

Bone has an elastic modulus in the range of 20-30 GPa, about 10 times that of collagen or synthetic resins, table 1 (1,2). This increase is achieved by reinforcement with 40-50 vol% of hydroxyapatite ribbons which have a thickness of a few nanometers and a high aspect ratio. Various theories can be used to predict composite moduli from composition (3) and bone seems to be in the range expected for the known component moduli, volume fraction and particle shape.

Particle-filled composites generally show a decrease in strength and a rapid decrease in toughness with increasing particle content. Fiber-filled composites get

0097–6156/95/0585–0006$12.00/0

Table 1: Comparative Properties of Bone and Synthetic Composites

	Volume Fraction, %	Tensile Modulus GPa	Tensile Strength MPa	Strain to break %	Work of Fracture J m^{-2}
Sheet Molding compound (resin + filler + fiber)	50, 15-30% fiber	15	86-230	1.3-1.7	
Polybutyleneterephthalate/ glass beads	25	4.9	95		
Polyethyleneterephthalate + short glass fiber	35	20	165	1	3200
Bone (bovine femur)	41	20	220	10	1700
Collagen (tendon)	0	3	100		
Hydroxyapatite	100	110	100		
Polyethyleneterephthalate	0	3.3	60	275	7300
E glass	100	70	3000		

stronger but also less tough (*4*). Properties are very dependent on the aspect ratio of the fibers which is generally limited to a range from 10-50 by the processing conditions. Glass fibers are typically 10 μm in diameter and standard processing equipment will break fibers with lengths in the millimeter range. Finer fibers are undesirable because they would be a health hazard during handling. In addition normal mixing procedures, which randomize fiber orientations, limit the fiber volume fraction to the percolation threshold. This is around 40 vol% at an aspect ratio of 10 and 10 vol% at 50 (*5*).

Inorganic-organic hybrids have been studied for about 10 years by several groups and the general pattern of mechanical properties is now becoming clear. The addition of the inorganic phase stiffens glassy polymers in line with composite theory. The change in modulus does depend on morphology in that a material formed with a continuous inorganic network is stiffer than one with an continuous organic phase and inorganic particles. There is a much more marked increase in modulus above the glass transition (Figure 1).

The mechanical strength in compression does increase, as for many particulate composites, but there is generally a decrease in tensile strength and in toughness, especially at high particle contents (*6,7*). To date there is no clear

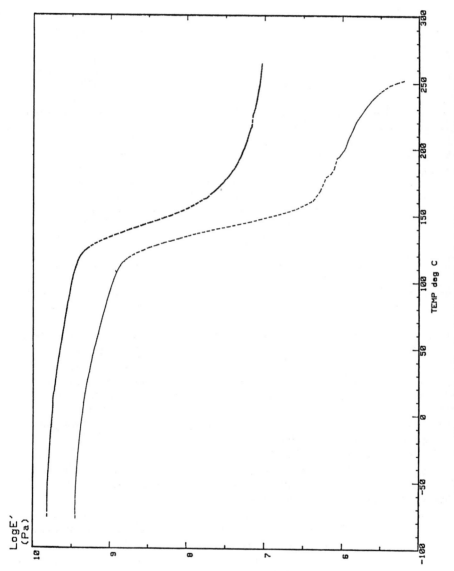

Figure 1. Storage Moduli of a PMMA-Silica hybrid composite and of PMMA

sign of special properties arising from the nanometer scale of the inorganic filler except the effect on rubber modulus (Figure 1) and an increase in degradation temperature for acrylic polymers (*8*).

Based on current knowledge of organic-inorganic hybrids and the preceding discussion of bone, there are two obvious routes for achieving better nanoscale composites.   Firstly the reinforcing particles should have an elongated shape rather than being spherical. This needs to be achieved without paying a penalty in packing density.   Secondly weak planes should be built into the structure to control failure; a similar mechanism is being tested by incorporating weak interfaces in ceramics (*9*).   The possibility of building into the matrix special structures with high work of fracture should also be explored.

One approach to control the particle morphology is to use a crystalline polymer matrix rather than an amorphous polymer. The lamellar morphology of crystalline polymers should redistribute the alkoxide to the interlamellar amorphous regions during crystallization.   Subsequent hydrolysis should lead to titania particles in layers between the polymer crystals.   In addition the toughness of crystalline polymers is related to the ability of chains to pull out of the lamellae during yielding and this toughness may be carried over into the composites.

To investigate a system of this type, polypropylene was melt blended with titanium n-butoxide and extruded.   After extrusion the film was treated with boiling water for 24 hours to precipitate the titania

Experimental

Escorene Polypropylene PP-5022 in pellet form was obtained from Exxon Chemicals.   Polypropylene was dissolved in boiling xylene over a 45 minute period. A white polypropylene 'sponge' was solvent precipitated by pouring the polypropylene/xylene solution into methanol. This material was broken into small lumps, compressed to remove bulk solvent, and left to dry for 24 hours. The dried material was then ground to a fine powder over a 2 minute period using a commercial coffee grinder. To obtain an extremely fine powder, the material was sieved using a 60-mesh sieve (0.246mm opening). The powder was then placed in a vacuum oven at 55°C for 48 hours to completely remove residual solvent.

Potassium Benzoate was considered to be an ideal nucleating agent (*10,11*), and was added (1 wt%) to some batches of polypropylene powder. The addition of a nucleating agent appears to reduce the spherulite diameter to a few microns and improve the strength of the spherulite boundaries.   The composite containing no nucleating agent shows very large, TiO_2 covered, spherulites, whereas the composite containing a nucleating agent shows a very different fracture surface, with large areas of 'flat' fracture topography (Figure 2a,2b).   Composites prepared using the nucleating agent were much less brittle than non-nucleating systems.

A quantity of titanium n-butoxide was added to obtain a known weight percentage of titanium n-butoxide (10% 15%, 20%, 30%, and 40% batches were prepared).   Also 0.8 wt% Ken-React KR-TTS coupling agent, isopropyl

(a)

(b)

Figure 2.   Fracture surfaces of (a) 9.4 wt.% titania-polypropylene composite prepared without nucleating agent and (b) nucleated 7 wt% composite.

triisostearoyl titanate $(CH_3CH(CH_3)-O-Ti-(OOC_{17}H_{35})_3)$, was added. The mixture was then ground for approximately 1 minute to obtain a homogenous material. This was then placed in Mason-jars and sealed until used.

A home-built ram extruder was used to extrude 25 mm. wide x 3 mm. thick strips. Polymer samples were placed in the pre-heated barrel (180°C), and compressed. The ram was then retracted slightly to relieve pressure, and the material left to melt over a 1 hour period. The barrel and die temperature was then lowered to 170°C over a 30 minute period. To extrude samples containing 40 wt% alkoxide, it was necessary to lower the die temperature to around 158°C. Extrudate was allowed to cool (un-assisted) to room temperature on exit from the die.

Thin film material (less than 0.05mm thick) was produced using a commercial 0.25" single-screw extruder. Temperature zones within the extruder were programmed to approximate the conditions in the ram-extruder. Extruded film was fed onto a chill roll at 70°C. Extruded strips were hydrolyzed by placing in boiling water for 12-14 hours. Extruded film was boiled for 4-6 hours. The samples were then placed in a vacuum oven for 24 hours at 50°C to dry.

Dynamic mechanical analysis was performed on extruded strip. Samples were machined flat to an accuracy of better than 0.01mm. Experiments were run in triplicate. Samples were tested using a Du Pont Thermal Analysis system with a model 983 dynamic mechanical analyzer. Samples were run with an oscillation amplitude of 0.2mm. Heating rates were 3°C/min (30°C to 140°C) with a fixed frequency of 1Hz. Sample dimensions were typically 21mm x 14mm x 0.7mm (L x W x T). Values for the flex loss modulus (E"), flex storage modulus (E'), and shear moduli (G' and G") were obtained.

Tensile tests were performed using an Instron model 1011. Films were tested using low mass clamps (10g max force), flat-faced grips with a 10g load-cell at a strain rate of 0.1 mm/min. Ram-extruded strips were examined using serrated edge grips with a 1000lb load-cell at a strain rate of 0.1 mm/min. ASTM standard test D 882-88 was followed as closely as possible, however due to limitations in sample size, it was not possible to follow recommended sample dimensions, and gauge lengths. Films were generally 0.05 mm thick and 12 mm or 18 mm wide. Strip samples were generally 0.8 mm thick and 12 mm wide. A 25 mm gauge length was used for both strip and film samples.

Differential scanning calorimetry was performed using a Du Pont model 910 differential scanning calorimeter. Samples were run in nitrogen and ramped at 30°C/min from room temperature to 220°C. A value for % crystallinity was calculated using a value of 207 Jg^{-1} for ΔH_{fusion} for isotactic polypropylene, as determined by Gee and Melia (*12*). SEM analysis was performed on fractured samples. Cross sections of strip samples were prepared by fracturing in liquid nitrogen. Fracture analysis was also performed on the cross-sections obtained from tensile breaks during tensile tests.

Results

During ram extrusion it was noticed that samples containing more than 15 wt% titanium n-butoxide exhibited markedly different solidification behavior to samples with lower concentrations of titanium n-butoxide. It was noticed that with composites containing up to 15 wt% titanium n-butoxide (0%, 2%, 5%, 10%, and 15% samples were compared), the extruded composite appeared clear on first emerging from the die, rapidly turning opalescent. With composites containing more than 15 wt% titanium n-butoxide, the behavior was similar, except that after several minutes of cooling, a further whitening of the material occurred, spreading throughout the sample at a rate of about 1mm/sec. Prior to the onset of whitening, the extrudate had cooled substantially and was slightly opaque indicating that crystallization of the polypropylene matrix had already occurred. At the whitening point, the temperature of the material had dropped to around 70°C-80°C.

Figure 3 shows titania particles and voiding at the spherulite boundaries in an unnucleated sample. Individual spherulites appear joined by polymer fibrils, which tend to also appear meshed with TiO_2 particulates on the surface of the spherulites. Composites prepared with more than 15 wt% titanium n-butoxide, and no nucleating or coupling agents, were mechanically fragile.

Rigorous hydrolysis conditions employed to allow complete hydrolysis of the alkoxide to an amorphous TiO_2 precipitate. TGA of the treated (hydrolyzed and dried) polypropylene composite, originally containing 30 wt% titanium n-butoxide, showed no evidence of release of water, butanol, or unreacted titanium n-butoxide. X-ray diffraction showed no evidence for crystalline (anatase or rutile) TiO_2. We concluded that our polypropylene composites contained amorphous TiO_2 derived from the full and complete hydrolysis of titanium n-butoxide.

Micrographs show no evidence of large titania particles. This is confirmed by small angle x-ray scattering studies which show the presence of sub-micron particles (D.W.Schaefer and P.D.Calvert, to be published). These very fine particles would not be expected to be responsible for extensive whitening of the composite, which is more likely to reflect refractive index variations on the scale of the spherulites.

Elastic Properties. Dynamic mechanical analysis was performed on ram-extruded strip, and the results are shown in Figure 4(a,b). Figure 4(a) shows the flexural storage modulus E' as a function of temperature. Figure 4(b) shows the loss modulus E''. The modulus increase is in line with that seen in the polymethylmethacrylate-silica system shown in Figure 1.

An increase in concentration of TiO_2 precipitate results in an increase in the dynamic modulus. This is in agreement with other workers, and the general finding is that an increase in filler content or a decrease in filler size results in an increase in the dynamic modulus (*13,14,15,16,17,18*).

Figure 4(b) shows that the loss modulus is little changed at 2.3 wt% TiO_2, but drops sharply with further increases in filler. This is believed to be due to hardening of the spherulite boundaries.

The effect of the coupling agent on the bonding between the titania and the polypropylene can be analyzed by measuring the shear modulus, and fitting data to established theories relating component viscosities to filler volume fraction in two component systems. The Einstein model was used with and without

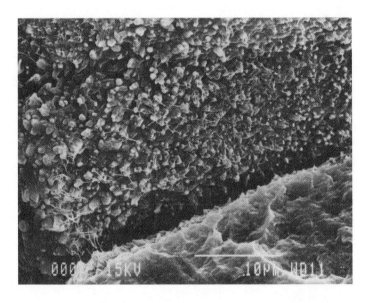

Figure 3. Spherulite boundary of 7 wt.% titania, un-nucleated composite.

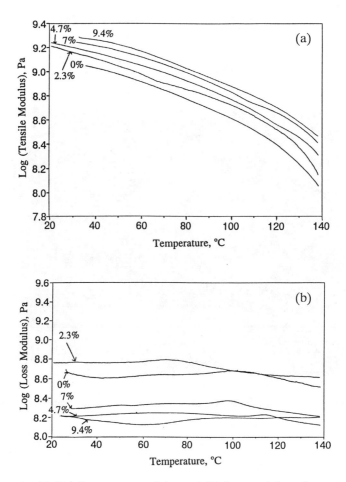

Figure 4.   (a) Tensile storage modulus and (b) loss modulus of composites.

modifications for adhesion between the two phases. Figure 5 shows results calculated from experimental data using the Einstein model assuming no adhesion, with three different density values for amorphous TiO$_2$. Theoretical data is also presented using the Einstein model with and without adhesion between the amorphous TiO$_2$ and polypropylene matrix, equations 1 and 2 respectively, where G'$_c$ and G'$_p$ are the shear moduli of the composite and polymer and Φ_F is the volume fraction of filler.

$$Gc'/Gp' = (1 + \Phi_F) \tag{1}$$

$$Gc'/Gp' = (1 + 2.5\Phi_F) \tag{2}$$

As the volume fraction increases, both models predict higher values, although the experimental data with a density of 3 for the amorphous TiO$_2$ tends to fit the Einstein no-adhesion model quite closely. Judging from the comparison of experimental and theoretical data, it is clear that the isopropyl triisostearoyl titanate coupling agent is not producing a high level of interaction between the TiO$_2$ particles and polypropylene matrix. At very low volume fractions, the experimental data tends to fall between the Einstein model with adhesion and Einstein model without adhesion. Possibly, the positive effects of the coupling agent are masked at higher volume fractions by the decrease in crystallinity of the polypropylene matrix.

Tensile Properties. Stress-strain analysis was performed on both ram-extruded strip, and-screw-extruded film. Figure 6 shows that the yield strength of strip is level up to up to 2.3 wt% TiO$_2$, and then drops sharply. The screw extruded film behaves similarly.

The addition of particulate fillers to tough thermoplastics such as polypropylene tends always to reduce the toughness of the material. This behavior is seen as a decrease in the area under the stress-strain curve.

Some further information can be extracted by analyzing the fracture surfaces following tensile testing. At up to 4.7% titania the fracture surface shows extensive yielding. At 7% titania and above, fracture becomes clearly brittle with interspherulitic failure.

Discussion

A conventional hybrid of amorphous polymer and silica will phase separate on a scale which depends on the chemical kinetics of the polymerization or condensation of the oxide and on the thermodynamics of phase separation (*19,20,21*). The distribution of the two phases should be essentially random.

In the case of a crystalline polymer, the development of polymer morphology will interact with the development of the inorganic particles. In the present system, the titanium n-butoxide forms a homogeneous melt with polypropylene. On cooling, phase separation of the alkoxide may precede or follow the crystallization of the polymer. We believe that phase separation occurs

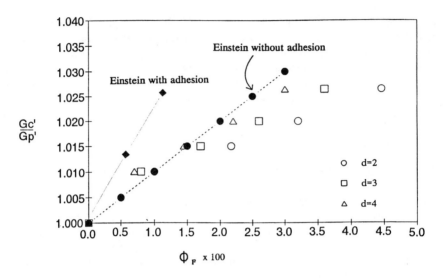

Figure 5. Comparison of composite moduli with Einstein's equation.

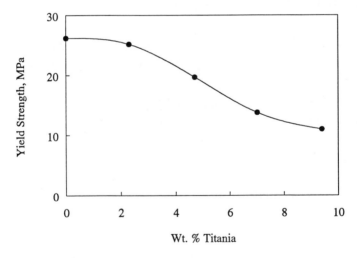

Figure 6. Tensile yield strength of composite strips prepared by ram extrusion.

at the higher alkoxide contents when the sample becomes much more opaque, at about 80°C. At lower alkoxide contents the system may remain locally homogeneous down to room temperature.

If the alkoxide remains miscible with the amorphous polypropylene down to room temperature, there will nonetheless be segregation. A soluble diluent in a crystallizing polymer will tend to be rejected from growing spherulites and will accumulate at the spherulite boundaries (22,23). The extent of this segregation will depend on the dimensionless number given by the diffusion coefficient divided by the growth rate and spherulite radius. Slow diffusion will allow little segregation. In this case the appearance of large amounts of titania at the spherulite boundaries suggests that there is extensive segregation.

In the resulting microstructure, the effects of structure at the scale of the individual lamellae will be essentially hidden by the effects of large amounts of brittle ceramic at the spherulite boundaries. A material with tough polymer bonding brittle ceramic grains would be most desirable, this structure is the opposite. The brittleness and low loss modulus can probably be attributed to this ceramic boundary. The implication is that a lamellar, but non-spherulitic system would be more desirable.

We have demonstrated that hybrid materials can be formed by an extrusion process. We have not focussed on the hydrolysis reaction, which we envisage as occurring as a continuous bath treatment following the extrusion. The time required for this is a function of the sample thickness and diffusion coefficient of water in the polymer. A reasonable value for the diffusion coefficient of water in polypropylene at 100°C is 10^{-6} cm^{2}sec^{-1}, which would give a treatment time of about 10 minutes for a thickness of 1mm (24). It is clear that larger parts would require some form of reaction injection molding.

Polypropylene is not an ideal matrix because it is hard to achieve good bonding to the titania particles. However, many other polymers are unsuitable as a result of reaction with the alkoxide. Polyesters, for instance, are rapidly degraded by titanium alkoxides in the presence of traces of moisture.

Conclusions

We have shown that it is possible to extrude a polypropylene-titanium n-butoxide compatible melt, and that by completing a hydrolysis step, extruded polypropylene-titanium n-butoxide composite strips and films could be efficiently converted to polypropylene-TiO$_2$ composites containing sub-micron TiO$_2$ particles and that the process allowed these particles to be formed *in-situ*.

Initial experiments with in-situ precipitated TiO$_2$ composites showed that to obtain materials with reasonable mechanical properties, it was necessary to incorporate a coupling agent to increase the interfacial adhesion between TiO$_2$ filler and surrounding matrix, and that it was also necessary to reduce the polypropylene spherulite diameter by adding a nucleating agent.

In-situ growth of TiO$_2$ in a polypropylene matrix results in an increase in the modulus, while tensile, yield strengths and toughness decrease. These results were similar to those obtained for conventionally filled polypropylene (25). At low TiO$_2$ concentrations ductile failure is clearly evident. At intermediate TiO$_2$ concentrations, a transition from ductile to brittle failure produces a fracture surface which exhibits some yield, but matrix void formation around TiO$_2$ eventually causes catastrophic failure resulting in a fibrillated fracture surface. At high TiO$_2$ concentrations, failure is purely brittle.

Improvement of the properties of these composites will depend on achieving better particle-polymer bonding and on reducing the degree of segregation of alkoxide to the spherulite boundaries.

Acknowledgements

We would like to thank Elf Atochem, Hughes Missile Systems Co. and the Army Research Office for support of this work.

Literature Cited

1.   *International Encyclopedia of Composites*. Lee, S.M. (Ed.), VCH Publishers, Berlin, 1991
2.   Currey, J. *The Mechanical Adaptations of Bones*; Princeton University Press: Princeton NJ, 1984.
3.   Utracki, L.A.; Vu Khanh, T. In *Multicomponent Polymer Systems*; Miles, I.S.; Rostami, S. (Eds), Longman, London,1992
4.   Kardos, J.L. In *International Encyclopedia of Composites, Volume 5*. Lee, S.M. (Ed.), VCH Publishers, Berlin, 1991
5.   Milewski, J.V.; Katz, H.S. *Handbook of Reinforcements for Plastics*, Van Nostrand Reinhold, New York NY 1987
6.   Novak, B.M.; Ellsworth, M.W.; Verrier, C. *Amer. Chem. Soc. PMSE Preprints* **1994** *70* 266
7.   Landry, C.J.T.; Coltrain, B.K.; Brady, B.K. *Polymer* **1992**, *33*, 1486
8.   Mackenzie, J.D. *Amer. Chem. Soc. PMSE Preprints* **1994**, *70*, 380
9.   Baskaran, S.; Nunn, S.D.; Popovic, D.; Halloran, J.W. *J. Amer. Ceram. Soc.* **1993**, *76*, 2209
10.   Beck, H.N.; Ledbetter, H.D. *J. Appl. Polym. Sci.* **1965**, *9*, 2131.
11.   Beck, H.N. *J. Appl. Polym. Sci.,* **1967**, *11*, 673.
12.   Gee, D.R.; Melia, T.P. *Die Makromolekulare Chemie,* **1970**, *132*, 195.
13.   Jancar, J. *J. Mater. Sci.,* **1989**, *24*, 4268.
14.   Jancar, J. *J. Mater. Sci.,* **1991**, *26*, 4123.
15.   Vollenberg, P.H.T.; Heikens, D. *Polymer*, **1989**, *30*, 1656.
16.   Sumita, M.; Tsukihi, H.; Miyasaka, K.; Ishikawa, K. *J. Appl. Polym. Sci.,***1984**, *29*, 1523.
17.   Maito, S.N.; Mahapatro, P.K. *J. Appl, Polym. Sci.,* **1989**, *37*, 1889.
18.   Maito, S.N.; Mahapatro, P.K. *J. Appl. Polym. Sci.,* **1991**, *42*, 3101.
19.   David, I.A.; Scherer, G.W. *Polymer Preprints* **1991** *32(3)* 530.
20.   Novak, B.M. *Advanced Mater.* **1993** *5* 422.
21.   Schaefer, D.W.; Mark, J.E.; Jian, L.; Sun, C.-C.; McCarthy, D.; Jiang, C.-Y.; Ning, Y.-P., Spooner, S. In *Ultrastructure Processing of Advanced Materials*, Uhlmann, D.R.; Ulrich, D.R. (Eds.) Wiley, New York, NY, 1992.
22.   Calvert, P.D.; Ryan, T.G. *Polymer* **1984** *25* 921
23.   Padden, F.J.Jr; Keith, H.D. *J. Appl. Phys.* **1973**, *35*, 1270,1286.
24.   Van Krevelen, D.W. *Properties of Polymers* Elsevier, Amsterdam, 1976.
25.   Friedrich, K.; Karsch, U.A. *J. Appl. Polym. Sci.* **1981**, *16*, 2167

RECEIVED December 1, 1994

Chapter 3

Inorganic–Protein Interactions in the Synthesis of a Ferrimagnetic Nanocomposite

T. Douglas[1,6], J. W. M. Bulte[2], D. P. E. Dickson[3], R. B. Frankel[4], Q. A. Pankhurst[3], B. M. Moskowitz[5], and S. Mann[1]

[1]School of Chemistry, University of Bath, Bath BA2 7AY, United Kingdom
[2]National Institutes of Health, Bethesda, MD 20892
[3]Department of Physics, University of Liverpool, Liverpool L69 3BX, United Kingdom
[4]Department of Physics, California Polytechnic State University, San Luis Obispo, CA 93407
[5]Department of Geology and Geophysics, University of Minnesota, Minneapolis, MN 55455

The iron-storage protein ferritin provides a spatially constrained reaction environment within a catalytically active bio-polymer. These properties have facilitated the synthesis of a ferrimagnetic iron oxide-protein composite, formed by tailoring conditions for the synthesis of magnetite. The controlled partial oxidation of Fe(II), at high pH and elevated temperature, in the presence of apo-ferritin resulted in the formation of a colloidal composite with a narrow particle size distribution. This material combines characteristics of both the protein and the inorganic phase. Direct magnetic measurements indicated 13,100 Bohr magnetons per particle. Electron and x-ray diffraction data indicated the presence of a cubic iron oxide mineral phase but could not distinguish between magnetite and maghemite. Data from Mössbauer spectroscopy, measured both in the presence and absence of an applied field as well as at low temperature, suggested that the predominant mineral phase was maghemite rather than magnetite.

The protein ferritin is ubiquitous in biological systems. Ferritin stabilizes and stores colloidal iron oxide in the form of the hydrated mineral ferrihydrite (1), ($5Fe_2O_3.9H_2O$ with variable amounts of phosphate). The protein comprises 24 subunits, of two types (H (heavy) and L (light), that self-assemble forming a central cavity, approximately 80Å in diameter, in which the iron oxide is mineralized. This arrangement gives rise to a highly symmetrical spherical protein, 120Å in diameter, intersected by channels through which the ions move. As its biological function suggests, the protein has high specificity for iron but this has not prevented the synthetic formation of other, non-ferrous minerals, such as a

[6]Current address: Department of Chemistry, Ithaca College, Ithaca, NY 14850

manganese oxyhydroxide (2) and a uranium oxyhydroxide, within the protein cavity (3). The mechanism for iron biomineralization in ferritin is thought to involve two distinct clusters of amino acids (on each subunit) that act as ferroxidase centers (H subunit only) and crystal nucleation sites (both H and L subunits) respectively (4).

As a bio-polymer matrix ferritin combines a number of properties that make it ideal for the synthesis of nanocomposite materials: a) a well defined, constrained reaction environment b) stable protein matrix (pH4 to 9, and up to 90°C) c) catalytic oxidation site and crystal nucleation sites d) non-immunogenic protein (high sequence homology amongst ferritins from varying sources) e) biodegradable, non-toxic polymer. This work reports the characterization of a magnetic core synthesized within the cavity of the ferritin protein.

Synthesis of Magnetoferritin

Chemical reduction of the native ferrihydrite followed by chelation and removal of the ferrous ion leaves the apo-protein intact and the central cavity available for remineralization (5). The synthesis of magnetoferritin (6) was carried out anaerobically at elevated temperature (60-65°C) and pH (8.6) by the stepwise addition of Fe(II) to a solution of apoferritin followed by partial oxidation until an average loading of 2500 Fe atoms per ferritin had been achieved. The reaction product was a homogeneous black-brown solution, unlike the blood red color of the native ferritin, which could be precipitated by the application of a high gradient magnetic field.

Characterization of Magnetoferritin

Transmission Electron Microscopy (TEM). The magnetically isolated magnetoferritin cores were imaged by TEM (Figure 1a.) and revealed an average particle size of 73Å diameter with a homogeneous size distribution ($\sigma = 14$Å) (Figure 2). Electron diffraction confirmed a face-centered cubic mineral phase (Figure 1b.), different from the native ferrihydrite, but it was not possible to distinguish between magnetite, Fe_3O_4, and maghemite, γ-Fe_2O_3 (Table I). Control experiments using 0.1M NaCl in place of apoferritin produced a black, extensively aggregated material (figure 3), which exhibited the same electron diffraction pattern as for the magnetoferritin core. The lattice image (figure 4) of a magnetoferritin core approximately 80Å in diameter, obtained by high resolution transmission electron microscopy (HRTEM), clearly shows two sets of lattice fringes (d =

Table I. Electron diffraction of magnetite, maghemite and magnetoferritin

Fe_3O_4/γ-Fe_2O_3 d spacing (Å)	(hkl)	Magnetoferritin measured d spacing (Å)	Control reaction measured d spacing (Å)
4.85	111	4.80	4.85
2.97	220	3.01	3.11
2.53	311	2.54	2.53
2.099	400	2.11	2.11
1.715	422	1.73	1.73
1.616	511	1.65	1.60
1.485	440	1.51	1.50

(magnetite: space group $Fd3m$, a=8.39Å; maghemite: a=8.34Å)

(a)

(b)

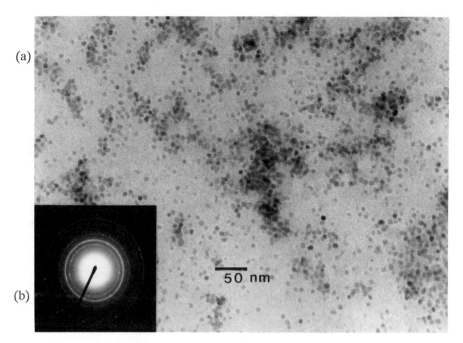

Figure 1. (a) TEM image of magnetoferritin cores. Scale bar is 50 nm.
(b) Electron diffraction from magnetoferritin cores.

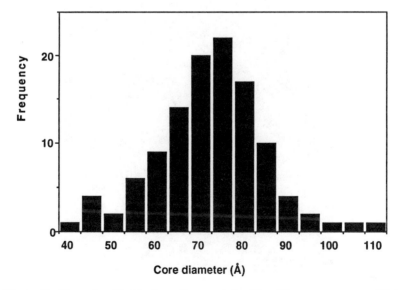

Figure 2. Core size distribution of magnetoferritin. The cores average 73 ±
14Å in size.

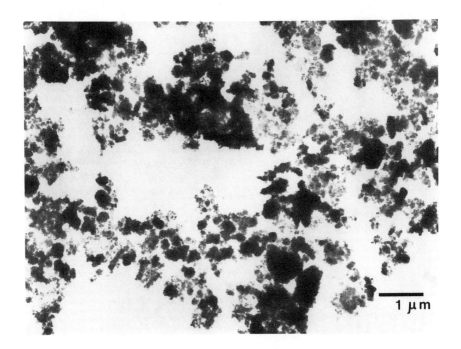

Figure 3. TEM image of the precipitate formed in the absence of ferritin.

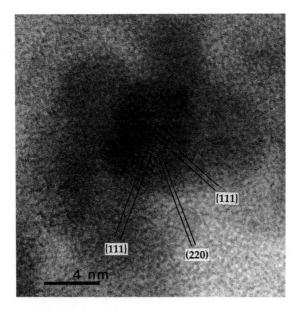

Figure 4. High resolution TEM (lattice image) of the magnetic core showing the {111} and (220) lattice fringes.

4.85Å) 70° apart, as well as a another lattice fringe (d = 2.25Å) at 35° to these. These are consistent with the {111} and (220) lattice planes respectively of a mineral having cubic symmetry and lattice parameters of 8.4Å.

Mössbauer spectroscopy. In an attempt to identify the magnetic iron oxide phase present, the ^{57}Fe Mössbauer spectrum of magnetoferritin was measured (7). Spectra (Figure 5a.) were recorded both in the presence and absence (8) of an applied magnetic field and compared with those of crystalline ferrihydrite, a poorly crystalline ferrihydrite, magnetite and maghemite samples (Figure 5b.). At 100K applied magnetic field and compared with those of crystalline ferrihydrite, a poorly crystalline ferrihydrite, magnetite and maghemite samples (Figure 5b.). At 100K the zero field Mössbauer spectrum is clearly different from native ferritin (9), although the Fe^{3+} component of the spectrum could indicate the presence of some ferrihydrite-like material. At 4.2K the zero field spectrum shows a sextet with broad asymmetric lines. The lineshapes are consistent with the spectrum of maghemite but the line broadening precludes any definitive distinction between magnetite and maghemite. In addition, the value of the hyperfine field (50 T at 4.2K) is consistent with both magnetite and maghemite (10).

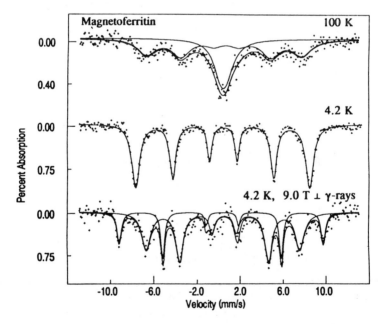

Figure 5a. Mössbauer spectra of the magnetoferritin sample at 100 K, and at 4.2K in zero field and with a magnetic field of 9T field applied perpendicular to the γ–ray beam.

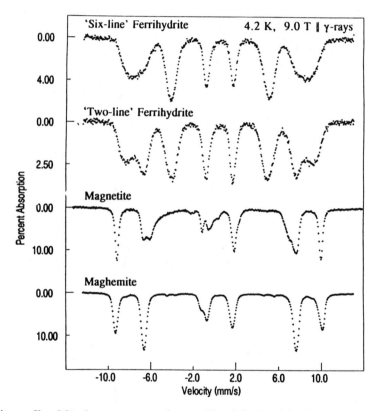

Figure 5b. Mössbauer spectra of crystalline "six line" ferrihydrite, poorly crystalline "two line" ferrihydrite, magnetite, and maghemite recorded at 4.2K in a magnetic field of 9T applied parallel to the direction of the γ–ray beam.

In the presence of a 9T field, applied perpendicular to the γ−ray beam, the Mössbauer spectrum consists of two six line subspectra (Figure 5b.), one with relatively sharp lines, the other with broader lines (relative areas 69% and 31%). In the subspectrum with sharper lines the magnetic splitting is 58.8T, while the subspectrum with broader lines has magnetic splitting of 44.2T. The applied field spectrum of magnetoferritin closely resembles that of maghemite, particularly in the relative intensities of the two lines on the right hand side of the spectrum. This characteristic is apparent in the spectrum of maghemite and is the opposite of the intensity ratio in the magnetite spectrum. The relative areas of the two subspectra of maghemite are 37.5% and 62.5% which are close to those observed in magnetoferritin. As a result, the Mössbauer spectrum of magnetoferritin resembles that of maghemite more than that of magnetite.

Magnetic Measurements. The field dependent magnetization of magnetoferritin was measured at 300 K with a SQUID magnetometer. The magnetization saturated but showed no hysteresis (Figure 6), consistent with superparamagetic behavior. The magnetic moment per molecule was calculated from a fit to the data using the Langevin magnetization function:

$$M(B_0) = M_{sat}L(\mu B_0/kT) + \alpha B_0$$

where $M(Bo)$ is the magnetization of the sample, M_{sat} is the saturation magnetization, Bo is the applied field, μ is magnetic dipole moment per molecule, k is Boltzmann's constant, T is the absolute temperature, and α is a constant. $L(\mu Bo/kT) = \coth(\mu Bo/kT) - (kT/\mu Bo)$. The results of the fit, shown in Figure 6, indicated 13,100 Bohr magnetons per molecule. If we assume the core material is maghemite (based on the Mössbauer results) this corresponds to a particle of 8.5 nm diameter, containing 12,000 Fe atoms.

Electrophoresis. Polyacrylamide gel electrophoresis (PAGE) of magnetoferritin, under non-denaturing conditions, revealed a monomeric protein of roughly 450kDa, essentially the same as the native horse spleen ferritin (HS-ferritin) (Figure 7a.). Both HS-ferritin and magnetoferritin showed a trace of oligomeric protein (dimers or trimers). However, in the magnetoferritin sample a high molecular weight fraction could be detected (not present in HS-ferritin) which contained a relatively large amount of iron.

Under denaturing conditions (4-20%SDS) both HS-ferritin and magnetoferritin showed two bands at 22-24kDa which are the subunits L and H. The cores, no longer encaged by the protein, showed similar size distribution for both HS-ferritin and magnetoferritin.

Isoelectric focusing of magnetoferritin showed that the pI is in the range 4.3-4.6, the same as that observed for HS-ferritin (Figure 7b.).

Discussion. The results indicate that magnetoferritin is a superparamagnetic nanocomposite exhibiting the combined properties of the bio-polymer and the inorganic colloid.

Electron diffraction and lattice imaging data identified the mineral core as an iron oxide of cubic symmetry, either magnetite or maghemite. Mössbauer spectroscopy confirmed the phase as maghemite. It is not known whether the maghemite was formed *in situ* or by oxidation of a precursor magnetite core. There is a small discrepancy between measured (TEM) and calculated (SQUID) core sizes. The magnetic data indicates a particle of maghemite approximately 85Å

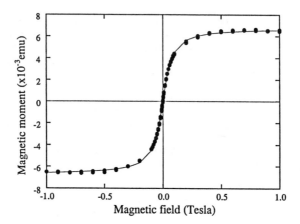

Figure 6. Paramagnetic moment of magnetoferritin in water at room temperature, plotted as a function of applied magnetic field. The solid line is a fit of the Langevin function to the data, yielding a magnetic dipole moment of 13,100 μ_B per molecule.

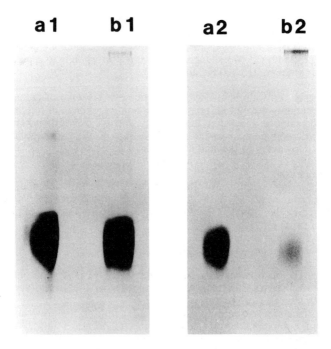

Figure 7a. Non-denaturing polyacrylamide gel electrophoresis of horse spleen ferritin (lanes a1 and a2) and magnetoferritin (lanes b1 and b2). The gel is stained for protein using Coomassie Blue (lanes a1 and b1) and for iron using Prussian Blue (lanes a2 and b2).

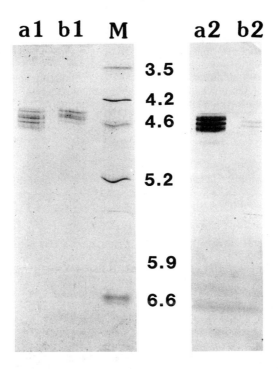

Figure 7b. Isoelectric focusing gel of horse spleen ferritin (lanes a1 and a2) and magnetoferritin (lanes b1 and b2) and markers (M). The gel is stained for protein using Coomassie Blue (lanes a1, b1 and M) and for iron using Prussian Blue (lanes a2 and b2).

diameter, containing on the order of 12,000 Fe atoms, whereas the average particle size (measured by TEM) of 73Å corresponds to approximately 7,500 Fe atoms per particle.

The measured core size is larger than expected from theoretical Fe/protein ratios. This might be mechanistically insightful, implying an "all or nothing" mechanism for mineralization within of the protein. It has been suggested that adsorption of Fe(II) onto a ferric oxyhydroxide can bring about the transformation to the spinel phase by a process of dissolution and reprecipitation (11). It is conceivable that the magnetoferritin core is formed in this way; i.e the addition of Fe(II) to an initially formed ferrihydrite-like core thus transforming it to the spinel phase. We speculate that core formation in magnetoferritin is dominated by catalytic oxidation at the mineral surface with little or no involvement of the ferroxidase center of the protein, except perhaps at a very early stage in the reaction. This mechanism of reconstitution favors crystal growth over crystal nucleation, the result of which is a protein containing cores of maximum size (~80Å), and presumably leaving many proteins unmineralized.

The pI of magnetoferritin is essentially unchanged from that of native horse spleen ferritin indicating that the surface charge of the monomeric protein is unaffected by the transformation. Gel electrophoresis indicates that the subunits of magnetoferritin are also unchanged. A high molecular weight fraction, containing iron, in the gel electrophoresis indicates that there is some aggregation in the magnetoferritin preparation, possibly due to iron mediated disulfide crosslinking.

Conclusions. Magnetoferritin exhibits the magnetic properties of a ferrimagnetic inorganic colloid while possessing the unaltered features of the protective ferritin protein. These properties make this nano-composite material ideal for biological and medical applications such as magnetic resonance imaging and work is ongoing in the study of this material as a bio-compatible nmr imaging contrast agent (12). The elucidation of the mechanism of maghemite core formation, as well as the synthesis of other mineral phases within the protein matrix are areas of current investigation.

References

1. *Biomineralization: Chemical and Biochemical Perspectives*; Mann, S.; Webb, J.; Williams, R.J.P.,Eds., VCH, Weinheim: **1989**.
2. Mackle, P.; Charnock, J.M.; Garner, C.D.; Meldrum, F.C.; Mann, S. *JACS* **1993**, *115*, 8471-8472.
3. Meldrum, F.C.; Wade, V.J.; Nimmo, D.L.; Heywood, B.R.; Mann, S. *Nature* **1991**, *349*, 684-687. Hainfeld, J.F. *Proc. Natl. Acad. Sci. USA* **1992**, *89*, 11064-11068.
4. Lawson, D.M.; Treffry, A.; Artymiuk, P.J.; Harrison, P.M.; Yewdall, S.J.; Luzzago, A.; Cesareni, G.; Levi, S.; Arosio, P. *FEBS Lett.* **1989**, *254*, 207.
5. Meldrum F.C.; Wade V.J.; Nimmo D.L.; Heywood B.R.; Mann S. *Nature* **1991**, *349*, 684-687.
6. Meldrum, F.C.; Heywood, B.R.; Mann, S. *Science* **1992**, *257*, 522-523.
7. Pankhurst, Q.A.; Betteridge, S.; Dickson, D.P.E.; Douglas, T.; Mann, S.; Frankel, R.B. *Hyperfine Interactions* **1994**, in press.
8. Pankhurst, Q.A.; Pollard, R.J. *J. Phys. Chem.* **1992**, *2*, 7329-7337.
9. St. Pierre, T.G.; Webb, J.; Mann, S. in *Biomineralization: Chemical and Biochemical Perspectives*; Mann, S.; Webb, J.; Williams, R.J.P. Eds. VCH, Weinheim: **1989**, .
10. Pankhurst, Q.A., Pollard, R.J., *Clays Clay Min.*, **1992**, *40*, 268-277.
11. Tronc, E.; Belleville, P.; Jolivet, J-P.; Livage, J. *Langmuir* **1992**, *8*, 313-319. Mann, S.; Sparks, N.H.C.; Couling, S.B.; Larcombe, M.C.; Frankel, R.B. *J. Chem. Soc., Faraday Trans.* **1989**, *85*, 3033-3044.
12. Bulte, J.W.M.; Douglas, T.; Mann, S.; Frankel, R.B.; Moskowitz, B.M.; Brooks, R.A.; Baumgarner, C.D.; Vymazal, J.; Strub, M-P.; Frank, J.A. *Journal of Magnetic Resonance Imaging* **1994**, *4*, 497-505.

RECEIVED October 25, 1994

Chapter 4

Ion-Exchange Intercalation into the MPS₃ Layered Compounds

Design of Nanocomposites with Unusual Magnetic, Electrical, and Nonlinear Optical Properties

R. Clément

Laboratoire de Chimie Inorganique, Centre National de la Recherche Scientifique, Unité de Recherche Associée 420, Université Paris Sud, 91405 Orsay Cedex, France

Cation exchange intercalation into the layered $M^{II}PS_3$ compounds allows the synthesis of many air stable intercalates $Mn_{1-x}PS_3(G)_{2x}(solv)_y$ (G stands here for a guest monocation) containing intralayer metallic vacancies. These nanocomposites exhibit various unusual physical properties: *(i)* Intercalation deeply modifies the magnetic properties of the pure layered materials. Whereas pure MPS_3 phases are antiferromagnets, many intercalates become (ferri)magnets below a Curie temperature that can be as high as 92 K; *(ii)* Insertion of tetrathiafulvalene species gives a new type of organic-inorganic metal due to partial host - guest electron transfer; *(iii)* The MPS_3 host lattice is capable of enforcing spontaneous poling of guest chromophores having high hyper-polarizability. Intercalates of dimethylamino-N-methylstilbazolium have been obtained (M= Mn, Cd) which possess an efficiency for second harmonic generation about 750 times larger than the efficiency of urea. The manganese intercalate is the first material to associate both strong NLO effects and spontaneous magnetization.

The MPS₃ Layered Compounds

In the past decades, there has been an increasing interest in low dimensional compounds and intercalation chemistry (1, 2, 3). Research in the field has been stimulated by many important discoveries such as enhancement of the critical temperature T_c for the onset of superconductivity, design of lithium batteries, heterogeneous catalysis. Several reviews are now available. Much less work has been done concerning the magnetic and optical properties of insulating or semiconducting intercalation compounds. One reason is probably that there is quite a small number of intercalation compounds which are at the same time insulating, optically transparent, air stable, well crystallized and which contain atoms with localized spins. It is the purpose of this article to show how the intercalation chemistry of the MPS₃ compounds can provide new directions for the synthesis of novel hybrid organic inorganic nanocomposites. The topic will be illustrated by describing the synthesis of the first nanocomposite that combines non linear optical properties due to the inserted chromophores with ferrimagnetic properties due to the host lattice.

0097–6156/95/0585–0029$12.00/0

The MPS3 compounds, where M stands for a metal in the +2 oxidation state, form a class of lamellar semiconductors (Figure 1) with the $CdCl_2$ structural type (4, 5). The electronic structure and properties of these materials have received considerable attention (for a review, see the articles by Brec (6) and O'Hare(7)). Several of them are optically transparent (M= Mn, Cd, Zn). These compounds can be described as broad band semiconductors, but this picture has been refined on the basis of tight binding band structure calculations (8).

Because of their structural analogy with the dichalcogenides MX_2, the MPS3 compounds were in the late seventies considered merely as an exotic appendix of this class. and their reactivity was expected to be similar. Indeed, the early chemistry carried out supported this analogy. Thus, several MPS3 compounds (M= Fe, Ni) were shown to react with electron-donor species (butyllithium (9), cobaltocene (10)) to form intercalation compounds according to a redox, guest to host electron transfer process. NiPS3 actually proved to be a particularly efficient cathodic material in rechargable lithium batteries (11).

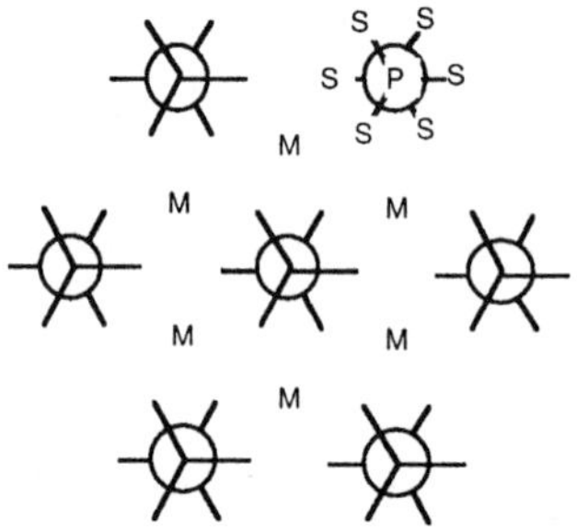

Figure 1. Schematic top view of a MPS3 slab, array of M^{2+} cations coordinated to $P_2S_6^{4-}$ bridging ligands.

In 1980, the author showed that MnPS3 exhibited a topochemical reactivity of a different type, based on ion transfer rather than on electron transfer. This article focus on this new type of intercalation chemistry and gives an updated account of the physical properties studied in the author's group on the derived intercalation compounds.

Cation Transfer Intercalation Chemistry. The basic chemical feature is the spontaneous reaction, at room temperature, of MnPS3 with aqueous ≈ 1 M solutions of a number of salts, such as KCl, NH_4Cl and salts of organometallic cations such as cobalticenium. These reactions lead to compounds $Mn_{1-x}PS_3(G)_{2x}(H_2O)_y$, where the positive charge of the entering guest cations G^+ is counterbalanced by the departure of an equivalent amount of intralayer Mn^{2+} cations from the solid (12).

$$MnPS_3 \text{ (s)} + 2x \ Me_4NCl \longrightarrow Mn_{1-x}PS_3[Me_4N]_{2x}(H_2O)_y \text{ (s)} + x \ MnCl_2 \quad (x \approx 0.2)$$

Reaction durations range from a few minutes (pyridinium or tetramethylammonium chlorides) to several hours (cobalticenium chloride). The material remains solid throughout the reaction, and therefore such a cation exchange process is highly unusual, even unique, as it implies that manganese cations are able to leave their **intra**lamellar sites to pass into the solution. Therefore, a central point is that metallic vacancies are created in the host lattice during the insertion process. The guest species are solvated by water molecules. This is particularly obvious in the case of small alkali metal ions, where the interlamellar distance is determined by the Van der Waals thickness of the water molecules: for example, potassium cations retain a single layer of solvating water molecules ($y \approx 1$, $d = 9.3$ Å) whereas sodium cations retain two layers of water molecules ($y = 4$, $d \approx 12$Å). The composites are well crystallized, as their X-ray powder diffraction patterns exhibit sharp *hkl* reflections. These reflections can be indexed using a monoclinic unit cell closely related to that of pristine MnPS$_3$ or a superstructure (see details below). When the reaction is not complete, the samples contain diphasic mixtures of MPS$_3$ and of the fully intercalated composite. Infrared spectra (13, 14) in the range 4000 cm^{-1} - 200 cm^{-1} exhibit the absorption bands of the guest species and those of the (S$_3$P-PS$_3$) groups. The v(PS$_3$) asymmetric stretching, which occurs at 570 cm^{-1} in pure MnPS$_3$, is always split into two components around 605 cm^{-1} and 555 cm^{-1} in the intercalated materials. This large splitting, very useful from an analytical point of view, is a consequence of the presence of intralamellar metallic vacancies which renders P-S bonds inequivalent.

Insertion often fails when the size of the guest species is too large. However, many bulky species can nevertheless be inserted in two steps: (i) insertion of alkali metal, pyridinium or tetramethylammonium ions; (ii) exchange of the "pre-inserted" ions with the bulky species (15). Small hydrated guest species are actually quite mobile (as in many other layered materials) and can be further exchanged by a broad range of cations, whatever their size (for example species as big as Ru(2,2'bipy)$_3^{2+}$ or FeCp(CO)diphos$^+$).

Several other MPS$_3$ phases of the series (M = Mn, Cd, Zn, Fe) give rise to a similar insertion process but the reaction may require an assist, which can be provided by a complexing agent such as EDTA, to help the M^{2+} cations to leave the host lattice (16). Each host material in the series has its own particularities. ZnPS$_3$ is very sensitive to hydrolysis and should be handled in ethanol rather than in water. CdPS$_3$ reacts very rapidly with Me$_4$NCl in ethanol but not in water (probably due to the formation of CdCl$_4^{2-}$ complexes which help Cd^{2+} departure). Deintercalation can be achieved when the guest species are mobile. For instance, when Mn$_{0.8}$PS$_3$K$_{0.4}$(H$_2$O)$_y$ is treated with an aqueous solution of NiCl$_2$, the solvated K$^+$ ions are exchanged by Ni(H$_2$O)$_6^{2+}$ hexaaqua species in a first step. Upon drying, the nickel ions then lose their aqua ligands and move into the intralayer vacancies, and pure MnPS$_3$ is reformed (17).

A "Destruction-Reconstruction" Intercalation Mechanism Involving Heterogeneous Equilibria: The MPS$_3$ Materials as Coordination Compounds. The high mobility of the intralamellar M^{2+} cations is difficult to understand in terms of diffusion in the solid state, which must be very slow at room temperature. Such a process has never been observed so far in any other layered material. We have suggested that the cation exchange and intercalation process proceed via destruction (local microdissolution) and subsequent reconstruction (microrecrystallization) of the host lattice. Despite the structural analogy with the transition metal dichalcogenides, the chemical nature of the MPS$_3$ compounds is different because of the ionicity of the metal - sulfur bonds and of the stability both the M^{2+} and P$_2$S$_6^{4-}$ in protic solution. The MPS$_3$ compounds should be rather considered as polynuclear coordination compounds, that is each slab should be seen as an array of M^{2+} ions assembled together by P$_2$S$_6^{4-}$ bridging ligands. We have brought arguments in favor of the occurence of an heterogeneous equilibrium between solid MPS$_3$ and the solvated species M^{2+}aq and P$_2$S$_6^{4-}$aq (14). In the presence of suitable intercalatable species, the ions then rearrange themselves to give the intercalate. It should be emphasized that P$_2$S$_6^{4-}$ is a real molecular anion that is stable in solution.

$$Mn_2P_2S_6\ (s) \rightleftharpoons Mn^{2+}{}_{aq} + P_2S_6^{4-}{}_{aq} \xrightarrow{+\ 4x\ G^+} Mn_{2-2x}P_2S_6G_{4x}(H_2O)_y\ (s)$$

Indeed, it is well known that coordination compounds of the first transition series are generally labile when the metal is in the +2 oxidation state. The most reactive MPS_3 materials are precisely those where the M^{2+} ions have a d^5 or d^{10} electronic structure, i.e. those in which the $[MS_6]$ chromophores have no crystal field stabilization energy. In support of this scheme, it has been possible to synthesize MPS_3 materials by the usual techniques of coordination chemistry, i.e. by precipitating aqueous solutions of the sodium salt $Na_4P_2S_6, 6H_2O$ by metallic cations M^{2+} (16, 18, 19).

It is quite staightforward to explain why this unusual reactivity is not commonly encountered in the chemistry of other lamellar materials such as the dichalcogenides or oxides. Such materials are lamellar because of the existence of strong intralayer covalent M-X bonds. The covalent character of these bonds requires a high oxidation state of the metallic cations and consequently such M-X bonds are not labile. In contrast, the lamellar structure of the MPS_3 materials is due to the geometry of the P_2S_6 ligand. As the metallic ions are in a 2+ oxidation state, the metal -ligand bonds can be thermodynamically and kinetically quite easily broken.

This new chemistry gives rise to a large number of intercalates which are air stable, often transparent, and reasonably well crystallized. The loss of intralayer metallic cations has been confirmed by several authors (20, 21, 22). Unless specific species are inserted (see below the case of TTF), they are semiconductors or insulators. In addition, depending on the nature of the guest species, the intercalates display a whole range of physical properties. Thus, intercalation has enabled us to synthesize molecular insulating magnets with high Curie temperature (by the standards of this field), a new type of molecular metal, and materials for non linear optics (second harmonic generation). In addition, because these materials are nanocomoposites, intercalation provide opportunities to combine different properties.

A Recent Development: Intercalation of Soluble Polymers (PEO). In the perspective of obtaining new "supported" solid electrolytes, we have attempted to insert polymeric ethers into MPS_3 host lattices (23). For example, potassium intercalate $Mn_{1-x}PS_3K_{2x}(H_2O)_y$ was treated with solutions of polyethyleneglycol (PEG) oligomers of average molecular weight 1500, 6000 and also with solutions of polyethyleneoxide (PEO) of average molecular weight 100000. In a typical experiment, ≈ 500 mg of the polymer was dissolved in 10 ml of water or methanol, and ≈ 200 mg of the potassium intercalate was then introduced and stirred for about 12h at room temperature. In the case of PEG 6000 and PEO, additional heating at 50°C for 1h was carried out.

The X ray diffraction patterns of the powders obtained exhibit sharp 001 reflections (up to 006 or more) which enable one to calculate an interlamellar distance close to 15.2 Å. The typical formula of the materials is $Mn_{0.88}PS_{3.00}K_{0.20}(CH_2CH_2O)_{2.14}$. Insertion of PEG's and PEO into the $Mn_{1-x}PS_3Na_{2x}(H_2O)_y$ compound and $Cd_{1-x}PS_3A_{2x}(H_2O)_y$ (A = K, Na) was also carried out under similar conditions as above. The PEO and PEG intercalates display a basal spacing value close to 15Å, regardless of the chain length (compared to 6.5 Å for $MnPS_3$, 9.3 Å for $Mn_{1-x}PS_3K_{2x}(H_2O)_y$). The polymer chains therefore run parallel to the layers, but different arrangements can be envisaged (double layer of zig zag chains, helical structure (24)). High resolution ^{13}C NMR spectra of PEO and PEG inserted in $CdPS_3$ show a very sharp resonance line at a chemical shift of 70.5 ppm, which suggest a helicoidal arrangement of the inserted polymer chains. On the other hand the NMR confirms that the polymer is much better organized in the interlamellar space than it is in the pure state (25). It is also reasonable to assume that solvation of the guest alkali ions by the oxygen atoms of the polyether is the driving force of the polymer insertion process. These results are currently being successfully extended to MPS_3 intercalates containing other hydrated metallic guest cations, such as lithium and rare earth cations. The guest cations trapped in the polymer intercalates retain some mobility and can still be exchanged, so that these polymer intercalates could themselves act as new organic - inorganic host lattices.

Another interest is that polymer insertion allows processing the intercalates as transparent thin films. Thus the lithium intercalate $Mn_{1-x}PS_3Li_{2x}(H_2O)_y$ disperses when washed with water, giving a colloidal pale green solution. Insertion of polymer at this stage stabilizes the colloidal solution. If the latter is spread onto a glass plate and dried, a transparent pale green thin films is formed which is able to take up very rapidly large cationic molecules (for instance "crystal violet " dye).

Intercalates as Insulating Magnets with "High" T_c

In recent years, considerable effort has been devoted to the synthesis of insulating or semiconducting, molecular based materials exhibiting bulk ferro or ferrimagnetism (26, 27, 28, 29).The strategy used so far consists either in trying to achieve stacking of proper molecular species or to build heterometallic ordered chains. A large number of "molecular" magnets have been obtained, but Curie temperatures are always very low, typically in the range 1-15 K, the main reason being that intermolecular magnetic interactions between adjacent chains or molecular species are very weak.

An alternative approach to the synthesis of semiconducting "molecular" magnets is to modify antiferromagnetic layered materials to induce bulk magnetization. The major interest of this approach is that magnetic interactions can be strong along two dimensions within the slabs, so that higher ordering temperatures can be expected. We have shown in the past years that a spontaneous magnetization occured in several MPS_3 intercalates (M = Mn, Fe), even though the pure MPS_3 host orders antiferromagnetically (30). One purpose of this paper is to describe the basic features concerning the magnetic properties of the $MnPS_3$ intercalates, focussing on the tetramethylammonium - $MnPS_3$ intercalate which we have studied in great detail.

The Magnetic Properties of the MPS_3 Intercalates. The temperature dependence of the magnetic susceptibility of pure $MnPS_3$ and of the $Mn_{0.85}PS_3(Me_4N)_{0.3}(H_2O)_y$ intercalate are shown in figure 2 as $1/\chi = f(T)$. At low temperature, the susceptibility of the intercalate considerably increases and becomes field dependant (31, 32). In contrast to pure $MnPS_3$, which orders antiferromagnetically below a Neel temperature of 78 K (33), the intercalate orders magnetically in such a way that bulk spontaneous magnetization occurs below ~35K.

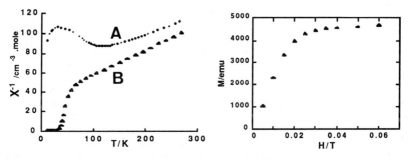

Figure 2. left: Temperature dependance of the inverse of the magnetic susceptibility of pure $MnPS_3$.(trace A) and of intercalate $Mn_{0.85}PS_3(Me_4N)_{0.3}(H_2O)_y$ (trace B); right: Magnetization versus applied magnetic field H of a powdered sample of the intercalate $Mn_{0.85}PS_3(Me_4N)_{0.3}(H_2O)_y$ (adapted from ref. 2.)

This is confirmed by the dependance of the bulk magnetization M versus applied magnetic field (fig. 2) The magnetization rapidly increases, then saturates. The saturation value (around 4000 emu per mole of Mn) is a fraction ($\approx 15\%$) of the value NgbS that would be reached if all manganese spins were parallel inside each Weiss domain (34).

Intercalation of Me_4N^+ cations into a monocrystalline platelet of $MnPS_3$ can be carried out within a few minutes. A monocristalline platelet of the intercalate was therefore synthesized to study the anisotropy of the magnetization below the Curie temperature and deduce the orientation of the magnetization with respect to the host layers. The dependence of the magnetization at 10 K versus applied magnetic field is shown in Figure 3 for two orientations (parallel and perpendicular) of the field with respect to the plane of the platelet (that is also the plane of the slabs). It is clear that the axis of easy magnetization is essentially perpendicular to the plane of the platelets, which means that the spins in the magnetically ordered state are essentially perpendicular to the plane of the slabs, as in pure $MnPS_3$ (35).

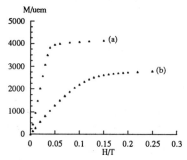

Figure 3. Magnetization of a platelet of the $Mn_{0.85}PS_3(Me_4N)_{0.3}(H_2O)_y$ intercalate versus magnetic field H: (a: H perpendicular to the slabs; b: H parallel to the slabs).

X-ray powder diffraction spectra of the $Mn_{0.85}PS_3(Me_4N)_{0.3}(H_2O)_y$ intercalate were studied using an asymmetric Seeman Bohlin camera. This method is well adapted to materials giving many reflections at low angles because it affords a good resolution. The spectrum of $Mn_{0.85}PS_3(Me_4N)_{0.3}(H_2O)_y$ displays reflections which are practically as sharp as those of $MnPS_3$, indicating a high degree of order. However, the intercalate exhibits many more reflections at low angles than pure $MnPS_3$, suggesting the occurence of a superlattice. Analysis of the spectra shows that all reflections can be indexed in a unit cell having the a parameter tripled with respect to the a parameter of the unit cell of the parent $MnPS_3$ (Figure 4). Therefore the structure of the intercalate along the plane of the slabs is a superlattice of the pure host. Recent X-ray and neutron diffraction experiments carried out in collaboration with O'Hare and Evans on monocrystals of the same intercalate have confirmed these results and demonstrated that the superstructure was due to the ordering of the intralayer vacancies (J. Evans, Thesis, 1994, University of Oxford, U.K., and J. Evans, D. O'Hare, R. Clément, to be published). It is interesting to note that spontaneous ordering of metallic vacancies had already been evidenced in the pure $In_{2/3}PS_3$ compound (36).

A Ferrimagnetic Model. The ordering of the intralayer metallic vacancies in the tetramethyl intercalate is an important feature, because it provides a straightforward and reasonable mechanism to explain the occurence of a spontaneous magnetization. The basic idea is that ordering of the vacancies destroys the balance between the antiferromagnetically coupled spins and hence causes ferrimagnetism to appear in the intercalate, as detailed below.

The magnetic structure of pure $MnPS_3$ below 78 K consists of spins 5/2 located at the corners of a honeycomb lattice and ordered antiferromagnetically (37, 38, 39). Each spin "up" is surrounded by three spins "down" and *vice versa*, so that there is an equal number of spins in each direction and hence there is no net macroscopic magnetization.

If we assume that the intralayer vacancies in the intercalate are ordered on the same sublattice, the amount of spins in each direction will be different and a net macroscopic magnetization will appear, due to non compensation. A tentative scheme of a possible arrangement is represented in Figure 4. The phenomenon therefore resembles <u>ferri</u>magnetism, and it is consistent with the value of the magnetization found experimentally: As the stoichiometry of the intercalated host lattice is close to $Mn_{5/6}PS_3$, an initial set of 6 manganese atoms in pure $MnPS_3$ will give 3 spins up for only 2 spins down plus a vacancy in the intercalate. Therefore the difference accounts for a magnetization equal to one sixth of the maximum value $M = Ng\beta S$ that would occur if all the manganese spins were parallel (i.e. if they were ferromagnetically coupled). The experimental value, ≈ 4000 emu per mole of manganese, is actually very close to $1/6[Ng\beta S]$ (≈ 24000 emu if g = 2).

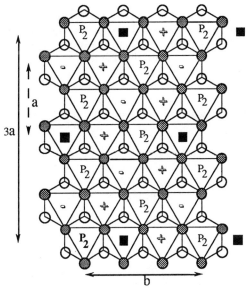

Figure 4. Tentative scheme of the magnetic structure of a slab (top view) of the $Mn_{0.85}PS_3(Me_4N)_{0.3}(H_2O)_y$ intercalate below the Curie temperature. The symbols ⊹ and ⊖ represent the up and down orientations of the Mn^{2+} spins. The black squares represent metallic vacancies. The shaded and open circles represent sulfur atoms.

The discussion so far has been focused on the tetramethylammonium intercalate. Actually a large number of intercalates show up the same general behavior, that is they become magnets at low temperature. However the ordering temperature Tc and the value of the magnetization at saturation depend on the inserted species (32). A few compounds are compared in Table I.

T_c decreases as the amount of intralayer vacancies increases. This is probably due to dilution of the interacting spins. We have synthesized a sodium intercalate $Mn_{0.5}PS_3 (Na)_{1.0} (H_2O)_{\sim 4}$ treating the above potassium intercalate with a concentrated solution of NaCl in the presence of EDTA. This intercalate is still a layer material with a basal spacing close to 12.5 Å (two layers of solvating water molecules), but it no longer exhibits any magnetic transition.Therefore there seems to be a "percolation threshold" for the magnetic ordering around the composition $Mn_{0.7}PS_3$.

TABLE I. Saturated Magnetization (per mole of Mn) at 10 K for some MnPS$_3$ Composites

Composite	Spacing (Å)	Magnetization (emu / mol)	T_C / K
$Mn_{0.81}PS_3$ (K)$_{0.38}$ (H$_2$O)$_{\sim 1}$	9.37	$\approx$ 1000	$\approx$ 20
$Mn_{0.80}PS_3$ (NH$_4$)$_{0.4}$ (H$_2$O)$_{\sim 1}$	9.38	3100	$\approx$ 15
$Mn_{0.89}PS_3$ (n-octyl NH$_3$)$_{0.22}$ (H$_2$O)$_{0.5}$	10.38	1150	$\approx$ 45
$Mn_{0.85}PS_3$ (CoCp$_2$)$_{0.30}$(H$_2$O)$_{\sim 1}$	12.15	$\approx$ 4200	$\approx$ 35
$Mn_{0.86}PS_3$ (pyH)$_{0.28}$ (H$_2$O)$_{0.7}$	9.65	80	$\approx$ 35
$Mn0.83PS_3$ (pyH + py)$_{0.42}$ (H$_2$O)$_{0.7}$	12.36	$\approx$ 1500	$\approx$ 35

It is interesting to note that the ordering temperatures of the nanocomposites obtained are quite high by the standards of the field of "molecular ferromagnets. This is obviously due to the two-dimensional character of the materials. In the search for magnets with higher T_C's, it is worth mentioning that we have managed to synthesize a pyridine - pyridinium intercalate $Fe_{0.88}PS_3$ (pyH + py)$_{0.36}$ (H$_2$O)$_{\approx 0.5}$. having its T_C well above liquid nitrogen temperature. (Cell dimensions: a = 6.001 Å, c = 10.527 Å, b = 10.361 Å, b = 112.60° (40)).

The temperature dependence of the inverse of the magnetic susceptibility of the latter material is shown in Figure 5. Whereas pristine FePS$_3$ orders antiferromagnetically below $\approx$120 K, the intercalate undergoes a rather abrupt transition between 90 K and 100 K. The following simple experiment is quite spectacular: When a pyrex tube containing $Fe_{0.88}PS_3$ (pyH + py)$_{0.36}$ (H$_2$O)$_{\approx 0.5}$ is placed in liquid nitrogen on top of a standard magnetic stirrer, the powder "flows " in the tube and follows the rotating rod; the phenomenon stops upon reheating .

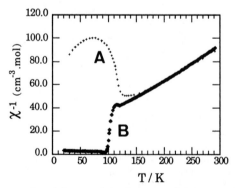

Figure 5. Temperature dependence of the inverse of the magnetic susceptibility of FePS$_3$ (trace A) and of the composite $Fe_{0.88}PS_3$ (pyH + py)$_{0.36}$ (H$_2$O)$_{\approx 0.5}$ (trace B).

Study of the magnetization of $Fe_{0.88}PS_3$ (pyH + py)$_{0.36}$ (H$_2$O)$_{\approx 0.5}$ as a function of the applied magnetic field has been carried out at 85 K. The saturation value of M (around 500 emu per mole of iron) is only a weak fraction of the value expected

(~22000 cgs.emu per mole of Fe) if all spins were ordered parallel. The M(H) behavior at lower temperatures is more complicated and indicates some interlayer antiferromagnetic interaction. It is not clear why the magnetization is weaker than in the case of $MnPS_3$ intercalates. The magnetic structure of $FePS_3$ is very different from that of $MnPS_3$. and might be one of the reasons. Studies on other $FePS_3$ intercalates are in progress in our group and will be published elsewhere.

In conclusion of this section, it should be emphasized that the surprising modification of the magnetic properties of $MnPS_3$ ultimately appears to be a consequence of the creation of intralayer metallic vacancies (loss of M^{2+} ions) during intercalation. The vacancies cause non compensation between the antiferromagnetically coupled spins. To our knowledge, this is a novel feature in the field of intercalation science. The fact that the intensity of the magnetization varies from one intercalate to another is probably related to various ways for the vacancies to order (or imperfect ordering), but additional studies are necessary to establish this point, as well as to establish whether the magnetic properties of the $FePS_3$ intercalates can be interpreted in a similar way .

Beside the competition for high T_c's, another interest for synthesizing 2D nanocomposite magnets is that the magnetic properties of the host lattice can be hopefully combined with other properties, such as optical properties (see below).

Intercalation as an Approach to Molecular Metals

A possible approach to the synthesis of low dimensional organic metals is to use a layered material as a macroionic host that could be able to accept electrons from(or donate to) a suitable guest species and therefore allow mixed valency, regardless of the fact that the guest species can be initially inserted by ion exchange.

Intercalation of TTF^+ species into MPS_3 (M=Mn, Fe,Cd)was achieved (41, 42) by treating a tetraethylammonium intercalate $Fe_{1-x}PS_3(Et_4N)_{2x}(solv)_y$ ($x \approx 0.14$) with a solution of $(TTF)_3(BF_4)_2$ in dry acetonitrile for 2 days at 50°C. The material is well crystallized and is stable in air over a long period. Its monoclinic unit cell is closely related to that of pristine $FePS_3$ (Cell Dimensions a= 5.919 Å, b = 10.348 Å, c = 12.642 Å, β = 107.45°). The interlamellar distance increases by ~ 5.65 Å upon intercalation, which strongly suggests that the TTF species stand "edge on" with respect to the slabs, with the C=C binary axis parallel to the layers, as in the FeOCl intercalates (43- 46). Elemental analysis of the intercalate leads to a formula $Fe_{0.82}PS_3(TTF)_{0.38}$ close to the scheme $Fe_{1-x}PS_3(TTF)_{2x}$ ($x \sim 0.18$).

The *d.c.* electrical conductivity σ of $Fe_{0.82}PS_3(TTF)_{0.38}$ along the plane of the layers was measured in the range 110 K - 370 K on a monocristalline platelet fitted with gold electrodes deposited by evaporation under vacuum. The conductivity is quite large ($\sigma \approx 3\ \Omega^{-1}\ cm^{-1}$ at 25°C) and is not thermally activated (Figure 6). Indeed, σ

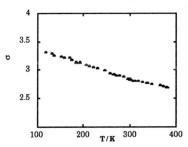

Figure 6. Temperature dependence of the conductivity σ of $Fe_{0.82}PS_3(TTF)_{0.38}$ (adapted from ref. 42)

increases as the temperature is lowered, a behaviour which indicates a metallic character .

The metallic properties of $Fe_{0.82}PS_3(TTF)_{0.38}$ contrast with the semiconducting properties of most MPS_3 intercalates, including $Mn_{0.83}PS_3(TTF)_{0.36}$ (47). The conductivity of the latter ($\sigma \approx 10^{-2} \Omega^{-1} cm^{-1}$ à 298 K) is still thermally activated (activation energy 0.17 eV). The metallic properties of the $Fe_{0.82}PS_3(TTF)_{0.38}$ intercalate are therefore certainly related to the ability of the intralayer Fe(II) cations to be partially oxidized by the guest tetrathiafulvalenium cationic species. Such partial charge transfer results in mixed valency on both the host and the guest sublattices. This interpretation is supported by ^{57}Fe Mossbauer spectra which provide evidence for the presence in $Fe_{0.82}PS_3(TTF)_{0.38}$ of about 14% of Fe(III) cations (Francis, A.H.and Dunham, R., The University of Michigan at Ann Arbor, unpublished data). Many other organic cations are likely to behave in a similar way as the TTF^+ species. It has been claimed that charge transfer also takes place in the $Mn_{0.83}PS_3(TTF)_{0.36}$ intercalate, although this compound remains semiconducting (48). We have recently inserted the BEDT-TTF monocation into $MnPS_3$, and the study of this material is currently in progress.

Optical Properties of MPS_3 Intercalates

Insertion of luminescent centers. Some members of the MPS_3 family are optically transparent or pale colored (M= Mn, Cd, Zn) and are therefore suitable for optical studies. Luminescent centers (Eu^{3+} ions, $Ru(bipy)_3^{2+}$) have been inserted into $MnPS_3$ and $CdPS_3$ following the cation exchange route described above. Photoluminesceence and photoexcitation studies have demonstrated the possibility for inserted Eu^{3+} ions to shift from the interlamellar space to the intralamellar space, and lattice photoluminescence was found to be activated by energy transfer from the intercalate (49). In contrast, photoexcitation spectra of the $Cd_{1-x}PS_3[Ru(bipy)_3]_{2x}$ intercalation compound have shown that energy can migrate from the host lattice towards the guest species (50).

Non Linear Optical Properties of MPS_3 Intercalates. Tremendous research has been directed over the last three decades towards the synthesis and processing of organic materials exhibiting second order nonlinear optical (NLO) properties (51, 52, 53) and many chromophores with large hyperpolarizability have been synthesized. However, the chromophores must be arranged in a noncentrosymmetric structure if the molecular hyperpolarizability (b) is to contribute to an observable bulk effect (χ^2). Various strategies have been reported to engineer molecules into acentric arrangements, in the realm of single crystals or following the well documented poled polymer approach (54) in which the polarizable chromophores are oriented by a strong electric field at temperatures above the glass transition of the films.

The fact that such promising materials did not fully reach the level of marketing is closely related to problems in their alignment stability over time, hence the need for alternative routes to permanently stabilize the polar order of chromophores. Several attempts have been already performed to use the incorporation of chromophores into organic (54-56) and inorganic (57-62) hosts so as to generate spontaneous poling. Efficiencies of several times that of urea were achieved even when neither the guest nor the host exhibited second harmonic generation. We have recently shown that intercalation dimethylamino-N-methyl stilbazolium cation ($DAMS^+$) into MPS_3 provides a new strategy to build up nanocomposites having very large NLO responses due to spontaneous poling of the inserted chromophores. The DAMS cation actually has been reported to exhibit one of the largest known efficiencies for second harmonic generation (63).

DAMS$^+$

We first found that DAMS$^+$ species could be successfully intercalated by ion exchange with a potassium preintercalate $M_{1-x}PS3 K_{2x}$ ($M = Mn^{II}$, Cd^{II}). However the resulting materials were found to give very small SHG responses, of the order of that of urea (64). This was quite disapointing, as DAMS chromophores can be a thousand times more efficient than urea. To increase the cristallinity of the materials, we targeted an optimized intercalation process in one step by direct reaction at 130°C of MPS3 powder with an ethanolic solution of DAMS iodide in the presence of pyridinium chloride. The role of pyridinium chloride is to generate *in situ* an intermediate pyridinium intercalate that undergoes rapid exchange with the DAMS species. Eventually, we synthezised two compounds of formula $Cd_{0.86}PS3(DAMS)_{0.28}$ and $Mn_{0.86}PS3(DAMS)_{0.28}$. These intercalates are very similar to the former ones but they are much better crystallized and do not contain any residual solvent.

The SHG properties were recorded by the Kurtz powder technique (using a 1.34µm laser beam as the irradiation source (Q-switched Nd^{3+}:YAG laser) (65). The manganese derivative exhibits an efficiency 300 times that of urea, and the cadmium derivative reaches 750 times that of urea (66), which is among the largest powder efficiencies ever reported for stilbazolium based chromophores.

Very efficient materials are also obtained if a $Cd_{1-x}PS3(pyridinium)_{2x}$ preintercalate is exchanged with an ethanolic solution of DAMS at 120°C.

The strong NLO effects recorded imply that the DAMS chromophores are inserted with their dipole mments fairly well aligned. It is not obvious why such spontaneous poling occurs along with intercalation. The structure of pure CdPS3 and MnPS3 is centrosymmetric and should not tend to generate acentric arrangement of the guest chromophores. However, as pointed out earlier in this paper, several features indicate that symmetry lowering accompanies intercalation. Recent Infrared and Raman studies have shown that the center of symetry of pristine MnPS3 is removed upon intercalating pyridine (22). Analysis of the luminescence spectra of intercalated rare earth cations show strong Stark effects due to very low local symmetry (67). Superstructures due to intralayer vacancy ordering have been evidenced. We also have recently obtained AFM images which bring some characterization on the atomic scale of the modified structure of a $Mn_{1-x}PS3(K)_{2x}$ intercalate (68). The most salient feature is that the PS3 pyramids are tilted along the same direction (Figure 7). As a result, the

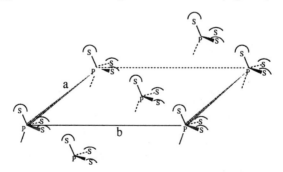

Figure 7. Scheme of a layer of the $Mn_{1-x}PS3(K)_{2x}$ intercalate according to our AFM study (68)

dipole moments of the PS3 pyramids must become tilted all along the same direction, so that they add up along this specific direction. This process might generate an "internal" electric field along a specific direction in the close vicinity of the layer. If the same modifications occur in the DAMS intercalate (no AFM image of this material has been obtained so far), it can be tentatively suggested that the PS_3 tilting might be responsible for poling the dipolar chromophores.

Annealing the samples at 140°C for several months (under vacuum) has no effect on their efficiency for second harmonic generation, and hence it seems that the main factor responsible for spontaneous poling has a thermodynamic origin rather than a kinetic one. This is a crucial point in the search for suitable SHG materials. However, in order to be appropriate in optical devices, such materials must be available as single crystals or thin films. It is too early to predict whether or not this goal will be achieved, but efforts are in progress towards processing the MPS_3 intercalates as thin films, taking advantage of the aqueous chemistry of the MPS_3 phases and of their ability to insert polymers such as polyethyleneoxide.

$Mn_{0.86}PS_3(DAMS)_{0.28}$: The First Magnet that Exhibits Strong SHG

Associating different types of physical properties in a single material and seeking synergy is certainly one of the important goals in the fields of molecular materials and hybrid organic-inorganic nanocomposites. The intercalation chemistry of the transparent $MnPS_3$ described in this article could possibly bring such an opprtunity: the host lattice may provide spontaneous magnetization once intercalated, the guest organic chromophores generate second harmonic radiation once oriented by the host lattice, and therefore it can be expected that the $Mn_{0.86}PS_3(DAMS)_{0.28}$ nanocomposite would act as a magnet able to generate strong SHG.

We have shown recently that this compound actually becomes a magnet below about 40 K (66). The magnetization at saturation strongly depends on the synthesis route. For example, starting with a pyridinium preintercalate eventually affords a DAMS intercalate with a magnetization around 400 emu / mole of manganese, whereas using a tétramethylammonium pre-intercalate gives a DAMS intercalate with a magnetization around 1000 emu / mole of manganese. Much work remains to be done on these nanocomposites, in particular to study whether the magnetic and NLO phenomena can be coupled.This composite, which appears to be the first material that possesses both strong spontaneous NLO properties and spontaneous magnetization up to a quite high temperature ($T_c \approx 40$ K), might be an interesting compound for physics. We note that a compound exhibiting weak NLO properties and magnetization at extremely low temperature was also recently reported (69).

Conclusion

The MPS_3 layered materials appear to open new directions in the science of intercalation and bring perspectives in the field of molecular materials, in particular to associate different properties in a single material. Cation exchange intercalation proves to have numerous consequences: high T_c insulating magnets, new organic molecular metals, photoluminescent materials, spontaneous poling of dipolar molecules, and non linear optics. The results described in this article show that intercalation can provide an alternative to other methods, such as sol-gel techniques (70) for the synthesis of hybrid organic-inorganic nanocomposites, bringing the advantage of a better crystallinity, a factor which can be of crucial importance when cooperative interactions are required. The ability of the MPS_3 to insert polymers still increases their potential chemical versatility and may help for developing processing of layered materials. Nanochemistry utilizing assemblies of molecules grafted on MPS_3 slabs might also give rise to multidisciplinary developments in the future.

Acknowledgement. The author is grateful to the MRT, CNRS (Ultimatec) and Conseil General de l'Essonne for their support. International collaboration was greatly assisted by awards from the NATO Research Programme and the British French Alliance Programme.

References

1. *Intercalation Chemistry*, Whittingham, M.S., Jacobson, A.J., Eds.; Academic Press, New York **1982**.
2. *Chemical Physics of Intercalation II*, Bernier, P., Fisher, J.E., Roth, S.; Solin, S.A., Eds., NATO ASI Series B, Plenum New York **1993**, Physics Vol.305.
3. *Chemical Physics of Intercalation* , Legrand, .A P.; Flandrois, S. Ed., NATO ASI Series B, Plenum New York,**1987** Physics Vol.172.
4. Klingen, W.; Ott, R.; Hahn, H. *Z. Anorg. Allg. Chem.* **1973** , *396*, 271.
5. Klingen, W.; Eulenberger, G.; Hahn, H. *Z. Anorg. Allg. Chem.* **1973**, *401*, 97.
6. Brec, R. *Solid State Ionics* **1986**, *22*, 3 and references therein.
7. O'Hare, D. in *Inorganic Materials*,   Bruce D.; O'Hare, D. eds, John Wiley NY1992. p165.
8. Whangbo, M.H.; Brec, R.; Ouvrard, G.; Rouxel, *Inorg. Chem..* **1985**, *24*, 2459.
9. Brec, R.; Scleidh, D.; Ouvrard, G.; Loiuisy,A.; Rouxel, J. *Inorg. Chem.* **1979**, *18*, 1814
10. Clément, R.; Green, M.L.H. *J. Chem. Soc., Dalton Trans.,* **1979**, 1566.
11. Le Méhauté, A.; Brec, R.; Ouvrard, G.; Rouxel, J. *Mat. Res. Bull.* **1977**, *12*, 1191. Thompson, A.H.; Whittingham, M.S. *Mat. Res. Bull.* **1977**, *12*, 741.
12. Clément, R. *J. Chem. Soc., Chem. Commun.* **1980**, 647.
13. Mathey, Y.; Clément, R.; .Sourisseau, C.; Lucazeau, G. *Inorg.Chem.*,**1982**, *6*, 2773.
14. Sourisseau, C.; Forgerit, J.P.; Mathey, Y. *J. Solid State Chem.*, **1983**, *49*, 134.
15. Clément, R. *J. Amer. Chem. Soc.* **1981**, *103*, 6998.
16. Clément, R.; Garnier, O.; Jegoudez, J. *Inorg. Chem.* **1986**, *25*, 1404.
17. Clément, R.; Michalowicz, A. *Rev. Chim. Min.* **1984**, *21*, 426.
18. Clément, R.; Doeuff, M.; Gledel, C. *J. Chim. Phys.* **1988**, *85*, 1053.
19. Foot, P.J.S.; Nevett, B.A. *J. Chem. Soc., Chem. Commun.* **1987**, 380
20. Cleary, D.A.; Groh, J.; Lifshitz, E.; Francis, A.H. *J. Phys. Chem.* **1988**, *92*, 551.
21. Read, J.A.; Chick, C., Francis, A.H. *J. Phys. Chem.* **1992**, *96*, 2010.
22. Joy, P.A.; Vasudevan, S. *J. Am. Chem. Soc.* **1992**, *114*, 7792.
23. Lagadic, I.; Léaustic, A.; Clément, R. *J. Chem. Soc., Chem. Commun.*,**1992**, 1396.
24. Shriver, D.F.; Ratner, M.A. *Chem. Rev.* **1988**, *88*, 109.
25. Ruiz - Hitzky, E.; Aranda, P. *Adv. Mater.* **1990**, *2*, 545.
26. Day, P. *Acc. Chem. Res.* **1979**, *12*, 236.
27. Miller, J.S.; Epstein, A.J.; Reiff, W.M. *Acc. Chem. Res.* **1988**, *21*, 114
28. Kahn, O.; Pei, Y.; Journaux, Y. in *Inorganic Materials*, edited by D. Bruce and D. O'Hare, John Wiley NY, **1992**. p 59.
29. Kahn, O.; *Molecular Magnetism*, VCH ed., New York, **1993**.
30. Le Flem, G.; Brec, R.; Ouvrard, G.; Loiuisy,A.; Segransan, P. *J. Phys. Chem. Solids*, **1982**, *43*, 455.
31. Clément, R.; Girerd, J.J.; Morgenstern , *I. Inorg. Chem.* **1980**,*19*, 2852.
32. Clément, R.; Audière, J.P.; Renard, J.P. *Rev. Chim. Miner.* **1982** ,*19*, 560.
33. Berthier, C.; Chabre, Y.; Minier, M. *Solid State Commun.* **1978**, *28*, 327.
34. Clément, R. et al.*in Inorganic and Organometallic Polymers with Specials Properties, NATO ASI,* Laine, R.M. Ed., Kluwer Acad. Publishers, 1992, p.115.
35. Clément, R.; Lagadic, I.;. Léaustic, A.; ,Audière, J.P., Lomas, L., in *Chemical Physics of Intercalation II , NATO ASI* , Bernier, P. et al Ed., Plenum Press, New York,1993, 315
36. Soled, S.; Wold, A. *Mater. Res. Bull.*, **1976**, *11*, 657.

37 Kurowasa, K.; Saito S.; Yamaguchi, Y. *J. Phys. Soc. Japan* **1983**, *52*, 3919.
38 Jernberg, P.; Bjarman, S.; Wappling, R. *J. Magn. & Magn. Mater.* **1984**, *46*, 178.
39 Wiedenmann, A.; Rossat-Mignod, J.; Louisy, A.; Brec, R.; Rouxel, J. *Solid State Commun.*, **1981**, *40*, 1067.
40 Clément, R.; Lomas, L.; Audière, J.P. *Chem. Mater.* **1990**, *2*, 641.
41 Lacroix, P.; Audière, JP.; Clément, R. *J. Chem. Soc., Chem. Commun.*, **1989**, 537.
42 Lomas, L.; Lacroix, P.; Audière, JP.; Clément, R. *J. Mater. Chem.* **1991**, *1*, 475.
43 Kauzlarich, S.M.; Teo, B.K.; Averill, B.A. *Inorg. Chem.*, **1986**, *25*, 1209.
44 Kauzlarich, S.M.; Stanton, J.L.; Fabre, J.; Averill, B.A. *J. Am. Chem. Soc.* **1986**, *108*, 7946.
45 Bringley, J.F.; Fabre, J.M.; Averill, B.A. *J. Am. Chem. Soc.*, **1990**, *112*, 4577.
46 Bringley, J.F.; Fabre, J.M.; Averill, B.A. *Chem. Mater.*, **1992**, *4*, 522.
47 Ichimura, K.; Miyazaki, T.; Matsuzaki, S.; Sano, M. *Mater. Sci. Forum* ., **1992**, *91-93*, 505.
48 Miyazaki, T.; Matsuzaki, S.; Ichimura, K.; Sano, M. *Solid.State Commun.* **1993**, *85*, 949.
49 Clément, R.; Léaustic, A.; Francis, A.H. *J Phys. Chem.* **1991**, *95*, 5405.
50 Lifshitz, E.; Clément, R.; Yu-Hallada, L.C.; Francis, A.H. *J. Phys. Chem. Solids* **1991**, *52*, 1081.
51 Williams, D.J. *Angew. Chem. Int. Ed. Engl.* **1984**, *23*, 690.
52 Eaton, D.F. *Science*, **1991**, *253*, 281 .
53 Zyss, J.; Ledoux, I.; Nicoud, J.F. in *Molecular non linear optics*, J. Zyss Ed. (Academic Press, Boston, **1993**).
54 Eich, M.; Bjorklund, G.C.; Yoon, D.Y.*Polymers Adv. Tech.* **1990**, *1*, 189.
55 Tomaru, S.; Zembutsu, S.; Kawachi, M.; Kobayashi, M. *J. Chem. Soc., Chem. Commun.* **1984**, 1207 .
56 Eaton, D.F.; Anderson, A.G.; Tam, W.; Wang, Y. *J. Am. Chem. Soc.* **1987**, *109*, 1886.
57 Tam, W. *et al.*, *Chem. Mater.* **1989**, *1*, 128 .
58 Weissbuch, I.; Lahav, M.; Leiserowitz, L.; Meredith, G.R.; Vanherzeele, H. *Chem. Mater.* **1989**, *1*, 114 .
59 Pécaut, J. ; Le Fur, Y.; Levy, J.P.; Masse, R. *J. Mater. Chem.* **1993**, *3*, 333 .
60 Masse, R.; Zyss, J. *Mol. Engineer..* **1991**, *1*, 141 .
61 Cox, S.D.; Gier, T.E.; Stucky, G.D. *Chem. Mater.* **1990**, *2*, 609 .
62 Kotler, Z.; Hieile, R.; Josse, D.; Zyss, J.; Masse, R. *J. Opt. Soc. Am .* **1992**, *B9*, 534
63 Marder, S.R.; Perry, J.W.; Schaefer, W.P. *Science* **1989,** *245*, 626 .
64 Lacroix, P.G.; Veret Lemarinier, A.V.; Clément, R.; Nakatani, K.; Delaire, J. *J. Mater. Chem.* **1993**, *3*, 499.
65 Kurtz , S.K.; Perry, T.T. *J. Appl. Phys.* **1968**, *39*, 3798 .
66 Lacroix, P.G.; Clément ,R.; Nakatani, K.; Zyss, J.; Ledoux, I. *Science* , **1994**, *263*, 658.
67 Clément, R.; Léaustic, A.; Marney, K.; Francis, A.H. *J. Phys. Chem. Solids*, **1994**, in press.
68 Lagadic, I. ; Clément, R. *Microsc. Microanal. Microstruct.* **1993**, *4*, 453.
69 Nicoud, J.F.; Serbutoviez, C.; Puccetti, G.; Ledoux, I.; Zyss, J.*Chem. Phys. Lett.* **1990**, *175*, 257.
70 Griesmar, P.; Sanchez, C.; Puccetti, G.; Ledoux, I.; Zyss, J. *Mol. Engineering* **1991**, *1*, 205 .

RECEIVED October 12, 1994

Chapter 5

Preparation and Characterization of Nanocomposites of Poly(ethylene oxide) with Layered Solids

John P. Lemmon[1], Jinghe Wu, and Michael M. Lerner

Department of Chemistry and Center for Advanced Materials Research, Oregon State University, Corvallis, OR 97331–4003

Single-phase nanocomposites containing montmorillonite, MoS_2, MoO_3 or TiS_2 with poly(ethylene oxide) are obtained by the exfoliation of the layered solid, adsorption of polymer, and subsequent precipitation of solid product. Aqueous solutions can be employed for all syntheses except PEO/TiS_2, which is prepared from $LiTiS_2$ in an N-methyl formamide (NMF) solution. X-ray diffraction indicates that the resulting solids increase in basal-plane repeat by approximately 4 or 8 Å, consistent with the incorporation of single or double layers of polymer between sheets. Reaction stoichiometry and elemental analyses provide compositions for the single-phase products, and thermal and electrical properties are examined.

The incorporation of poly(ethylene glycols) or poly(ethylene oxide) by smectite clays has been known for several decades (1). More recently, novel materials derived from polyethers and other low-dimensional solids, including oxide (2-5), chalcogenide (6-8), and MPS_3 (9) lattices have also been described. In each case, a significant increase in the interlayer spacing is observed, indicating that oligomers or polymers are incorporated within the galleries and that a new phase is produced. The appellation "nanocomposite" underlines the unusual architecture of materials which incorporate structural elements as diverse as polymer chains and two-dimensional lattices. Interest in the incorporation of polymers within inorganic hosts stems from the potential mechanical, structural, and electrical properties of organic/inorganic nanocomposites (10-13). This work will describe nanocomposites derived from PEO with a variety of layered structures, as summarized in Table I.

---

[1]Current address: Osram/Sylvania, Hawes Street, Tawanda, PA 18848

0097–6156/95/0585–0043$12.00/0

Methods for Generating Polymer-Containing Nanocomposites

Topotactic Methods. The intercalation of water or polar organic molecules into layered transition metal dichalcogenides has been widely studied since the 1970's (*14-18*). In general, layered ternary chalcogenides of the transition metals, A_xMCh_2 (A = alkali metal; M = Ti, Nb, Ta; Ch = S, Se, Te) spontaneously incorporate water or polar organic

Table I. Layered nanocomposites containing PEO

PEO Nanocomposite	Host	Basal Spacing (Å) Nanocomposite	Expansion
PEO/Na-montmorillonite	9.6	**17.7**[a]	**8.1**
		13.7	4.1
$Li_{0.1}(PEO)_{1.0-1.3}MoS_2$[b]	6.5	**14.5**	**8.0**
$M_{0.3}(PEO)_{0.8}TiS_2$	6.0	**14.2**	**8.2**
$Li_{0.3}(PEO)_{0.9}MoO_3$	8.1	**16.0**	**7.9**
		12.7	4.6
$Li_x(PEO)_yMoSe_2$	7.3	**15.2**	**7.9**

[a]Boldface indicates excess polymer employed in synthesis.
[b]Stoichiometry of PEO indicates moles of monomer repeat (C_2H_4O)

molecules to form stable compounds with solvated cations:

$$A_xMCh_2 \ + \ n\,L \ \rightarrow \ A_x(L)_nMCh_2 \tag{1}$$

Exchange reactions can also be utilized to expand the chemistry of these materials. An example is the incorporation of the oligoether poly(ethylene glycol) by exchange of water in the hydrated lithium salt:

$$A_x(H_2O)_yMS_2 \ + \ z\,PEG \ \rightarrow \ A_x(PEG)_zMS_2 \ + \ y\,H_2O \tag{2}$$

A topotactic mechanism for the above reactions, in which the intersheet galleries expand or contract but the solid retains its two-dimensional character should usually be ineffective for the introduction of high-molecular-weight polymers due to the slow diffusion of macromolecules into the galleries. It should be noted, however, that recent work utilizing elevated temperatures shows some promise for this method of obtaining nanocomposites (*5*).

An alternate route to polymer-containing nanocomposites involves the *in situ* polymerization of monomeric intercalants, and the preparation of materials such as poly(pyrrole) / FeOCl (*19*), poly(styrene) / MoS_2 (*20*), poly(aniline) / V_2O_5 (*21*), and poly(aniline) / MoO_3 (*22*) have been reported by this method. As a variant of this method, Nazar and coworkers (*23*) have reported the incorporation of a water-soluble precursor into MoO_3 by the exfoliation / adsorption method, followed by the *in situ* thermal conversion into poly(*p*-phenylenevinylene).

Exfoliation / Adsorption Method. Murphy, *et. al.* (*24*) and Lerf and Schöllhorn (*25*) observed that, under appropriate conditions, single-sheet colloids can be obtained by chemical oxidation of the lithiated metal disulfides. Morrison and co-workers (*20,26-28*), and other groups (*29*), have demonstrated that a colloidal suspension of single-sheet MoS_2 can incorporate molecular organics, organometallic complexes, or other complex cations when the single-sheets are restacked. The general method is simple: layered compounds are reduced (usually by chemical lithiation), rapidly re-oxidized by hydrolysis to form a stable colloid, and then interacted with a soluble polymer and the solid product precipitated from the solution.

PEO / Na-Montmorillonite Nanocomposites

Polymer composites with smectite clays are produced by the exfoliation / adsorption method, although no special techniques are required to prepare the colloid (which is directly obtained by the dissolution of the clay). X-ray diffraction data of the products obtained by the interaction of colloidal montmorillonite with PEO show that the montmorillonite galleries are expanded from the 9.6 Å basal plane spacing of anhydrous Na-montmorillonite. The diffraction indicates a regular lattice spacing of 17.7 Å, an expansion of the clay galleries by 8.1 Å.

Stoichiometry and Structure. The variation of polymer / clay stoichiometry is indicated in Figure 1. At polymer/montmorillonite ratios of 0.15 and 0.30 g/g, single-phase products exhibit relatively sharp diffraction peaks and several higher-order (00ℓ) reflections. Preferred orientation in the pressed pellets reduces the intensity of all but (00l) reflections; therefore, little useful information on the in-plane structure can be deduced from these data. Peak widths indicate ordered domains of 150 Å (approximately 10 layers) along the stacking direction for these products. Least-squares fits to these data yield lattice spacings of 13.73(4) Å and 17.65(4) Å. At stoichiometries richer in polymer than 0.30 g/g an admixture of the 17.7 Å phase and unreacted, crystalline PEO is obtained. The incorporation of polymer by the method employed is therefore limited to 0.30 g/g.

The polymer conformation in these nanocomposites raises significant fundamental issues and will also govern important properties such as ion transport and mechanical strength. Two models which may be considered are a helical conformation similar to that observed in crystalline PEO, or an adsorbed polymer layer on the clay surfaces. Although evidence suggesting a helical polymer conformation for these materials has been presented (*2*), the formation of a 13.6 Å phase involves a gallery expansion of 4.0 Å, which sterically limits the polymer conformation to approximate a single adsorbed layer between the clay surfaces. The 17.7 Å phase displays twice the gallery expansion and also twice the polymer content of 13.6 Å phase, and therefore is consistent with the incorporation of two such polymer layers. This architecture can be derived from the coalescence of colloidal clay sheets with single adsorbed polymer layers on both surfaces. The similarity of complexes prepared from PEO and an amorphous copolymer PEM, $[OCH_2(OCH_2CH_2)_m]_n$, is also significant in that the helical conformation is disrupted in PEM by methylene linkages in the polymer chain.

The 4 Å and 8 Å gallery expansions are also similar with those obtained when oligomeric (PEG) or small-molecule ethers such as ethylene glycol are incorporated into montmorillonite. This coincidence is reasonable if each of these species adopts an adsorbed-layer conformation within the galleries.

Nanocomposites prepared with stoichiometries between 0.15 and 0.30 g/g do not produce a sharp (001) peak, but show highly asymmetric peak profiles. When physical mixtures of the single-layer (0.15 g/g) and double-layer (0.30 g/g) phase are ground together at ambient temperature, two sharp peaks corresponding to the discreet phases are obtained, but a broad diffraction profile of intermediate repeat spacing appears upon annealing this mixture at 100 °C for two days. These patterns therefore appear to correspond to homogenous nanocomposites of intermediate composition (0.15 - 0.30 g/g) which are a solid solution of the single and double-layer phases, that is, they contain a homogenous distribution of these expanded galleries rather than a uniform basal plane spacing. Computer modelling studies indicate that the X-ray profiles of this interphase region are closely simulated by nanocomposite structures with ordered domains of approximately 100 Å containing both single and double polymer layers.

Thermal and Electrical Measurements. TGA and DTA traces indicate a single, irreversible, endothermic event at 360 °C corresponding to polymer decomposition. Samples maintained briefly above this temperature are discolored to brown or black, which is consistent with the expected carbonization of the organic component. The total weight loss recorded by 400 °C is 23 %, and corresponds to the initial stoichiometry of polymer in the nanocomposite. The melting transition of crystalline PEO (a strong endotherm at 67 °C), is absent in these profiles. The lack of either glass or melting transitions indicates the absence of a discrete polymer phase, and again demonstrates that these two-dimensional polymer layers are not structurally similar to the native polymers.

An Arrhenius plot for the nanocomposite is provided in Figure 2. An increase in conductivity of several orders of magnitude, relative to that of the native anhydrous Na-montmorillonite, is obtained. The highest values, near 10^{-5} Scm^{-1}, are less than those reported previously for PEO - salt complexes (which exhibit maximum conductivities of 10^{-3} - 10^{-4} Scm^{-1} [13]), and occur at higher temperatures. It should be noted that the conductivity may be anisotropic in these low-dimensional materials, and observed values may reflect dimensional averaging. The ionic conductivity is assumed to arise from an enhanced mobility for Na$^+$ within the polymer-containing galleries.

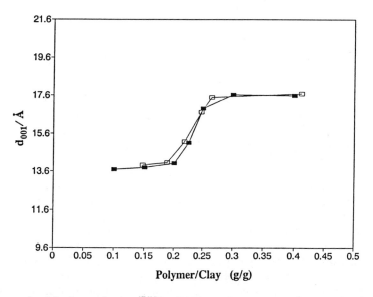

Figure 1. Maximum in the (001) reflection of nanocomposites vs PEO/Na-montmorillonite ratio (g/g) for (■) PEO and (□) PEM.

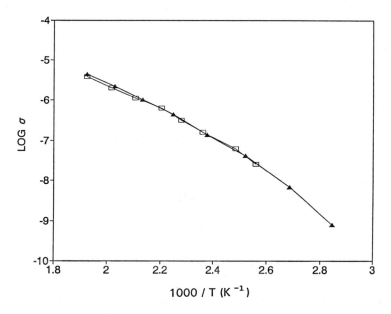

Figure 2. Arrhenius plot for PEO/Na-montmorillonite nanocomposite (0.30 g/g). Conductivities are reported as log [σ (Scm^{-1})]. Data were recorded for sample heating from 350 to 520 K (▲) and subsequent cooling (□). Time between readings is approximately 1 hour.

$Li_xPEO_yMoS_2$

The powder diffraction data from the nancomposites prepared via exfoliation / adsorption in aqueous media using excess PEO indicate single-phase products. Basal-repeat spacings obtained increase slightly with higher polymer content, and range from 14.2 to 14.6 Å. Peak widths indicate stacking coherence lengths in the range of approximately 200 - 400 Å (20 - 25 unit repeats). At low polymer stoichiometries, a broad peak associated with restacked MoS_2 appears.

The distance between Mo planes in these nanocomposites is 8.0 - 8.3 Å greater than in $LiMoS_2$. This expansion, similar to that obtained by insertion of poly(ethylene glycol) or PEO into MS_2 or montmorillonite, suggests that a polymer bilayer is incorporated into the intersheet region. There is no evidence of a greater expansion, indicating a phase containing more than a double layer of polymer.

The polymer conformation within the galleries cannot be determined directly from diffraction studies, and a helical polymer conformation, as in crystalline PEO, cannot be excluded by steric arguments. Since the favorable energy of polymer adsorption will decrease dramatically once monolayer coverage of the sheet surfaces is complete, a bilayer within the galleries (resulting from monolayer coverage of each sheet face) is a reasonable expectation from this exfoliation-adsorption method.

Elemental analyses for C, H, and Li for reactions indicate that the polymer stoichiometry for single-phase products can vary between 1.0 and 1.3 moles of monomer repeat to MoS_2. The polymer / MoS_2 ratio found in the product depends on the mixing stoichiometries utilized in the nanocomposite syntheses. The change in polymer content found for these products indicates a higher packing density of polymer within the disulfide galleries. This explanation is consistent with the reproducible change in relative peak intensities and slight increase in c-repeat distance with polymer content.

The overall stoichiometry was found to be close to $Li_{0.12}(PEO)_xMoS_2$ for samples with a range of x, which reflects a negative charge on MoS_2 sheets similar to that present in the hydrated lithium salts. The lithium stoichiometry is consistently similar to that obtained for hydrates prepared under similar conditions. Electrical measurements indicate a semimetallic conductivity and thermal response, which are again consistent with negatively-charged MoS_2 layers. The presence of hydroxide has also been suggested to account for the negative charge associated with the MoS_2 layers (28), but we have observed that similar colloids and restacked solids can be generated in non-aqueous solvents, polar organics such as DMSO.

Thermal Analyses. A simultaneous DTA / TGA scan of the nanocomposite is provided in Figure 3. The absence of a melting endotherm at 60 °C indicates that no crystalline PEO phase occurs in the product. An exotherm near 310 °C arises from degradation of the polymer and the associated weight loss of 27 %, due to volatilization of the decomposition products, closely corresponds to the polymer content within the nanocomposite. X-ray diffraction of the nanocomposite after heating to 450 °C shows only a restacked MoS_2 phase. For comparison, the decomposition of pure PEO is observed at 360 °C in an inert atmosphere.

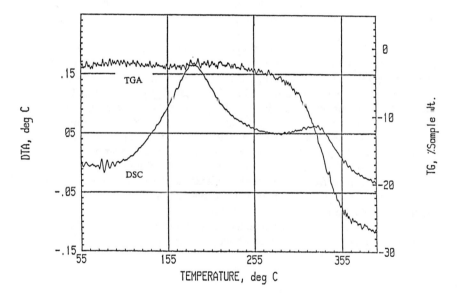

Figure 3. Simultaneous DTA / TGA scan obtained for $Li_{0.12}PEO_{1.34}MoS_2$.

An additional exotherm between 125 and 255 °C is not associated with any significant sample weight loss.   PEM / MoS_2 nanocomposites have also been examined, and show similar behavior.   In order to further explore this transition, ambient-temperature X-ray diffraction pattern of a single pellet of the nanocomposite following heating at 200 °C in an Ar atmosphere for 3 - 6 hrs show an irreversible loss of sample order.   After six hours, the material appears entirely disordered by X-ray diffraction.   Subsequent heating of the pellet above 400 °C under inert conditions for several hours to remove the polymer component regenerates crystalline MoS_2.   The disordered phase must therefore contain MoS_2 sheets without a coherent stacking arrangement.   The generation of the disordered state requires a large rearrangement in the structure.   The driving energy for this process may be provided by the thermally-activated rearrangement of adsorbed polymer.   The potentially high electroactive area for the disordered phase make it an attractive candidate for electrochemical applications.

Chromatography.   GPC data on polymer extracted from the nanocomposites indicates that the polymeric component of these materials is not degraded into low-molecular-weight fragments, although the results of crosslinking and scission are evident.   The extracted polymer from $Li_{0.12}PEO_{1.34}MoS_2$ (prepared with 95,000 Da PEO) contains significant molecular weight components centered at 450,000 and 20,000 Da.   After heat treatment to form the unstacked nanocomposite, the extracted polymer has a single peak centered at 45,000 Da.   In these samples, only 10 - 50% of the total polymer component can be extracted into an organic solution for analysis.   Low-molecular-weight and other highly-soluble products should be more readily desorbed from the disulfide surfaces, so the data obtained may not be representative of the polymer component of the sample.   As no low-molecular-weight fragments are observed, however, the gross degradation of PEO does not occur.   The details of polymer interaction with the disulfide lamella, and the mechanism of crosslink formation and bond cleavage, are of considerable interest and worthy of further study.

$M_xPEO_yTiS_2$

The exfoliation / adsorption process can probably be extended to other layered disulfides (TaS_2, WS_2, and NbS_2) or dichalcogenides.   For example, we have recently obtained results with $MoSe_2$ indicating single-phase nanocomposites are readily prepared by methods described above.

Some target compounds, however, are not amenable to processing in aqueous solution.   A notable example is TiS_2-containing nanocomposites; TiS_2 is susceptible to hydrolysis under acidic conditions, and forms a stable hydrate in aqueous base. The charge-storage properties of TiS_2, generally as the anode in a solid-state Li cell, naturally leads to interest in the preparation and characterization of a single-phase polymer/TiS_2 nanocomposite.

A recent communication notes that the addition of a solution of PEO with sodium perchlorate in acetonitrile to an aqueous colloid of TiS_2 will produce a nanocomposite (*7*). Our results indicate that the PEO/TiS$_2$ nanocomposite can indeed be obtained from aqueous media, but a single-phase product is not obtained.

Exfoliation in N-Methyl Formamide (NMF). The exfoliation of TiS$_2$ and other disulfides in NMF has been known for some time (*24,25*). Two phases can be obtained by the interaction of MTiS$_2$ with NMF; M$_x$(NMF)$_y$TiS$_2$ (M = Li, Na), with c-repeat $\approx$ 13.7 Å, and (RNH$_2$)$_x$(NMF)$_y$TiS$_2$, containing CH$_3$NH$_2$CHO$^+$ (RNH$_2$$^+$) cations, with c $\approx$ 19.5 Å. The relative abundance of the metal and RNH$_2$$^+$ - containing products is pH dependent, with the latter predominating when the reacton is carried out under acidic conditions. Similarly, the insertion of NMF has been reported for other lithium or sodium transition metal disulfides (*14*). Optimal conditions for the synthesis of PEO / TiS$_2$ are found to be a basic NMF solution using excess PEO (which can be easily washed away from the desired product). In this way, single-phase products can be reproducibly prepared. A representative diffraction pattern is provide in Figure 4.

The products obtained in this manner do not contain a large component of the molecular solvent within the galleries. Elemental analysis allow for a maximum ratio of NMF to monomer repeat near 0.1 mol/mol. Unlike the NMF-intercalated products, these nanocomposites do not lose a significant fraction of weight during DSC scans through the boiling point of NMF (185 °C).

The powder diffraction data for the single-phase products obtained provide a basal-plane repeat of 14.2 Å. Peak widths for the (00l) set provide a stacking coherence length of 300 Å (approximately 20 repeat units). The absence of a diffraction peak at d = 5.7 Å indicates that the product is devoid of unreacted TiS$_2$, and the absence of a melting exotherm at 60 °C in the DSC trace shows that no crystalline PEO phase occurs in the nanocomposite.

The distance between Ti planes in the nanocomposite is 8.0 - 8.2 Å greater than in LiTiS$_2$. This expansion suggests that a polymer bilayer has been incorporated into the intersheet region. Elemental analyses indicate a product stoichiometry of Li$_{0.02}$Na$_{0.25}$(PEO)$_{0.79}$TiS$_2$·0.09NMF; the notation (PEO)$_n$ refers to n moles of the monomer repeat unit (CH$_2$CH$_2$O). The overall cationic content is consistent with other similar compounds which retain the composition A$_x$(solv)$_y$TiS$_2$ (x $\approx$ 0.3), (*14,15*) and reflects the partial re-oxidation of the TiS$_2$$^{1-}$ sheets through interaction with solvents. The polymer content is somewhat less than that obtained with PEO / MoS$_2$ nanocomposites (*6*).

Thermal and Electrical Measurements. Thermal analysis of the nanocomposite reveals a broad endotherm between 90 and 140 °C. This endotherm is also observed with unreacted TiS$_2$. A sharper endotherm is observed at 180 °C and could be associated with the minor NMF component in the product. Both TiS$_2$ and the nanocomposite are unstable to loss of sulfur above 220 °C.

Electrical measurements on pellets pressed from powders indicate a semimetallic conductivity and thermal response; the conductivity of the nanocomposite is somewhat less than the pure TiS$_2$. The bulk conductivity for the partially reduced sheets, TiS$_2$$^{0.3-}$, is expected to increase significantly relative to TiS$_2$

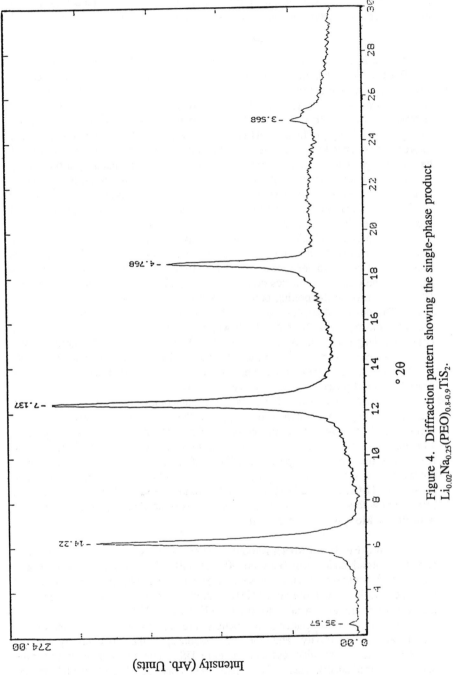

Figure 4. Diffraction pattern showing the single-phase product $Li_{0.02}Na_{0.25}(PEO)_{0.8-0.9}TiS_2$.

(*30*);   however, the reduced crystalline order in the nanocomposites and the incorporation of a large fraction of electronically insulating polymer between the sheets should decrease the observed conductivities of these pellets.

$Li_xPEO_yMoO_3$

Li_xMoO_3 rapidly exfoliates during sonication in aqueous solution, and the nanocomposite with PEO is easily obtained. The extent of lithiation in the Li_xMoO_3 employed (i.e. x in Li_xMoO_3) was not evaluated quantitatively in our studies, other studies indicate that the maximum value for x by these methods may be less than 0.5 (*31,32*). Complete exfoliation results only for the most highly reduced salts, when the anhydrous lithiated compounds show a basal-plane expansion of at least 1 Å relative to MoO_3. The nanocomposites prepared by the exfoliation/adsorption method expand by approximately 8 Å when excess PEO is employed: the stacking repeat of 8.2 Å in Li_xMoO_3 is increased to 15.9 - 16.5 Å. A 4.4 Å expansion is observed, however, when the PEO content is limited to 0.1 g/g Li_xMoO_3. These data are consistent with adsorbed bilayers or monolayers of PEO between MoO_3 sheets. The results obtained can be compared with the 5 - 6 Å expansion observed with inclusion of PPV or PAN, where aromatic rings are believed to be oriented perpendicular to the MoO_3 sheets (*33,34*).

 Elemental analysis of the products provides a Li/MoO_3 mole ratio of 0.25, indicating that a partial negative charge is retained by the sheets. The nanocomposite rapidly reforms a colloid when placed in aqueous solution, so the products cannot be washed to remove excess polymer. Some products therefore contain excess PEO; in this case scanning calorimetry of the products indicates a sharp exotherm near 60 °C corresponding to a bulk PEO melting transition. Single-phase products are obtained by careful control of the PEO / MoO_3 and solvent ratios. Elemental analyses for Li, C, and H in the single-phase products provides a stoichiometry of $Li_{0.2-0.3}PEO_{0.93}MoO_3 \cdot 0.06H_2O$. Thermal analyses of the products show an irreversible endotherm, associated with polymer decomposition, at 320 °C.

Acknowledgments. The authors gratefully acknowledge supporting grants from the National Science Foundation (DMR-9157005) and the AlliedSignal Foundation.

Literature Cited

1. *The Formation and Properties of Clay-Polymer Complexes*;   Theng, B. K. G., Ed.; Elsevier: New York, NY, 1979.

2. Aranda, P.; Ruiz-Hitsky, E. *Chem. Mater.* **1992**, *4*, 1395.

3. Wu, J.; Lerner, M. M. *Chem. Mater.* **1993**, *5*, 835.

4. Liu, Y.-J.; DeGroot, D.C.; Schindler, J.L.; Kannewurf, C.R.; Kanatzidis, M.G. *Chem. Mater.* **1991**, *3*, 992.

5. Vaia, R. A.; Ishii, H.; Giannelis, E. P. *Chem. Mater.* **1993**, *5*, 1694.

6. Lemmon, J.; Lerner, M. *Chem. Mater.* **1994**, *6*, 207.

7. Ruiz-Hitzky, E.; Jimenez, R.; Casal, B.; Manriquez, V.; Santa Ana, A.; Gonzalez, G. *Adv. Mater.* **1993**, *5*, 738.

8. Bissessur, R.; Kanatzidis, M. G.; Schindler, J. L.; Kannewurf, C. R. *J. Chem. Soc., Chem. Commun.* **1993**, 1582.

9. Lagadic, I.; Léaustic, A.; Clément R. *J. Chem. Soc., Chem. Commun.* **1992**, 1396.

10. Stucky, G. D. In *Progress in Inorganic Chemistry;* Lippert, S. J., Ed.; Wiley & Sons Inc.: New York, NY, 1992; Vol. 40.

11. Ozin, G. A. *Adv. Mater.* **1992**, *4*, 613.

12. Komarneni, S. *J. Mater. Chem.* **1992**, *2*, 1219.

13. Stein, A.; Keller, S. W.; Mallouk, T. E. *Science* **1993**, *259*, 1558.

14. Schöllhorn, R. *Angew. Chem. Int. Engl. Ed.* **1980**, *19*, 983.

15. *Intercalated Layered Materials;* Levy, F. Ed.; D. Reidel Publ.: Boston, MA, 1979.

16. Gamble, F. R.; DiSalvo, F. J.; Klemm, R. A.; Geballe, T. H. *Science* **1970**, *168*, 568.

17. Whittingham, M. S. *Mat. Res. Bull.* **1974**, *9*, 1681.

18. Murphy, D.W.; Hull, G.W. *J. Chem. Phys.* **1975**, *62*, 973.

19. Kanatzidis, M. G.; Tonge, L. M.; Marks, T. J.; Marcy, H. O.; Kannewurf, C. R., *J. Am. Chem. Soc.* **1987**, *109*, 3797.

20. Divigalpitiya, W. M.; Frindt, R. F.; Morrison, S. R. *J. Mater. Res.* **1991**, *6*, 1103.

21. Liu, Y.-J.; DeGroot, D.C.; Schindler, J.L.; Kannewurf, C.R.; Kanatzidis, M.G. *J. Chem. Soc., Chem. Commun.* **1993**, 593.

22. Bissessur, R.; DeGroot, D.; Schindler, J.; Kannewurf, C.; Kanatzidis, M. *J. Chem. Soc., Chem. Commun.* **1993**, 687.

23. Nazar, L. F.; Zhang, Z.; Zinkweg, D. *J. Am. Chem. Soc.*, **1992**, *114*, 6239.

24. Murphy, D. W.; DiSalvo, F. J.; Hull, G. W.; Waszczak, J. V. *Inorg. Chem.* **1976**, *15*, 17.

25. Lerf, A.; Schollhorn, R. *Inorg. Chem.* **1977**, *16*, 2950.

26. Divigalpitiya, W. M. R.; Frindt, R. F.; Morrison, S. R. *Science* **1989**, *246*, 369.

27. Gee, M. A.; Frindt, R. F.; Joensen, P.; Morrison, S. R. *Mat. Res. Bull.* **1986**, *21*, 543.

28. Miremadi, B. K.; Cowan, T.; Morrison, S. R. *J. Appl. Phys.* **1991**, *69*, 6373.

29. Tagaya, H.; Hashimoto, T.; Karasu, M.; Izumi, T.; Chiba, K. *Chem. Soc. Jap., Chem. Lett.* **1991**, 2113.

30. Uchida, T.; Kohiro, K.; Hinode, H.; Wakihara, M.; Taniguchi, M. *Mat. Res. Bull.* **1987**, *22*, 935.

31. Kuhlmann, R.; Schollhorn, R. *Mat. Res. Bull.* **1976**, *11*, 83.

32. Thomas, D.; McCarron, E. *Mat. Res. Bull.* **1986**, *21*, 945.

33. Nazar, L. F.; Liblong, S. W.; Yin, X. T. *J. Am. Chem. Soc.* **1991**, *113*, 5889.

34. Bissessur, R.; DeGroot, D. C.; Schindler, J. L.; Kannewurf, C. R.; Kanatzidis, M. G.; *J. Chem. Soc., Chem. Commun.* **1988**, 223.

RECEIVED October 15, 1994

Chapter 6

Polymer−Clay Hybrids

Akane Okada, Arimitsu Usuki, Toshio Kurauchi, and Osami Kamigaito

Toyota Central Research and Development Laboratories, Inc., Nagakute, Aichi 480−11, Japan

Polymer-clay hybrids, organic-inorganic molecular composites, were prepared. X-ray and TEM measurements revealed that each 10 Å template of clay mineral is dispersed in the polymer matrix and that the repeat unit increased from 12 Å in unintercalated material to more than 200 Å in the intercalated material. Thus, polymer-clay hybrids are "polymer based molecular composites" or "nanometer composites". They, when molded, show excellent mechanical properties compared to unfilled polymers and/or conventional composites. In the hybrids negatively charged silicate (clay mineral) and positively charged polymer-ends are directly bonded through ionic bonds. The mechanism of reinforcement is discussed with the results of CP-MAS NMR and pulsed NMR studies.

Polymer has been successfully reinforced by glass fiber or other inorganic materials. In these reinforced composites, the polymer and additives are not homogeneously dispersed on the microscopic level. If the dispersion could be achieved on the microscopic level, the mechanical properties would be expected to be further improved and/or new unexpected features might appear(1). Clay mineral is a potential candidate for the additive since it is composed of layered silicates, 10 Å thick, and undergoes intercalation with organic molecules(2). The lack of affinity between hydrophylic silicate and hydrophobic polymers makes it difficult to get homogeneously miscible with each other. Swelling of each template of silicate with organic molecules is a matter of vital importance to reach to this type of molecular composite

In this paper we present two polymer-clay hybrids; nylon 6-clay hybrid and rubber-clay hybrid. Swelling behavior, preparation and properties of hybrids, and interaction of the organic-inorganic surfaces are discussed.

Nylon 6-Clay Hybrid (NCH)

Nylon 6 (polycaploractam) has good mechanical properties and is a commonly used engineering polymer(3). We tried to prepare NCH by blending commercial nylon 6 and montmorillonite, a common clay mineral, in a twin screw extruder, which gave just a phase separated, conventional nylon 6-clay composite (termed as NCC). So, we tried polymerization of ε-caprolactam in the interlayer space of montmorillonite to disperse each template of silicate into nylon 6 matrix on the molecular level. Swelling of silicate by ε-caprolactam is of key importance. We found that montmorillonite ion-exchanged with 12-aminolauric acid can be swollen by ε-caprolactam to fulfill our purpose.

0097−6156/95/0585−0055$12.00/0

In the following, swelling behavior, preparation and properties of NCH and the mechanism of reinforcement are presented. NCH has been used in an automotive part and has been manufactured in a large scale.

Swelling   Behavior   of   Montmorillonite   with   ω-Amino   acid   by ε-Caprolactam(4).   Natural Na-montmorillonite is hydrophylic and not compatible with most organic molecules. Sodium cation in the interlayer space of montmorillonite can be exchanged with organic cations to yield organophilic montmorillonite. For the present purpose, polymerization in the interlayer space, ammonium cations of ω-amino acids were chosen as cations since they catalyze ring opening polymerization of ε-caprolactam.

In a 1000 mL beaker were placed 24mmol of ω-amino acid, 2.4 mL of concentrated hydrochloric acid and 200 mL of water at 80°C. The solution of the ω-amino acid was added into a dispersion composed of 10 g of montmorillonite and 1,000 mL of hot water, and this mixture was stirred vigorously for 10 min, giving a white precipitate. The product was filtered, washed with hot water, and freeze-dried. In this paper, we call the cation exchanged montmorillonites "n-montmorillonite", where n is the carbon number of ω-amino acid. Mixture of 0.5 g of the n-montmorillonite powder and 2.0 g of ε-caprolactam (mp=70 °C)was heated at 100 °C for swelling.

The degree of swelling was studied by means of X-ray powder diffraction (XRD) measurement using a Rigaku RAD-B diffractometer. Figure 1 shows XRD patterns of n-montmorillonites. The basal spacings (interlayer distance) of the samples were obtained from the peak position of XRD pattern as shown in Table I. The XRDs of the mixtures of the n-montmorillonites and ε-caprolactam were measured at 25 °C and 100 °C. They suggest that swelling did occur. Basal spacings of the swollen n-montmorillonites are also shown in Table I. The spacings of the specimens were equal at 25 °C and 100 °C for the n-montmorillonites, when n was less than 8. They corresponded with the sum of the molecular length and 10Å (template) at 25 °C. However, they exceeded the sum at 100 °C for longer n-montmorillonite. The schematic diagram is shown in Figure 2. For better swelling of ω-amino acid, n should be larger than 11. We chose 12-aminolauric acid to prepare NCH since it is the most available among the longer acids .

Table I. Basal Spacings of n-Montmorillonite in the Presence of ε-Caprolactam

ω-Amino Acid NH$_2$(CH$_2$)$_{n-1}$COOH		Spacing (Å)	Spacing in Caprolactam (Å)
n	Molecular length(Å)	25°C	100°C
2	6.7	12.7	14.3	14.4
3	8.1	13.1	19.3	19.7
4	9.8	13.2	19.3	19.9
5	11.0	13.2	20.3	20.4
6	12.2	13.2	23.3	23.4
8	14.7	13.4	26.2	26.4
11	18.5	17.4	30.2	35.7
12	19.7	17.2	31.5	38.7
18	27.3	28.2	43.8	71.2

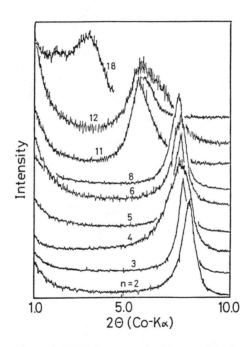

Figure 1. XRD Patterns of n-Montmorillonite

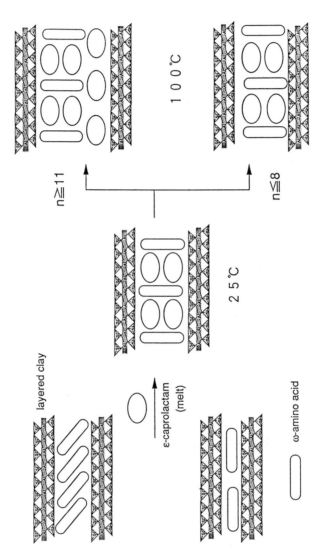

Figure 2.　Schematic Diagram of Intercalation of ε-Caprolactam

Preparation of NCH(5). In a vessel, 113 g of ε-caprolactam and 5.97 g of 12-montmorillonite were placed. The mixture was heated at 100 °C for 30 min. Then, it was heated at 250 °C for 48 h, yielding a polymeric product. After cooling, the product was mechanically crushed. The fine particles were washed with 2 L of water at 80 °C for 1 h. Thus we obtained NCH. Nylon 6 clay composites (termed as NCC) were prepared by blending commercial nylon 6 and montmorillonite in an extruder for comparison with NCH. These materials were injection-molded for various measurements. Basal spacing, *d*, was directly obtained in XRD. Figure 3 shows the transmission electron micrograph (TEM) of the section of molded NCH-15 measured on a JEOL-200CX TEM applying an acceleration voltage of 200V. The suffix of NCH means the amount of 12-montmorillonite used in polymerization. The dark lines are the intersection of the sheet silicate of 10 Å thickness and the spaces between the dark lines are interlayer spaces. Table II shows the basal spacing, *d*, obtained by XRD and TEM. The *d* values agree very well. It was found to be inversely proportional to the montmorillonite content. A maximum *d* of 214 Å was observed. The thickness of a layer of silicates is about 10 Å. This is of the order of molecular size and this layer can be thought to be an "inorganic macromolecule", so that, in NCH, the polymer and montmorillonite are mixed on the molecular level forming a "polymer based molecular composite". On the other hand, *d* in the NCC was 12 Å and it is unchanged from the pristine montmorillonite and therefore NCC is not a molecular composite.

Table II. Basal Spacings of NCHs

Specimen	Montmorillonite (wt %)	Spacing (X-ray) (Å)	Spacing(TEM) (Å)
NCH-5	4.2	150	214
NCH-10	9.0	121	115
NCH-15	14.5	64	62
NCH-30	25.0	51	50
NCC-5	5.0	12	

Properties(6). Mechanical properties of NCH-5 are shown in Table III together with nylon 6 and NCC-5 following ASTM. The tensile strength and tensile modulus of NCH were superior to others. The impact strength of NCH was identical with that of nylon 6. The most prominent effect was observed in heat distortion temperature (HDT). HDT of NCH-5 containing only 4 wt% of montmorillonite was 152 °C, which was 87 °C higher than that of nylon 6. This effect in NCH is attributed to drastic improvement in the quality of nylon 6. Resistance to water was also improved(7). The rate of water absorption in NCH was lowered by 40 % as compared to nylon 6 and NCC.

Table III. Properties of NCH-5(1)

Specimen	Monmorillonite (wt %)	Tensile Strength (MPa)	Tensile Modulus (GPa)	Charpy Impact Strength (KJ / m^2)
NCH-5	4.2	107	2.1	6.1
NCC-5	5.0	61	1.0	5.9
nylon 6	0	69	1.1	6.2

The molded specimen was found to be anisotropic. The coefficient of thermal expansion of NCH-5 in the flow direction was lower than half of that in the perpendicular direction. Nylon 6 was isotropic and NCC was intermediate. TEM studies revealed that sheets of silicate were parallel to the flow direction of the mold. The nylon molecules in NCH-5 were also oriented in the same direction. It seems that anisotropy of the thermal expansion results from the orientations of silicate and polymer chains.

Table III (continued). Properties of NCH-5(2)

Specimen	HDT at 18.5 kg/cm^2 ($°$C)	Rate of Water Absorption 23°C, 1 day (%)	Coefficient of Thermal Expansion	
			Flow Direction	Pependicular Direction
			(cm/cm $°$C x 10^5)	
NCH-5	152	0.51	6.3	13.1
NCC-5	89	0.90	10.3	13.4
nylon 6	65	0.87	11.7	11.8

Bond Character of Organic and Inorganic Surface. Excellent properties in NCH can be considered to have origin in an enormous surface area and ionic bonds between the organic polymer and inorganic silicate. Nylon 6 molecules have one $-NH_2$ and one $-COOH$ end group, so titration of nylon 6 with hydrochloric acid and NaOH aqueous solution gives the same value of concentration of both end groups. On the contrary, the titration study of NCH revealed that the concentration of $-COOH$ ends was much higher than that of $-NH_2$ ends. The results strongly indicate that substantial amount of nylon molecules in NCH are ion-bonded to silicate at $-NH_2$ end. Each silicate works as a crosslinker of them and restricts motion of them. In anticipation of better NCH, other clay minerals were examined such as synthetic mica, saponite and hectorite. The mechanical properties of new hybrids are shown in Table IV. Contrary to the expectation, among four clays, montmorillonite is the most effective on improvement of the properties. In order to evaluate the strength of the ionic bond at the interface between silicate and organic moiety, we studied ^{15}N CP-MAS NMR spectroscopy using a Brucker MSL 300 WB spectrometer operating 30.41 Hz and a magnetic field strength of 7.0 T. Nylon 6 in NCH has more than 100 mers, so that the terminal (surface) nitrogen is less than 1 % of the total nitrogens. We had to choose a simpler organic compound as a model of ion-bonding nylon 6.

Table IV. Mechanical Properties of NCHs

Specimen	Clay Mineral	Tensile Strength at 120 °C (MPa)	Tensile Modulus at 120 °C (GPa)	HDT at 18.5 kg/cm^2 (°C)
NCH-5	montmorillonite	32.3	0.61	152
NCHM-5	synthetic mica	30.2	0.52	145
NCHP-5	saponite	29.0	0.29	93
NCHH-5	hectorite	26.4	0.29	93
nylon 6	-	26.6	0.19	65

Thus, we prepared intercalation compounds of clay minerals and glycine since it is the simplest amino acid and its isotope-enriched reagent is commercially available. Hexamethylenediamine was chosen as a model of non-bonding nylon 6. The ^{15}N chemical shifts are summarized in Table V. As the positive charge on nitrogen increases, its chemical shift goes downfield. The results revealed that all glycines in interlayer space of minerals are partially ionized, and that the order of positive charge density (i.e., the order of ionic bond strength) agrees with that of mechanical properties in Table IV. It is clarified that the chemical bond between the inorganic silicate and organic polymers on the molecular level causes improvement of macroscopic mechanical properties of materials.

Table V. ^{15}N-NMR Chemical Shift

Specimen	^{15}N-NMR Chemical Shift (ppm)
$Cl\text{-}N^+H_3CH_2COOH$	15.6
Montmorillonite-$N^+H_3CH_2COOH$	11.2
Mica-$N^+H_2CH_2COOH$	9.4
Saponite-$N^+H_3CH_2COOH$	8.4
Hectrite-$N^+H_3CH_2COOH$	8.3
$NH_2(CH_2)_6NH_2$	7.0

Application. The timing belt cover of automotive engines are usually made of glass fiber reinforced nylon or polypropylene. Since NCH has a high modulus and a high distortion temperature as mentioned above, we tried to make it by injection-molding as shown in Figure 4. The belt cover showed good rigidity, excellent thermal stability and no warp. Also, weight saving has reached up to 25 % due to the small content of inorganic material in NCH compared with that in current thermoplastic composites[8]. It has already been in practical use since 1989. NCH is the first example of hybrid organic-inorganic composite that has been manufactured in large quantities.

Rubber-Clay Hybrid (RCH)

Nylon forms hydrogen bonding and has more affinity to inorganic molecules. On the contrary, rubber is more hydrophobic. So, it is interesting to examine validity of the method for rubber. Vulcanized rubbers are not usually used as pure rubbers but are reinforced by fillers to improve mechanical properties. Carbon blacks are typical reinforcing fillers for such purposes and inorganic minerals are used with some limitation. Although carbon blacks are excellent in reinforcement due to strong interaction with rubbers, they often decrease the processability of rubber compounds on account of high viscosity at high volume loading. On the other hand, minerals have a variety of shapes suitable for reinforcement such as needles and sheets but they have only poor interaction with rubbers[9]. We prepared montmorillonite ion-exchanged with a liquid rubber (LR), termed as LR-montmorillonite for usage of its favorable shape. Co-vulcanization of nitrile rubber was done with the LR-montmorillonite, giving a molecular composite, a rubber-clay hybrid termed as RCH.

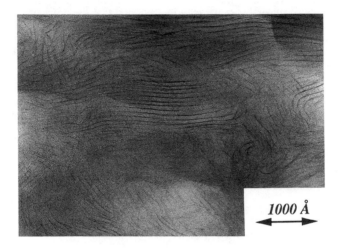

Figure 3.   TEM of the Section of Molded NCH-15

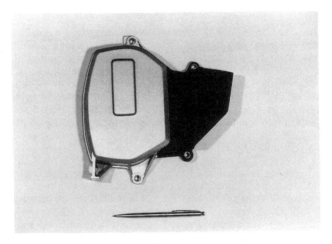

Figure 4.   Timing Belt Cover of Injection-Molded NCH

Preparation of RCH(*10*). Liquid polybutadiene, Hycar ATBN 1300*16 was supplied by Ube Industries LTD., which is a copolymer of butadiene and acrylonitrile (acrylonitrile (AN) content; 17%), and has two amino end groups and molecular weight of 3400. Nitrile rubber(NBR), Nipol 1042 (AN content; 33%) was supplied by Nippon Zeon Co.,LTD.

In a mixture of 30 mL of N,N-dimethylsulfoxide and 30 mL of ethanol, 9.2 g of ATBN was dissolved, and 5.4 mL of 1N hydrochloric acid was added. With an electric mixer, 4.2 g of montmorillonite was dispersed in 400mL of water, and the above ATBN solution was dropped to the dispersion. A rapid reaction occurred and gave a precipitate of LR-montmorillonite. The product was washed with 200 mL of ethanol and dried in vacuo at 60 °C.

By roll milling, 900 g of NBR was blended with 150 g of the above LR-montmorillonite, 15 g of sulfur, 30 g of zinc oxide and other additives. Vulcanized rubber sheets of 2 mm thickness were prepared by compression-molding at 160 °C for 15 min. The product is termed as RCH-5, the suffix of which means the content of mineral, parts per hundred of rubber (phr). For control experiments, NBR was blended with untreated clay mineral, giving a rubber-clay composite termed as RCC and with a conventional carbon filler, SRF;ASTM N770.

Properties of RCH. Properties are described for RCH before and after vulcanization.

Unvulcanized Rubbers. Mooney viscosity of unvulcanized rubber is an index of processability of unvulcanized rubber systems, and it was measured using a Shimazu Seisakusho Mooney Viscometer MV 200. The viscosity of carbon filled system (NBR-C) increases rapidly with the increase of the content of the filler, causing the lowering of processability. Surprisingly, the viscosity of the LR-montmorillonite system (RCH) decreases with the clay mineral content and was lower than even that of unfilled system (NBR). Thus, montmorillonite, if highly loaded, doesn't inhibit processability in sharp contrast with carbon black as summarized in Table VI.

The rate of cross-linking reaction was measured by a JSR curemeter. In this measurement, t-90 (min), time of 90 % curing, shows an index of the reactivity. The reaction rates of LR-montmorillonite systems were close to those of the carbon filled systems while they were lower than that of the unfilled system as shown in Table VI. The reaction of the clay composite system (RCC) retarded so remarkably that this system is not a practical use, maybe owing to reaction inhibition on mineral surface. Thus, this new system can improve processability.

Table VI. Properties of Unvulcanized RCH

Specimen	Filler	Content (phr)	Mooney Viscosity at 100 °C	t-90 (min)
RCH-5	montmorillonite	5	53	15.0
RCH-10	montmorillonite	10	47	16.0
RCC-10	montmorillonite	10	58	23.5
NBR-C20	carbon	20	64	14.0
NBR-C40	carbon	40	75	14.5
NBR	—	0	57	11.0

Vulcanized Rubbers. Transmission electron micrographic study indicates that sheets of silicate are dispersed in rubber on the molecular level. The storage modulus (E') was obtained at 10 Hz between −150 °C and 150 °C using an Iwamoto

Seisakusho VES-F Viscoelastometer. The E' values at 25 °C of RCH were shown in Table VII. In RCH, montmorillonite reinforces four times as much as carbon when added to the rubber as a filler. Thus the E' value of RCH-10 is almost identical with that of NBR-C40. Similar results were obtained also at 100 °C.

The tensile stress at 100% at 25 °C of RCH-10, containing 10 phr of clay mineral, is equal to that of the rubber containing 40 phr of carbon black. However, clay mineral doesn't have any effect of reinforcement in RCC, a conventional composite. Swelling of NBR reaches to equilibrium in benzene at 25 °C after 48h. The swelling ratio is defined as swollen volume of rubber matrix to unswollen volume of rubber matrix. Here, the degree of swelling of RCH-10 is identical with that of NBR-C40, too. The properties of vulcanized rubbers tested are summarized in Table VII. Thus the polymer in the vicinity of the silicate is restricted in mobility by the silicate so that the polymer is effectively reinforced and also restricted in swelling.

Table VII. Properties of Vulcanized RCH

Specimen	E' at 25 °C (MPa)	Stress at 100% Strain at 25 °C(MPa)	Swelling Ratio in Benzene
RCH-5	5.4	1.5	4.2
RCH-10	8.8	2.1	4.1
RCC-10	4.2	1.1	4.7
NBR-C20	5.6	1.4	4.4
NBR-C40	8.9	2.1	4.0
NBR	3.3	1.0	4.9

Mechanism of Reinforcement. These favorable properties of RCH (Table VII) can be considered to have origin in an enormous surface area and ionic bonds between the organic polymers and inorganic silicate sheets as in the case of NCH. If such strong interaction exists in rubber systems, one can expect some increase of fast component of proton spin-spin relaxation time, T_2. T_2 measurements were performed on a JEOL FSE 60Q broadline pulsed NMR spectrometer operating at 60 Hz for protons using the solid echo and CPMG (Carr-Purcell-Meiboon-Gill) pulse sequences. Figure 5 shows the T_2 signals for both liquid rubber (LR) and LR-montmorillonite. The T_2 signal for LR-montmorillonite was actually resolved into a fast decaying rigid component and a slowly decaying rubbery component (corresponding to the rubbery molecules in mobility). The rigid component which cannot be observed in LR is assigned to the motionally constrained regions of LR molecules in LR-montmorillonite through the rubber-filler interactions, and its fraction can be evaluated to be about 20%, which is equal to the fraction of tightly bound component in inextractable rubber-filler gel, so called "bound rubber", for carbon black filled systems(11). There is now general agreement that "bound rubber" is an important factor in reinforcement, depending on the characteristics of the filler particles such as particle size, surface area, structure, and surface activity. Thus, it is considered that there appears to be the strong rubber-filler interactions (comparable to those in carbon black filled systems) in LR-montmorillonite, in which negatively charged silicate layers are bonded to LR-molecules with positively charged terminal sites, forming "bound rubber".

Conclusion

Polymer-clay hybrids, organic-inorganic molecular composites, were prepared. X-ray and TEM measurements revealed that each 10 Å thick template of clay mineral was

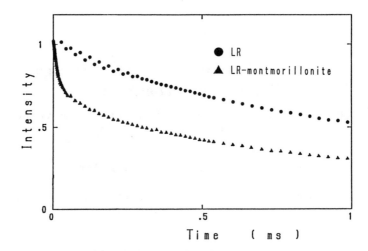

Figure 5. T$_2$ Signal for LR-Montmorillonite

dispersed in polymer matrix and that the repeat unit increased from 12 Å in unintercalated materials to more than 150 Å in the intercalated materials. Thus, polymer-clay hybrids are real molecular composites or nanocomposites. They, when molded, show excellent mechanical properties compared with unfilled polymers and/or conventional composites. In the hybrids, negatively charged silicates (clay mineral) and positively charged polymer ends are directly bonded through ionic bonds. The mechanism of reinforcement was explained using the results of ^{15}N CP-MAS and proton pulsed NMR studies.

Literatures Cited

1. Giannelis, E. P. *J.Minerals, Metals & Materials Soc.* **1992**, March, 28.
2. Jordan, J. M. *J. Phys. Colloid Chem.* **1950**, *53*, 245.
3. *Nylon Plastics;* Kohan, M. I.,Ed.; Interscience: New York, NY, 1973.
4. Usuki, A.; Kojima,Y.; Kawasumi, M.; Okada, A.; Kurauchi, T.; Kamigaito, O. *J. Mater. Res.* **1993**, *8*, 1174.
5. Usuki, A.; Kojima,Y.; Kawasumi, M.; Okada, A.; Kurauchi, T.; Fukushima, Y.; Kamigaito, O. *J. Mater. Res.* **1993**, *8*, 1179.
6. Kojima,Y.; Usuki, A.; Kawasumi, M.; Okada, A.; Kurauchi, T.; Fukushima, Y.; Kamigaito, O. *J. Mater. Res.* **1993**, *8*, 1185.
7. Okada, A.; Kawasumi, M.; Kurauchi, T.; Kamigaito,O. *Polym. Prepr.* **1987**, *28*, 447.
8. Kurauchi, T.; Okada, A.; Nomura, T.; Nishio,T.; Saegusa, S.; Deguchi, R. *SAE Paper Series 910584;* Soc. Automotive Engineers: Warrendale, PA, 1991.
9. Dannenberg, E. M. *Rubber Chem. Technol.* **1986**,*59*, 512.
10. Okada, A.; Fukumori, K.; Usuki, A.; Kojima, Y.; Kurauchi,T.; Kamigaito,O. *Polym. Prepr.* **1991**, *32*, 540; Fukumori, K.; Usuki, A.; Sato, N.; Okada, A.; Kurauchi, T. *Proc. 2nd Japan International SAMPE Symposium,* **1991**, 89.
11. Nishi, T. *J. polym. Sci. Polym. Phys. Ed.* **1974**, *12*, 685.

RECEIVED August 3, 1994

Chapter 7

(Perfluorosulfonate Ionomer)–(Inorganic Oxide) Nanocomposites

Organic Modification of Surfaces of Silicon Oxide Nanoparticles Grown In Situ

Q. Deng, K. A. Mauritz[1], and R. B. Moore

Department of Polymer Science, University of Southern Mississippi, Hattiesburg, MS 39406–0076

$SiO_{2[1-x/4]}(OH)_x$ nanoparticles were generated via *in situ* sol-gel reactions for tetraethoxysilane within polar clusters of perfluorosulfonic acid films. Then, residual SiOH groups on these nanoparticles were post-reacted with diethoxydimethylsilane (DEDMS) and ethoxytrimethylsilane (ETMS), resulting in organically-"shelled" and interknitted, and simply organically "shelled" nanoparticles, respectively. FT-IR and ^{29}Si solid state NMR spectroscopies established structural incorporation of the di- and tri-methylsilanes onto silicon oxide "cores" as well as degree of molecular connectivity within the inorganic phase. Post-reaction with DEDMS rendered the structure more linear. Mechanical tensile experiments indicated a ductile $\rightarrow$ brittle transformation upon DEDMS post-reaction, suggesting linking of silicon oxide nanoparticles. On the other hand, a measure of ductility persisted after ETMS post-reaction. Tensile strength was increased after both post-reaction schemes.

In earlier experiments, we utilized the polar clusters within perfluorosulfonate ionomer (PFSI) films as nanometers-in-scale reaction vessels in which $SiO_{2[1-x/4]}(OH)_x$ structures nucleate and grow via *in situ*, acid catalyzed sol-gel reactions for tetraethoxysilane (TEOS).[1-5] The pseudo-periodic array of clusters, with center-to-center spacings of 30-50 Å, offers an interactive polymerization template that directs the morphology of the resultant dry, solid silicon oxide phase. SO_3H groups in clusters catalyze alkoxide hydrolysis within these "nano-reactors," although various cation-exchanged forms are also useful.[6-8] We have characterized these unique hybrid materials for structure and properties using electron microscopy, SAXS, WAXD, solid state NMR, infrared and dielectric relaxation spectroscopies, DSC, TGA, mechanical tensile and gas permeation analyses.

---

[1]Corresponding author

0097–6156/95/0585–0066$12.00/0

This earlier work established the validity of our template hypothesis of *in situ* inorganic growth. SAXS, WAXD and DSC studies revealed that the clustered + semicrystalline morphology remains intact despite invasion by the silicon oxide component up to moderate filler levels.[4,5] Mechanical tensile tests, especially, indicated a percolation of the silicon oxide phase with increasing filler content,[1] and IR and ^{29}Si NMR spectroscopies[1,2,6] provided insight into molecular connectivity within the fractal-like silicon oxide phase.

Based on the success of these earlier efforts, we are synthesizing [inorganic oxide]/PFSI nanocomposites having greater compositional and structural complexity. The work reported here deals with creating an organic "shell" about existing silicon oxide nanoparticles via post-reactions involving ethoxymethylsilanes.

In the first step, $SiO_{2[1-x/4]}(OH)_x$ particles are grown within initially hydrated, alcohol-swollen Nafion-H$^+$ membranes via *in situ* sol-gel reactions for TEOS, as before. Our IR and ^{29}Si solid state NMR spectroscopic investigations[1,2,6] of such dried-annealed systems indicate large relative populations of uncondensed Si-OH groups mainly near nanoparticle surfaces, as illustrated in Fig. 1 (top). Of course, the concepts of "surface" and "shell" are somewhat vague within the context of particles of molecular dimensions. Perhaps these structures would be better described as being fractal-like in the inefficient manner in which they fill space. Also, in a strict sense, perhaps the term "*semi*-organic shell" would be more appropriate, since the chemical units added to nanoparticle surfaces contain silicon atoms so that the surface would have a composition that is "silicone-like." Our use of the word "organic" in this context simply refers to the situation wherein the methyl groups of these units shield the Si atoms to which they are bonded and are mainly the groups that are exposed to, and interact with the PFSI matrix. These residual SiOH groups are available for post-reaction with species as $R'_{4-x}Si(OR)_x$ (*e.g.* R = -$[CH_2]_nCH_3$), thereby producing organically, or hydrophobically "shelled" nanoparticles, as crudely illustrated in Fig. 1 (bottom). Diethoxydimethylsilane (DEDMS) and ethoxytrimethylsilane (ETMS) were chosen for the present study. DEDMS, by itself, can form linear oligomers or polydimethylsiloxane (PDMS). At the PFSI-nanoparticle interfacial zones, DEDMS can hypothetically link two favorably-positioned exterior SiOH groups either on the same nanoparticle or on two different nanoparticles that are sufficiently proximate. ETMS, on the other hand, can only react with one external SiOH group to produce, at best, a hydrocarbon "coating" on a silicon oxide phase.

Si-C bonds are stable against hydrolysis in an aqueous environment. It is anticipated that the unfavorable energetics of the [silanol]/[perfluorocarbon] interface can be rendered more favorable by this organic modification of nanoparticle surfaces. Consequently, mechanical, viscoelastic and thermal properties will be altered by tailoring the composition and structure of this interface.

Nanoparticle surfaces, after post-treatment, will display an increased interactive affinity for organic molecule (liquid or gas) permeants. The surface roughness, or porosity, of these particles will be modified in this way. Here, "roughness" or "porosity" refers to the fractal nature of these convoluted surfaces and there are implications with regard to size exclusion effects in molecular permeation through these materials when used a membrane in separation processes. Motivated by Yeager's discussion of how diffuse interfacial regions in PFSI materials affect ion

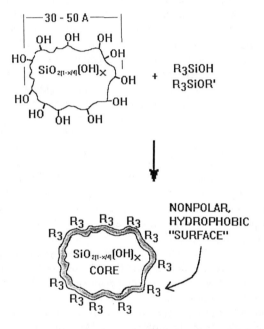

Figure 1. Silicon oxide nanoparticles imparted hydrocarbon "shells" by post-reaction of accessible SiOH groups with ethoxymethylsilanes.

permselectivity,[9,10] the utility of these nanocomposites when used as membranes in electrochemical cells is suggested.

These materials are expected to interact strongly and specifically with electromagnetic fields due to their unique heterophasic morphologies. As a dielectric, polarization at the nanoparticle/PFSI interface, owing to gradients of dielectric permittivity and conductivity across the phase boundaries, should be strong. Since the scale of phase separation is considerably smaller than wavelengths in the optical region of the electromagnetic spectrum, light, in transparent materials, should exhibit interesting behavior in this medium.

The particular goals and methods of the work reported here were: (1) To create organic/inorganic hybrids using a PFSI as a reactive morphological template for a two-step, *in situ* sol-gel reaction involving first tetra-, then di- and monofunctional alkoxysilanes to produce "core-shell" structures; (2) to verify structural incorporation of the post-reaction species, as well as probe the degree of molecular interconnection of the inorganic phase by IR and NMR spectroscopies; (3) to infer, using mechanical tensile studies, the relative energetics of nanoparticle/PFSI interfaces, as well as possible linking of silicon oxide nanoparticles via post-reactions with DEDMS.

Experimental Procedure

Materials. 1100 equivalent weight, 5 mil thick, PFSI membranes in the K^+ form ("Nafion 115") were supplied by the E.I DuPont Company. TEOS, DEDMS, ETMS, and methanol were obtained from the Aldrich Chemical Co. All water used was distilled/deionized.

Membrane Initialization. First, the membranes were converted to the sulfonic acid form by refluxing in 50% (v/v) HCl for 12 hr. Then, the membranes were refluxed for 6 hr in distilled/deionized water to leach out excess acid. Following this, a sequence of steps consisting of refluxing in methanol - then refluxing in distilled/deionized water - was repeated for 5 hr for each step until the pH of the solution reached 6.5 - 7.0. The total number of neccessary cycles in this sequence were ~6 - 8. The membranes were then dried at 100°C under vacuum for 24 hr. All membranes were reduced to this *standard initial state* prior to conducting the *in situ* sol-gel reaction for the purpose of acheiving maximum sample reproducibility. We have recently reported on the importance of membrane initialization.[11]

In Situ Sol-Gel Reaction. In covered glass jars, all initialized membranes were immersed in stirred solutions of 3:1 (v/v) methanol:water at 22°C for 20 hr. The sorbed water serves to initiated the hydrolysis of TEOS which was subsequently introduced in quantities such that H_2O:TEOS = 4:1 (mole/mole). Solutions of TEOS:methanol = 3:1 (v/v) were added to the containers (subsequently stoppered) while maintaining stirring. After specified times, the membranes were removed from these solutions. The time between the addition of TEOS and removal of the membranes from solution is referred to as "permeation time" rather than reaction time, since the hydrolysis-condensation reaction continues within the membrane after its removal from solution. Upon removal, the membranes were quickly soaked in methanol for 1-2 seconds to wash away excess surface reactants to reduce the possibility of a surface-attached silica layer. Finally, the samples were surface-blotted and dried at 100°C under vacuum for 24 hr to remove trapped volatiles as well as promote further condensation of residual SiOH groups. In each case, percent weight uptake was determined relative to the initial dry acid form weight. Samples were prepared to have silicon oxide uptakes in the range 5 - 7% and 11 - 14%, by varying permeation time. No surface-attached silica layers, as seen in earlier studies,[6,11,12] that might block post-reactions, were observed by light microscopy. At this point, and for these weight uptakes, silicon oxide "cores" are presumed, based on our earlier studies, to exist within the PFSI template.

Post-Reaction with DEDMS and ETMS . These dried-annealed SiO_2/PFSI hybrids were pre-swollen in 15 ml methanol at 22°C for 24 hr to enhance the permeation of DEDMS and ETMS and the delivery of these molecules to the interfacial zones. Afterwards, 12 ml of DEDMS or ETMS was added to the methanol bath (-in which the swollen membranes resided), with stirring at 22°C and the uptake of these reactive species proceeded. After selected times, the membranes were removed and washed for 1-2 seconds in pure methanol remove excess reactants from the surfaces, then surface-blotted dry.

The quantity of sorbed reactant for a given permeation time is expected to be a function of sample size, concentration in the methanol solution and temperature. In this set of experiments we have held these factors constant. The essential goal of this work was to demonstrate structure and property trends with reactant type and permeation time on a qualitative basis.

Afterwards, samples were placed on a Teflon-coated plate, transferred to a vacuum oven set at 40°C and thereafter heated to 100°C within 40 min (not under vacuum). During this heating, condensation between unreacted SiOH groups and the sorbed silanes proceeded. Then, the membranes were dried-annealed at 100°C under vacuum for 24 hr to remove volatiles as well as to promote further condensation. Afterward, all samples were kept in desiccator prior to structure-properties characterization.

One experimental series, for example, involved samples whose silicon oxide content was 13.4% before post-reaction. Such a film, approximately 50 x 50 mm in area, was divided into four pieces, one of which, untreated, served as a reference, the other three being post-reacted with DEDMS for three different times. Another series involved a similar chemical modification by ETMS in a membrane having 10.8% silicon content. Films containing other silicon oxide weight percents for the two post-reactants were prepared and are described in the following text.

Characterization Methods

Infrared Spectroscopy.   FT-IR/ATR spectra were obtained using a Bruker 88 Spectrometer set at a resolution of 4 cm[-1]. A thallium bromide-thallium iodide (KRS-5) crystal was used as the ATR plate with an angle of incidence of 45°. A total of 1000 interferograms was taken in each case.

As in our earlier similar IR investigations of these hybrids,[2] the reflectance mode must be used for these thicknesses, although it would certainly be more desirable to conduct investigations in transmission to obtain true bulk-averaged spectra. The penetration depth of an IR beam in ATR mode is considerable (0.2 - 4 μm) in relation to the size of ionic clusters in PFSIs (~30-50 Å), is a function of state of polarization, wavelength of the IR beam, incident angle and the indices of refraction of sample and plate. It was somewhat reassuring that in our earlier, albeit limited studies, ATR spectra were seen to be similar to those for transmission through thin films (≤ 1 mil thick). Nonetheless, the interpretation of FT-IR/ATR spectra for these hybrids should take into account the results of our scanning electron microscopy/EDS investigations that indicated the silicon oxide component is distributed nonuniformly across the film thickness, being most concentrated near the surfaces.[12]

^{29}Si *Solid State NMR Spectroscopy.* NMR spectra for the DEDMS-treated samples were obtained on a Bruker MSL-400 spectrometer operating at a frequency of 79.5 MHz for the silicon nucleus. A standard CP/MAS probe was used to acquire solid-state spectra. Samples were packed in fused zirconia rotors equipped with Kel-F caps and sample spinning rate was approximately 4 - 4.5 kHz. Spectra were acquired using high-power decoupling during acquisition only. Cross-polarization was not possible due to the paucity of protons in the samples. The 90° pulse width was 4.9 - 5.4 μs, and a relaxation delay of 10 s was used. The number of scans acquired per sample varied from 20,000 to 40,000. All chemical shifts were in reference to the downfield peak of tetrakis(trimethylsilyl)silane (-9.8 ppm with respect to TMS). All measurements were performed at 22°C.

Mechanical Tensile Measurements. Stress *vs* strain tests were performed at 22°C using an MTS 810 universal test machine at a strain rate of 0.2 mm/sec.

Results: DEDMS Post-Treatment

The following results refer to a hybrid of 13.4% SiO_2 uptake prior to post-treatment using DEDMS.

FT-IR Spectroscopy. Analysis of the inorganic phase of these hybrids by IR spectroscopy is rendered difficult because of the presence of numerous and complex bands characteristic of pure Nafion.[13] Therefore, subtraction of the spectrum of the unfilled dry Nafion film precursor from that of the hybrid is necessary. The very prominent SO_3^- stretching (~1200 cm^{-1}) and -CF_2-CF_2- symmetric stretching (~1140 cm^{-1}) vibrations were utilized as internal thickness standards to subtract the pure Nafion spectral contribution from that of the hybrids. The subtractions performed in this work were successful in uncovering the major peaks characteristic of molecular groups in the silicon oxide phase. In particular, the Si-O-Si asymmetric stretching vibration (~1000-1100 cm^{-1}), being the signature of bridging oxygens and completed condensation reactions, is split into two components arising from groups in linear and cyclic configurations as we pointed out in a previous report.[2] Comparison of the integrated intensities of the linear and cyclic components contributes to understanding the degree of molecular connectivity within the silicon oxide phase. Also of primary importance is the absorbance associated with Si-OH vibration (~940 cm^{-1}), a measure of degree of hydrolysis, or uncondensed silanol groups. The intensity ratio $I_{\upsilon(SiOH)}/I_{\upsilon(Si-O-Si)asym}$ measures degree of crosslinking within the silicon oxide phase. In this work, we will note trends in these spectral features on a qualitative basis.

Fig. 2 shows difference spectra (*i.e.* {silicon oxide containing + post-reacted film} - {dry, unfilled H^+ film}) for a time series (0 - 18 hr) of post-reactions with DEDMS. The silicon oxide content of this film was 13.4% prior to post-reaction. Since all four samples were cut from the same sheet, comparison of spectra is meaningful. All spectra have the same absorbance scale but have been vertically shifted for better visualization.

The distinct band observed at 1263 cm^{-1} for all post-reacted samples in Fig. 2 is a signature of CH_3 symmetric deformation in $Si-CH_3$ groups.[14-16] Also, two intense bands appear ~850 cm^{-1} and ~800 cm^{-1}. CH_3 rocking and Si-C stretching modes exist in the region ca. 870-750 cm^{-1} and it is logical to assume that the two peaks in this region in Fig. 2 correspond to these modes.[14-16]

Corresponding to the appearance of these bands is the suppression of the Si-OH stretching vibration which continues to diminish with increasing reaction time. This is the most direct IR indication of increasing degree of reaction of DEDMS with the pre-established silicon oxide "core" structure. On inspecting the trend of the two $\upsilon(Si-O-Si)_{asym}$ components, it is seen that post-reaction renders the structure more linear, which is reasonable since DEDMS has but two alkoxy groups. Although the cyclic component (~1080 cm^{-1}) becomes less intense than the linear peak (~1030 cm^{-1}) in proceeding from the sample containing only SiO_2 (non-DEDMS-reacted) to the sample having shortest post-reaction time (0.5 hr), the relative intensities of these two components do not change significantly thereafter. The low wavenumber, linear-associated peak shifts downward from ~1042 cm^{-1} to ~1030 cm^{-1} accompanied by an increase in peak width in proceeding from pure SiO_2-filled to the post-reacted samples; the shift is the same for all post-reaction times. A lesser downward shift occurs for the cyclic-associated peak. The peak width increase suggests a broader distribution of molecular environments of Si-O-Si groups in linear configurations, which is expected since the difunctional dimethylsilanes introduce a different structural unit within, or perhaps even spanning clusters. The peak shift to lower wavenumbers indicates an overall decrease of the Si-O force constant characteristic of dimethylsiloxane incorporation.

Differences in the three spectra of the post-reacted membranes are not profound, probably due to a condition wherein the reaction reached saturation after 0.5 hr. This idea is consistent with the fact that weight uptakes after post-reaction are around 4.5% for all three samples.

We should mention the existence of the symmetric stretching vibration of the Si-O-Si group, which occurs ~800 cm^{-1}. This vibrational mode, which appeared in IR spectra in earlier investigations of SiO_2/Nafion nanocomposites,[2] is theoretically inactive but nonetheless present, presumably owing to considerable distortion of coordinative symmetry of bonding to silicon atoms. We mention this here to avoid confusion between this band, which appears weak in the spectrum of non-postreacted, SiO_2-filled Nafion, and the strong Si-C bands that appear in the same region of the spectrum of the post-reacted hybrids.

Post-Reacted with (CH$_3$)$_2$Si(OC$_2$H$_5$)$_2$

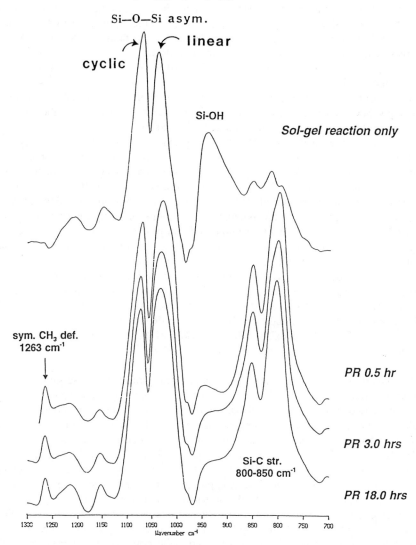

Figure 2. FT-IR/ATR difference spectra ({post-reacted SiO$_2$-containing PFSI} - {dry, unfilled PFSI[H$^+$]})for indicated DEDMS post-reaction times. SiO$_2$ content = 13.4% prior to post-reaction.

^{29}Si Solid State NMR Spectroscopy. Fig. 3 is the spectrum for a composition having 13.4 wt% silicon oxide and for the same composite, but post-reacted with DEDMS for 18h. While noise/signal is high in this spectrum, the chemical shift distribution in the Q-region quite noticeably shifts to reflect a greater degree of Si substitution about SiO_4 tetrahedra upon post-reaction because the signal in the Q^2 and Q^3 regions is greatly diminished. The symbols Q^2, Q^3 and Q^4 in this figure refer to di-, tri- and tetrasubstitution of Si. The approximate chemical shift (ppm, relative to $Si(Me)_4$) ranges of Q^n, taken from the literature, are as follows:[17]

Q^0: -60 to -82, Q^1: -68 to -83, Q^2: -74 to -93, Q^3: -91 to -101, Q^4: -106 to -120

The spectrum is not of such a quality to allow for conclusions regarding structural topology beyond nearest-neighbor Si atom placements, although the above-stated change in the chemical shift distribution is unmistakable.

The above result, coupled with the presence of strong $(CH_3)_2Si(O_{0.5})_2$ (so-called "D-unit") resonances[18-23] at the left of the spectrum (-19.2 ppm [shoulder ~17 ppm] , -21.8 ppm), reinforces the previous IR -based conclusion that dimethylsilane molecules are bonded onto SiO_2 cores. For reference, ^{29}Si NMR resonances for oligomeric and polymeric dimethylsiloxane chains occur from around -20 to -22 ppm, the latter corresponding to the polymer .[18-23] Chemical shifts for ring structures extend from -20 to -23 ppm, but we note that a ring of 3 D units occurs afar at -9.2 ppm. Pending future solid state NMR evidence, we are unable to comment on the length of D-chains or on the existence/size of incorporated rings.

While DEDMS molecules react with residual SiOH groups, increasing the average degree of Si atom coordination about SiO_4 groups on pre-existing silicon oxide structures, DEDMS can also react with itself so as to increase the overall linearity of the system, as indicated by the above IR studies.

To be sure, there exists a broad variety of Si atom environments in this sample.

Mechanical Tensile Studies. Tensile stress *vs* strain profiles for these samples (Fig. 4) show transition from ductility for the pure SiO_2 filled system, to near-brittleness, with increased tensile strength and decreased elongation-to-break, upon post-reaction with DEDMS. Interestingly, elongation-to-break is the same for all post-reaction times, a fact in harmony with the above-proposed concept of a system that has already reached DEDMS reaction saturation by 0.5 hr. 13.4% is known to be just below the threshold for percolation of the SiO_2 phase so that the intrinsically-ductile PFSI matrix-template is the load-bearing phase before post-reaction. However, since inter-cluster spacings are only ~10 Å,[24] it is a reasonable consideration, reinforced by the previous IR and NMR results that indicated increased linearity within the filler, that reactions of DEDMS form $[-O(CH_3)_2Si-O]_x$ bridges between adjacent SiO_2 cores so that the now-interpenetrating, glassy, inorganic component becomes the load-bearing phase as illustrated in Fig. 5.

It is noted that the initial modulus is the essentially the same for all hybrids, which implies that the resistance to small deformation is the same, regardless of treatment; thus, post-treatment essentially affects mechanical tensile properties associated with long range molecular motions.

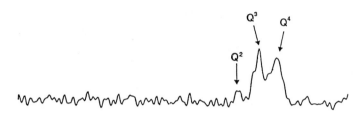

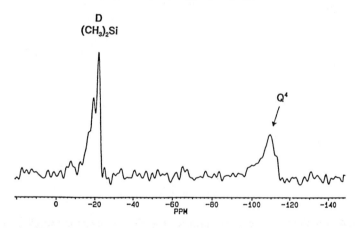

Figure 3. ^{29}Si solid state NMR spectrum for (top) a composition having 13.4 wt% SiO$_2$ and for (bottom) the same composite, but post-reacted with DEDMS for 18h.

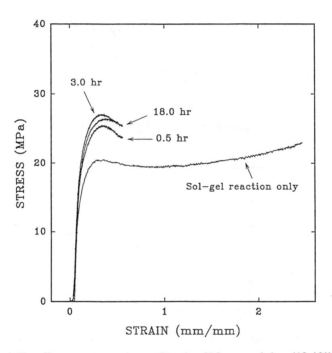

Figure 4. Tensile stress *vs* strain profiles for SiO_2-containing (13.4%) PFSI films that were post-reacted with DEDMS for indicated times.

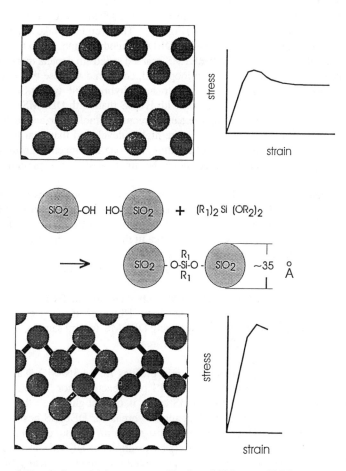

Figure 5. Depiction of (a) an array of isolated $SiO_{2[1-x/4]}(OH)_x$ nanoparticles and corresponding stress-strain curve (top), and (b) the same nanoparticles, but interknitted by $[-O-Si(R_1)_2]_n-O-$ bridges, and corresponding stress-strain curve (bottom). In the figure n = 1.

Results: ETMS Post-Treatment

FT-IR Spectroscopy. Spectra for 14.3% SiO$_2$/Nafion hybrids post-treated for 35 min (not shown) indicate that ETMS reaction is less complete than that for DEDMS, as revealed, for one, by considerably lesser suppression of the Si-OH stretch absorbance peak by ETMS. The same general conclusions are reached in a comparison of 35 min-reacted 6.5% SiO$_2$/Nafion hybrids (Fig. 6). The weight uptake after ETMS treatment is ~1%, compared with ~4.5% for DEDMS-modification. Also note in Fig. 6 that the DEDMS-, compared with the ETMS-reacted spectrum depicts more Si-O-Si bonds in linear fragments as opposed to loops and, compared to these two cases, the cyclic peak was the greatest for the un-postreacted hybrid. The two υ(Si-O-Si)$_{asym}$ bands are affected less upon post-reaction with ETMS than with DEDMS. In short, the IR evidence suggests that, given the same precursor, molecular connectivity in the filler in DEDMS-treated systems is more highly developed than that in ETMS-treated systems. Steric effects, less polar nature, and one less reactive group for the ETMS molecule are all factors that might be implicated in explaining these results.

    Absorbance in the Si-CH$_3$ stretching region is seen, but not with the same relative prominence as for the DEDMS-treated hybrid, showing again that ETMS post-reaction occurs to a lesser degree than that with DEDMS. This conclusion is further reinforced by noting the weak absorbance in the region of the symmetric CH$_3$ deformation. It must be allowed that, in addition to reaction with the silicon oxide phase, ETMS molecules might react with themselves to produce (CH$_3$)$_3$Si-O-Si(CH$_3$)$_3$ dimeric species.

    A ^{29}Si solid state NMR spectrum was obtained for an ETMS post-treated sample, but the signal/noise ratio was rather low. However, it was clear from this spectrum that, in contrast with the corresponding DEDMS-treated sample, the Q^2 and Q^3 regions were not severely affected by post-reaction.

Mechanical Tensile Studies. Stress *vs* strain profiles for 14.3% SiO$_2$/Nafion (Fig. 7) reveal that while both DEDMS and ETMS post-reactions for 35 min increase strength, elongation-to-break for the ETMS curve is only slightly less than that for pure SiO$_2$ filler. On the other hand, elongation-to-break for DEDMS post-treatment is reduced to about one-half the value for no post-reaction. This latter fact reinforces the concept of linking adjacent SiO$_2$ clusters by reactions of DEDMS as depicted in Fig. 5. However, ETMS treatment, while producing a more energetically-favorable interface that increases material strength, is not able to interknit the silicon oxide phase. It is worthy to note that all the curves exhibit yield following their respective elastic regions and that the moduli in these initial regions are not too different from each other. Therefore, from the mechanical tensile perspective, it might be said that these chemical modifications are most effective in altering ultimate properties, or long-range motions within the molecular architecture. Stress *vs* strain profiles for 6.5% SiO$_2$ (Fig. 8) indicate that both DEDMS and ETMS post-treatments conducted for 35 min enhance strength over the range of deformation to about the same extent, although elongation-to-break for the ETMS-reacted sample is somewhat lesser than that for DEDMS-treatment. Contrasted with the 14.3% SiO$_2$/Nafion systems, both post-treatments retain ductility almost as great as that for pure 6.5% SiO$_2$ filler.

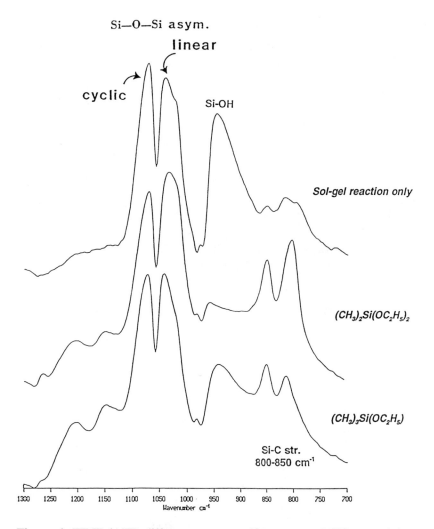

Figure 6. FT-IR/ATR difference spectra ({post-reacted SiO$_2$-containing PFSI} - {dry, unfilled PFSI[H$^+$]}) for DEDMS and ETMS post-reaction time = 35 min, as well as for the pure SiO$_2$-containing (top spectrum, content = 6.5%) precursor.

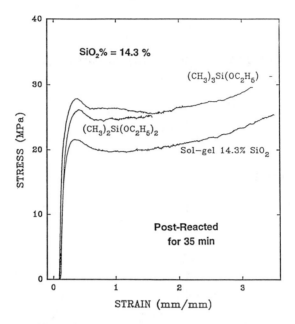

Figure 7. Tensile stress *vs* strain profiles for SiO_2-containing (14.3%) PFSI films that were post-reacted with DEDMS and ETMS for 35 min.

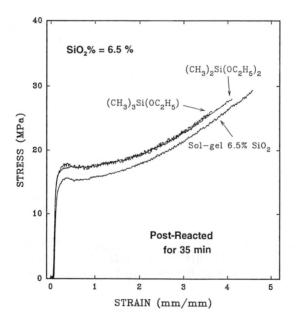

Figure 8. Tensile stress *vs* strain profiles for SiO_2-containing (6.5%) PFSI films that were post-reacted with DEDMS and ETMS for 35 min.

6.5% filler was earlier established as being considerably below the silicon oxide phase percolation threshold. Therefore, for this particle dispersion, it is reasonable to think that nanoparticle spacings are too large for cluster linkage by the DEDMS reaction, conducted under these conditions so that the PFSI template remains the load-bearing phase. However, the improvement of interfacial energetics by this organic modification of the surfaces of isolated nanoparticles does enhance mechanical strength perceptibly. 5.0% SiO_2 samples remain ductile when reacted with ETMS, at least up to 17h. The time dependence of the IR spectrum seen in Fig. 9 for 5.0% SiO_2 suggests that ETMS post-reaction with SiOH groups becomes essentially saturated, a result in harmony with the fact that post reaction for longer times causes little change in stress *vs* strain curve parameters.

Conclusions

First, we generated $SiO_{2[1 - x/4]}(OH)_x$ nanoparticles via *in situ* sol-gel reactions for tetraethoxysilane within hydrated Nafion perfluorosulfonic acid films that were dried after the reaction. Then, residual SiOH groups on these nanoparticles were post-reacted with diethoxydimethyl- and trimethylethoxysilane. This two step reaction resulted in interknitted and organically "shelled" nanoparticles, respectively. FT-IR and ^{29}Si solid state NMR spectroscopies established structural incorporation of these silanes onto silicon oxide "cores" and probed overall degree of molecular connectivity within the inorganic phase. The reaction of ETMS is less complete than that for DEDMS. Post-reaction with DEDMS renders the structure more linear. Mechanical tensile experiments indicated a ductile → brittle transformation upon DEDMS post-reaction, suggesting linking of previously-isolated silicon oxide nanoclusters. On the other hand, a measure of ductility persists after ETMS post-reaction. although tensile strength was increased after both post-reactions.

These results satisfy our previously-stated goals: (1) To create nanophase-separated organic/inorganic hybrids using PFSI films as reactive morphological templates for *in situ* sol-gel reactions involving tetra-, di- and monofunctional alkoxysilanes. We have shown by IR and NMR spectroscopies that the polar "surfaces" of silicon oxide nanoparticles can be chemically modified to be organic in nature by post-reaction with silanes having hydrocarbon groups; (2) to infer, on the basis of mechanical tensile studies, the relative energetics of nanoparticle/PFSI interfaces, as well as the linking of silicon oxide nanoparticles via post-reactions with a dialkoxysilane.

In the future, we will report on the behavior of thermal transitions and thermal degradation of these hybrids, as well as underlying molecular mechanisms, using dynamic mechanical analysis, differential scanning calorimetry and thermogravimetric analysis. Furthermore, small angle x-ray scattering studies will be conducted to probe inter-nanoparticle spacings and wide angle x-ray diffraction will investigate the nature of PFSI crystallinity after invasion by the inorganic phase.

Post-Reacted with (CH$_3$)$_3$Si(OC$_2$H$_5$)

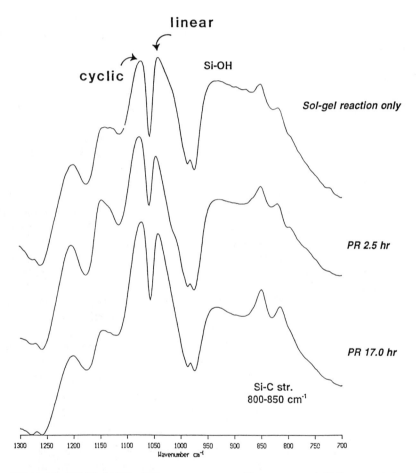

Figure 9. FT-IR/ATR difference spectra ({post-reacted SiO$_2$-containing PFSI} - {dry, unfilled PFSI[H$^+$]}) for indicated ETMS post-reaction times as well as for the pure SiO$_2$-containing precursor (top spectrum) having content = 5.0%.

Acknowledgments

This material is based partly upon work supported by a grant from the National Science Foundation/Electric Power Research Institute (Advanced Polymeric Materials: DMR-9211963). This work was also sponsored in part by the Air Force Office of Scientific Research, Air Force Systems Command, USAF, under grant number AFOSR F49620-93-1-0189. The U.S. Government is authorized to reproduce and distribute reprints for Governmental purposes notwithstanding any copyright notation thereon. We acknowledge the assistance of W. Jarrett, Department of Polymer Science, University of Southern Mississippi, in obtaining the ^{29}Si solid state NMR spectra. The donation of Nafion membranes by the E.I. duPont de Nemours & Co., through the efforts of J.T. Keating, is appreciated, as well.

Literature Cited

1.   Mauritz, K.A.; Storey, R.F.; Jones, C.K. in *Multiphase Polymer Materials: Blends,Ionomers and Interpenetrating Networks*, ACS Symp. Ser. No. 395, Ch. 16, L. A. Utracki and R. A. Weiss, Eds., 1989.
2.   Mauritz, K.A.; Warren, R.M. *Macromolecules*, **1989**, 22, 4483.
3.   Mauritz, K.A.; Stefanithis, I.D. *Macromolecules*, **1990**, 23, 1380.
4.   Stefanithis, I.D.; Mauritz, K.A. *Macromolecules* **1990**, 23, 2397.
5.   Mauritz, K.A.; Stefanithis, I.D.; Wilkes, G.L; Huang, Hao-Hsin, *ACS Div. Polym. Chem., Polym. Preprs.* **1991**, 32(1), 236.
6.   Davis, S.V.; Mauritz, K.A. *Am. Chem. Soc. Polymer Prepr.* **1992**, 33(2), 36.
7.   Davis, S.V.; Mauritz, K.A. *Am. Chem. Soc. Polymer Prepr.* **1993**, 34(1), 608.
8.   Bordayo, C.; Davis, S.V.; Moore, R.B.; Mauritz, K.A. *Am. Chem. Soc. PMSE Div. Prepr.* **1994**, 70, 230.
9.   Yeager, H.L.; Steck, A. *J. Electrochem. Soc.* **1981**, 128, 1880.
10. Yeager, H.L. in *Structure and Properties of Ionomers*, Pineri, M., Eisenberg, A., Eds.; D. Reidel: Dordrecht, **1987**, p377.
11. Davis, S.V; Mauritz, K.A.; Moore, R.B. *Am. Chem. Soc. Polymer Prepr.* **1994**, 35(1), 419.
12. Mauritz, K.A; Scheetz, R.W.; Pope, R.K.; Stefanithis, I.D. *ACS Div. Polym. Chem., Polym. Prepr.* **1991** 32(3), 528.
13. Falk, M. in *Perfluorinated Ionomer Membranes*, Eisenberg, A., Yeager, H.L., Eds., ACS Symp. Ser. 180, Amer. Chem. Soc.: D.C., **1982**, Ch. 8.
14. Smith, A.L. *Spectrochimica Acta* **1960**, 16, 87.
15. Wright, N; Hunter, M.J. *J. Amer. Chem. Soc.* **1947**, 69, 803.
16. Richards, R.E.; Thompson, H.W. *J. Chem. Soc. London* **1949**, 124.
17. Engelhardt, G.; Michel, D. *High Resolution Solid State NMR of Silicate and Zeolites* **1987**, Wiley: New York.
18. Babonneau, F.; Thorne, K.; MacKenzie, J.D. *Chemistry of Materials* **1989**, 1, 554.
19. Maciel, G.E.; Sindorf, D.W. *J. Am. Chem. Soc.* **1980**, 102, 7606.
20. Pouxviel, J.C.; Boilot, J.P.; Beloeil, J.C.; Lallemand, J.Y. *J. Non-Cryst. Solids* **1987**, 89, 345.

21.   Harris, R.K.; Kimber, B.J. *J. Organomet. Chem.* **1974**, <u>70</u>, 43.
22.   Harris, R.K.; Robbins, M.L. *Polymer* **1978**, <u>19</u>, 1123.
23.   Engelhardt, G.; Jancke, H.; Magi, M.; Pehk, T.; Lippmaa, E. *J. Organomet. Chem.* **1971**, <u>28</u>, 293.
24.   Gierke, T.D.; Munn, G.E.; Wilson, F.C. *J. Polym. Sci.: Polym. Phys. Ed.* **1981**, <u>19</u>, 1687.

RECEIVED November 21, 1994

New Techniques and Approaches

Chapter 8

Nanostructured Organic–Inorganic Hybrid Materials Synthesized Through Simultaneous Processes

Bruce M. Novak, Mark W. Ellsworth, and Celine Verrier

Department of Polymer Science and Engineering,
University of Massachusetts, Amherst, MA 01003

The sol-gel process, with its associated mild conditions, offers a new approach to the synthesis of composite materials with domain sizes approaching the molecular level. Both the organic and inorganic phases can be simultaneously formed through the synchronous polymerization of the organic monomer and the sol-gel precursors. Depending upon such factors as the structures of the organic and inorganic components, the phase morphology, the degree of interpenetration, and the presence of covalent bonds between the phases, the properties of these composites can vary greatly and range from elastomeric rubbers to high modulus materials. In an attempt to solve the truculent shrinkage problem historically associated with sol-gel derived glasses, we have synthesized tetraalkoxy orthosilicates possessing polymerizable alkoxide groups. The resultant alcohols, which are liberated during the hydrolysis and condensation steps associated with the formation of the inorganic component, are polymerized *in-situ* to form the organic phase of the composite material. The synthesis and mechanical properties of these nonshrinking hybrid materials will be discussed.

The properties of composite materials result from the rich interplay between the constituent components and are greatly influenced by the such factors as the phase's size and shape and the interfacial interactions (*1*). Tailoring properties by manipulating these important parameters particularly at very small length scales provides a myriad of challenges for the synthetic chemist. In an effort to control the phase domain size and uniformity in organic-inorganic nanocomposites, we have been investigating new *in situ* routes into their preparation. To this end, we, (*2*) as well as others, (*3*) have been interested in using the sol-gel process (*4*) as a means of forming hybrid materials which contain both inorganic and organic components commingled into intimate, new morphologies. The sizes of the phase domains are minimized using this technique (sometime approaching the molecular level), and highly transparent composite materials often result. The properties of these composites vary immensely and depend upon such factors as the inorganic to organic ratio, the structure of the inorganic component (highly condensed vs. lightly

0097–6156/95/0585–0086$12.00/0

ramified), and the nature of the organic polymer (amorphous vs. crystalline, thermal stability, T_g, T_m, etc.).

The Sol-Gel Process

Traditional sol-gel processes involve the hydrolysis and condensation of tetraalkyl orthosilicates to form amorphous silicate (SiO_2) networks (*4*). Although, similar reactions can be carried out using the corresponding orthotitanates and orthozirconates to form TiO_2 and ZrO_2, respectively, their very fast condensation rates make these reactions more difficult to control. The hydrolytic condensation reactions can be catalyzed using acid, base, or under neutral conditions using fluoride anion, and the choice of catalyst dramatically affects the structure of the inorganic phase (*5*). During the hydrolysis and condensation of tetraalkyl orthosilicates under basic conditions, the rate of condensation is fast relative to hydrolysis, resulting in the formation of dense, colloidal particles with highly condensed structures. At low pH, however, these relative rates reverse, and now the rate of condensation is slow relative to the rate of hydrolysis. As a result, highly ramified, low-fractal-dimension structures are formed. The most commonly used alkoxides are either methoxide (TMOS) or ethoxide (TEOS) because of their convenient hydrolysis rates and commercial availabilities. The nonpolar character of TEOS and TMOS requires the use of polar organic cosolvents (alcohols, THF, DMF, etc.) in order to mediate their miscibility with the aqueous phase. This overall process is shown in Equation 1.

$$Si(OCH_3)_4 \xrightarrow[\text{Cosolvents}]{H_2O/H^+} \qquad + \ 4\,CH_3OH \quad (1)$$

As the hydrolytic condensation process proceeds gellation occurs before complete conversion is reached, and a solvent-filled, three dimensional crosslinked network containing dangling alkoxide and silanol groups is formed. Slow condensation reactions continue to occur within the interior of these gels, and as a result, their properties continue to change as they are aged. Aging gels in typical pore solvents (water/alcohol mixtures) can increase the storage modulus, G', by as much as 4 orders of magnitude (from ≈ 0.5 MPa to $\approx 5,000$ MPa in 200 days) (*6*).

Once formed, the solvent-filled, inorganic gel must be dried, requiring the removal of excess water, cosolvent(s) and the liberated alcohol (typically ethanol or methanol). Simply drying under ambient conditions leads to rapid shrinking and cracking due to stresses generated by large capillary forces in the very small (< 10 nm) pores of the gel (*7*). If this drying processes is slowed down, crack-free monolithic samples can be obtained albeit still with shrinkage. One of the major obstacles to the wide-spread application of sol-gel techniques is this extraordinary shrinkage of the sample (shrinkages of $\geq 50\%$ are common). Minimizing shrinkage is particularly important in the fabrication of composites using sol-gel techniques as the shrinking tends to induce a high degree of stress within the material, particularly if the organic polymer is below its T_g, covalently bound to the inorganic network, or otherwise noncompliant. Described herein is our work on the formation an organic polymer within these sol-gel derived, inorganic structures to form mutually interpenetrating organic-inorganic networks. Among other issues we have been interested in arresting the large scale shrinkages associated with solvent removal during the drying process.

"Nonshrinking" Sol-Gel Composites

Synthesis.   As part of continuing efforts to circumvent the solubility limitation of preformed polymers, as well as to provide better homogeneity between the two phases, we began investigations into the formation of simultaneous interpenetrating networks (SIPN's) (*2a,b*) through the synchronous formation of the organic polymer and the inorganic glass network. Our approach is also designed to minimize the shrinkage and processing times of the sol-gel composite materials, and is centered around the concept of replacing the methoxide or ethoxide on the silicon orthoester precursors with polymerizable alkoxides.   By employing *in situ* organic polymerization catalysts, the alcohol liberated during the formation of the inorganic network is polymerized rather than being left to evaporate during the drying process. By using a stoichiometric quantity of water, and additional polymerizable alcohol as a cosolvent if needed, all components are converted into either the organic polymer or the inorganic network.   Because no evaporation is necessary, large scale shrinkages due to the evaporation of solvents are eliminated (Scheme I).

The glass content of materials derived from these precursors is governed by the stoichiometry of the precursors and are typically in the range of 10-18%. A further refinement of this nonshrinking motif that allows us to reach higher inorganic contents involves the formation of poly(silicic acid esters) elaborated with the same polymerizable alkoxides (*2c*).   These precursors can be readily prepared by allowing sodium metasilicate to hydrolyze and condense in acidic aqueous-organic solutions. Addition of salt forces the system to phase separate and the poly(silicic acid) partitions between the two phases with approximately 85% being found in the organic layer.   Alcohols can be added to the organic fraction and excess water removed through azeotrope distillation. During this process, varying numbers of the silanol groups are exchanged for alkoxides (DS $\approx$ 10 - 90% depending upon reaction times).

For characterization purposes the residual silanols can be capped with TMSCl or similar silylating agents to yield inert poly(silicic acid esters). Coupling agents can also be added to these reactions to yield particles with covalent carbon-silicon bonds. Some of the most successful composites result from incorporation of these coupling agents that act to modulate the interfacial interactions by forming bonds between the inorganic and organic phases. In addition to providing higher inorganic ratios, the precondensed structure of the poly(silicic acid esters) can be varied from wispy, fractal structures, to highly condensed particles by adjusting the pH of the condensation medium.

Nanocomposites can be formed from these poly(silicic acid esters) by dissolving them in typical organic monomers and then carrying out bulk polymerizations. Lined with these organic groups, these particulate precursors are soluble in a wide range of organic monomers including methacrylates, acrylates, acrylic acid, acrylamide and styrene. Addition of acid to these reactive solutions tends to scramble the inorganic structure, and the degree of this reorganization is dependent upon the scrambling rate relative to the organic polymerization rate. In the absence of major scrambling, the resulting composites have phase structures that reflect the structure of the poly(silicic acid) precursors. Therefore, at a fixed composition, a range of materials with different morphologies are accessible.

Structural Characterization. The techniques that can be used to characterize the structure of these nanocomposite materials are many and include solid state CP-MAS ^{13}C and ^{29}Si NMR, light scattering, x-ray scattering, and thermal mechanical analysis.

Silicon-29 NMR is particularly useful for the determining the average shape of the inorganic phase. Figure 1 compares the ^{29}Si NMR spectra of two poly(silicic acid) samples with differing structures: a) ramified, elongated structure, and b) dense particles. The degree of condensation within these particles is best evaluated by measuring percent Q^4 (a silicon bound to four other silicons through oxygen bridges) species in their structure. By varying the HCl concentration from 3.4 M to 6.0 M, the percent Q^4 can systematically be varied from *ca.* 40% to 72%.

A polymer's glass transition weakens and shifts to higher temperatures upon restricting the main chain's motion. Hence, thermal analysis can be used to help evaluate the degree of mixing between the two phases. The DSC spectra of a series of poly(2-hydroxy ethyl methacrylate) (HEMA)/SiO_2 composites are shown in Figure 2. The T_g of HEMA appearing at *ca.* 110 °C essentially disappears with loadings of SiO_2 as low as 10 % by weight ($\approx$ 6% by volume). This pronounced effect at these low loading levels indicates that both a high degree of mixing and strong interfacial interactions must be present in these materials. Direct evidence of residual covalent bonds between the SiO_2 and the alcohol polymer side chains of HEMA can be obtained through solid state CP-MAS ^{13}C NMR. The methylene carbon adjacent to the hydroxyl group is particularly sensitive to this bonding: the resonance for this carbon is shifted from 59 ppm to 63 ppm when its oxygen is covalently bound to silicon. Unfortunately, broadening and overlap with nearby resonances makes it impossible to quantitatively determine the number of these siloxane bridges present within a given sample using this technique.

Another thermal technique that can be used to probe structure at an even finer length scale is dynamic mechanical analysis (DMA). Figure 3 shows tan δ (the ratio of the loss modulus to the storage modulus) as a function of temperature for a series of HEMA/SiO_2 composites. At its T_g near 100 °C, HEMA shows a large increase in tan δ due to an increase in the loss modulus and a decrease in the storage modulus as the material traverses from the rubbery to the glassy state. Incorporating small amounts of SiO_2 within the polymer suppresses the T_g and virtually eliminates the tan δ peak. Hence, the use temperature of this polymer can be extended by *ca.* 100 °C through the incorporation of even small amounts of SiO_2.

A wide range of morphologies are assessable using these *in situ* synthetic techniques. At a constant composition, materials with phase structures varying from mutually interpenetrating, co-continuous networks to highly phase separated materials are possible. Two extremes within this continuum can be seen by the comparison of the SAXS profiles shown in Figure 4.

Properties. The properties of these materials depend heavily upon several factors including their compositional makeup, the domain size and shape, interfacial

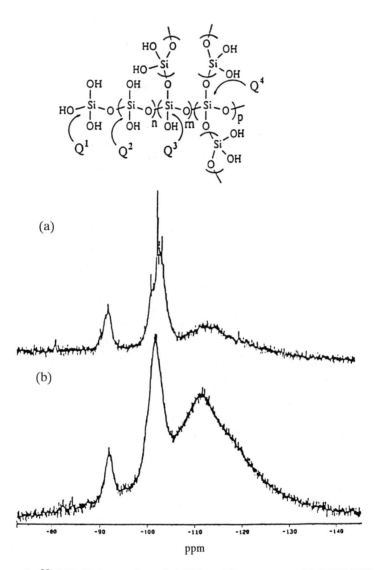

Figure 1: ^{29}Si NMR spectra for poly(silicic acid) precursors: (a) 3.0 M HCl, one hour reaction; (b) 6.0 M HCl, one hour reaction. Assignments: δ -81, Q^1; δ -92, Q^2; δ -103, Q^3, δ -112, Q^4.

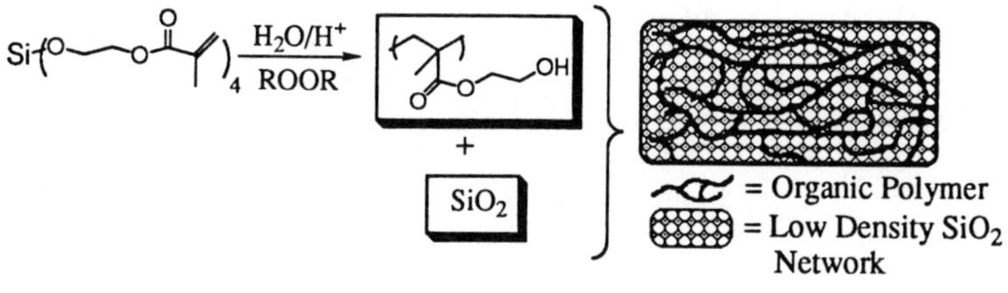

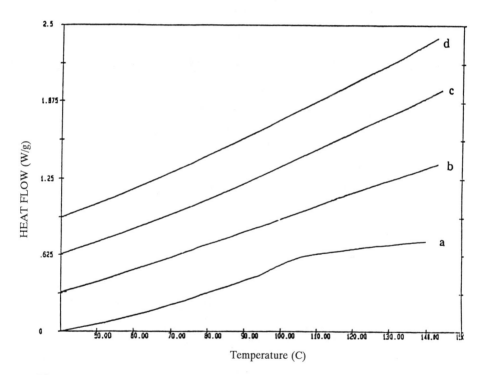

Scheme I

Figure 2: DSC scans for HEMA/SiO$_2$ nanocomposites. (a) 0% SiO$_2$ (poly(HEMA)); (b) 10% SiO$_2$; (c) 20% SiO$_2$; and (d) 40% SiO$_2$.

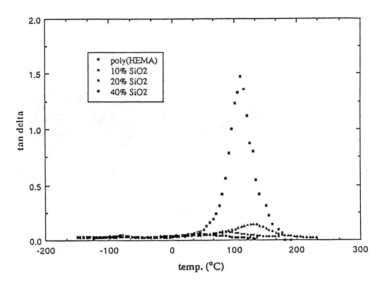

Figure 3: Tan δ plots for poly(HEMA) and HEMA/SiO$_2$ nanocomposites.

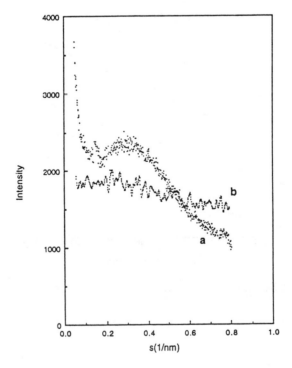

Figure 4: SAXS comparison of HEMA/SiO$_2$ nanocomposites formed through simultaneous processes: (a) particulate composite; (b) interpenetrating network. Both samples are 40% by weight SiO$_2$.

interactions, degree of interpenetration, interphase connectivities, the organic phase's T_g and level of intraphase crosslinking.

These *in situ* nanocomposites display excellent properties in comparison to the pure polymer. For example, intertwining the inorganic network into HEMA at a level of 27 volume percent increases the yield strength by over 835% (30 MPa vs. 281 MPa), and the elastic modulus by over 140% (900 MPa vs. 2.2 GPa). Furthermore, the mechanical properties of these materials compare very favorably with those of "conventional" glass/polymer composites. A direct comparison of these SIPNs to a chopped fiber-HEMA can be seen in Figure 5.

At any given composition and morphological structure, the properties of these composites will be highly dependent upon the properties of the organic polymer. Composites that span the continuum from elastomeric rubbers to high modulus materials are accessible through the use of different organic polymers. This can be illustrated by comparing the properties of composites prepared using HEMA ($T_g \approx$ 110 °C) with composite prepared using 2-hydroxyethyl acrylate (HEA) ($T_g \approx$ - 15 °C) (Figure 6).

HEMA
Tg = + 110 °C

HEA
Tg = - 15 °C

Using the *in situ* synthetic methodologies outlined above, materials possessing morphologies ranging from molecularly dispersed, interpenetrating networks to phase separated composites having controlled domain sizes. By holding all other variables constant, we can begin to explore properties as a function of composite morphology. In general, the mechanical properties of these composites improve as the phase domains decrease in size until a critical size is reached, after which a new "inverted" regime is entered and their properties begin to deteriorate. We have yet to pinpoint the critical size associated with this transition in all cases but it appears to be on the tens of nanometers length scale for the HEMA/SiO2 system. The properties in this inverted regime can be illustrated by comparing composites formed using the poly(silicic acid ester) precursors of different size. Unless extraordinary efforts are used, the inorganic phase of composites formed from the precursors retain their approximate original shape and size. Figures 7 and 8 compare the elastic modulii and yield strengths of two composites prepared using poly(silicic acid ester) precursors of different molecular weights (M_n = 3,000 and 15,000). The scattering profiles for the samples prepared for this high and low molecular weight precursors show maxima corresponding to an average repeat at approximately 35 and 55 Å, respectively. Although the difference in domain size between composites formed from these two precursors is relatively small, it has a measurable effect on their properties.

The properties of these composites are superior to analogous chopped glass composites which follows the size dependency trends that are valid in conventional particulate reinforced composites (*1,8*). However, as the domain size continues to decrease down to nanometer length dimensions, the composites prepared from the higher molecular weight precursors (i.e., larger phase sizes) now have the better properties. These differences are reflected in both the elastic modulus and the yield strengths of the composites and become more pronounced at higher loadings of SiO2.

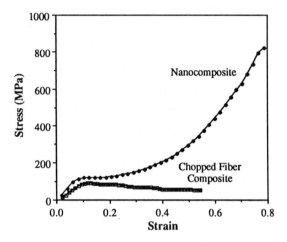

Figure 5: A comparison of the compression strength between a conventional HEMA/SiO$_2$ chopped fiber composite and a HEMA/SiO$_2$ nanocomposite.

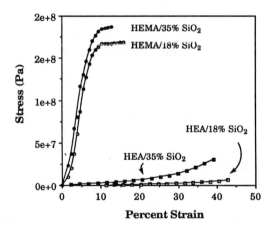

Figure 6: A comparison of the compression stress-strain curves for nanocomposites fashioned from HEMA and HEA.

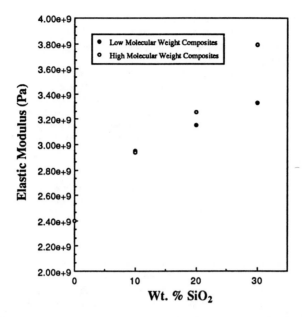

Figure 7: Elastic modulus versus weight percent SiO_2 for composites prepared from poly(silicic acid) precursors of different molecular weights (Low: M_n = 3,000; High: M_n = 15,000).

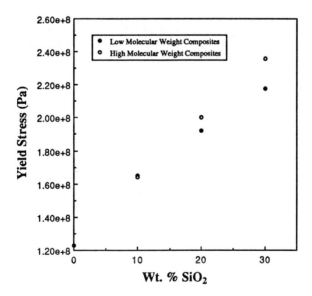

Figure 8: Yield strengths versus weight percent SiO_2 for composites prepared from poly(silicic acid) precursors of different molecular weights (Low: M_n = 3,000; High: M_n = 15,000).

Conclusions

The *in situ* method described herein provides a general route into a number composite materials possessing interesting new morphologies and phase structures. This inherent versatility allows for the formation of materials with properties ranging from high modulus materials to elastomeric rubbers. These diverse properties arise from a complex amalgamation of such factors as the material's composition, the domain sizes, domain shapes, interfacial interactions and intraphase connectivities. Attempts at deconvoluting these complicated structure-property relationships that exist in these nanocomposites are currently under investigation in a number of related systems.

Acknowledgments

The author gratefully acknowledges financial support for this work from the Office of Naval Research, the NSF Presidential Faculty Fellowship, the Alfred E. Sloan Foundation, Du Pont, Procter & Gamble, BFGoodrich, 3M and Kodak.

References

1.   (a) J. A. Manson, L. H. Sperling, *Polymer Blends and Composites*; Plenum Press: New York, 1976. (b) *Handbook of Composites,* Lubin, G., Ed., Van Nostrand Reinhold Co.: New York, 1982; Volumes 1 and 2.
2.   (a) Ellsworth, M. W.; Novak, B. M. *J. Am. Chem. Soc.* **1991**, *113*, 2756. (b) Novak, B. M.; Davies, C. *Macromolecules* **1991**, *24*, 5481. (c) Ellsworth, M. W.; Novak, B. M. *Chem. Mater.* **1993**, *5*, 839.
3.   Novak, B. M. *Ad. Mater.* **1993**, *5*, 422 and references therein.
4.   Brinker, C. J.; Scherer, G. W. *Sol-Gel Science*; Academic Press: New York, 1990.
5.   Schaefer, D. W.; Keefer, K. D. *Proc., Mater. Res. Soc. Symp. 32,* (1984) 1.
6.   Scherer, G. W. *J. Non-Cryst. Solids,* 109, (1989) 183.
7.   Zarzycki, J. In *Ultrastructure Processing of Ceramics, Glasses, and Composites;* Hench, L. L.; Ulrich, D. R. Eds.; Wiley: New York, 1984, p. 27.
8.   Richardson, W. M. O. *Polymer Engineering Composites,* Applied Science: London, 1977.

RECEIVED October 11, 1994

Chapter 9

Multiple Size Scale Structures in Silica–Siloxane Composites Studied by Small-Angle Scattering

G. Beaucage[1], T. A. Ulibarri, E. P. Black, and D. W. Schaefer

Sandia National Laboratories, Albuquerque, NM 87185

The physical properties of *in situ* produced composites, such as the TEOS-polydimethylsiloxane hybrids, are related to the complex interaction of structural features from the nano- to macroscopic scales. The nature of these structural interactions is important to understanding and controlling mechanical properties. We believed that the smallest scale structures, in the nanometer range, correlate with properties such as the modulus while large-scale structures in the micron range affect failure in these materials. In this paper we discuss techniques for the analysis of structural features and the interrelation of these features over a wide range of length scales using small-angle light, x-ray and neutron scattering (SALS, SAXS, SANS). The combination of data from a number of different instruments along with a new unified data analysis approach allows for characterization of the interaction between these different structural features.

Hybrid polymer/ceramic materials present a formidable challenge to morphologists. These systems often display multiple structural levels on length scales ranging from Angstroms to millimeters. The combination of small-angle scattering data with fractal approaches has lead to some relief in dealing with these systems. Fractal approaches describe power-law regimes often observed in measured scattering profiles. However, interpretation of such power-law functions is dangerous when self-similar structures exist in narrow size limits (*1*). All real systems are expected to show structural limits to these power-laws. For example, the mass-fractal power-law scattering from a polymer coil displays limits at the overall radius of gyration for the polymer coil and at the persistence length of the polymer chain.

By combining absolute intensity measurements from small-angle light scattering (SALS), ultra-low angle x-ray scattering (*2*) and conventional SAXS (*3*) or SANS (*4*), with diffraction data, morphologies are described from microns to Angstroms. Analysis of such extensive data requires a new approach that accounts for the interaction of structural features at a number of size levels. Although several semi-

---

[1]Current address: Department of Materials Science and Engineering, University of Cincinnati, Cincinnati, OH 45221–0012

0097–6156/95/0585–0097$12.00/0

empirical equations have been developed to fit length-scale-limited power-law scattering (5), none are adequate over five decades in length-scale. Beaucage (6), on the other hand, has developed a general equation that is capable of describing scattering functions containing multiple length-scales (Guinier regimes) separated by power-law scaling regimes. This unified approach is applied to scattering data from *in situ*-produced silica/siloxane composites in this paper. The approach is quite general, as can be seen in the examples given from this system, and can be easily applied to other hybrid polymer/ceramic or organic/ceramic systems.

A UNIFIED EQUATION FOR SMALL-ANGLE SCATTERING.

Beaucage (6), describes how Guinier's law in exponential-form and structurally limited power-laws can be derived from mutually exclusive scattering events. In the simplest case, the observed scattering is a sum of two components.

$$I(q) \approx G \exp\left(\frac{-q^2 R_g^2}{3}\right) + B \left[\frac{(erf(q R_g/\sqrt{6}))^3}{q}\right]^P \tag{1}$$

The first term describes an exponential decay in scattering at a characteristic size, R_g, for one structural level in a material such as phase-separated domains. The second term describes a power-law decay in scattering which follows the exponential regime. The power-law might be due to surface scattering from phase-separated domains whose radius of gyration is described by the first term. Such surface scattering has a limit at low-q described by the error function. $G=N_p \rho^2 V^2$ is the classic Guinier prefactor and B is a prefactor specific to the type of power-law scattering, specified by the regime in which the exponent, P, falls. The momentum transfer, q, has the units (length)$^{-1}$ so large-q scattering probes small length-scales. For Porod's law (scattering from sharp interfaces), for example, P=4 and $B=2\pi\rho^2 S$. For a Gaussian polymer, P=2, and $B = \frac{2 G}{R_g^2}$ (6). For polymeric mass-fractals of arbitrary mass fractal dimension, d_f, Beaucage (6) has shown that B is given by $B = \frac{d_f G \Gamma(d_f/2)}{R_g^{d_f}}$. This is termed a polymeric constraint (6) in analogy to the Debye equation for polymer coils. Generally, for surface fractals 4>P>3, while for mass fractals P<3, and for diffuse interfaces P>4 (1). The error function (erf) is available in a number of fitting programs (7) or can be calculated using an asymptotic expansion (8).

Since the crossover error function incorporates the same parameter (the Guinier radius, R_g), as the exponential term, no new parameters are generated by equation 1 compared to local Guinier and power-law fits. Equation 1 is a good approximation for randomly distributed model structures (spheres, disks, rods, and polymer coils) that display a single length-scale. In order to consider multiple-size-scale structures, however, the high-q, small-size limit to power-law scattering must also be considered.

Accounting for Multiple Structural Levels Using the Unified Approach.

All real power-law scattering regimes are limited by structural features in the material at both the small and large size limits. The large size (low-q) limit is described using the error function term in equation 1. We have used an exponential prefactor to describe the high-q, small-size, structural limit for fractal scaling (6). The two size limits to power-law scattering ($qR_{gi} \approx 1$ and $qR_{g(i+1)} \approx 1$) can be described through an extension of equation 1,

$$I(q) \approx \sum_{i=1}^{n} G_i \exp\left(\frac{-q^2 R_{gi}^2}{3}\right) + B_i \exp\left(\frac{-q^2 R_{g(i+1)}^2}{3}\right)\left[\frac{(\mathrm{erf}(q\,R_{gi}/\sqrt{6}))^3}{q}\right]^{P_i} \quad (2)$$

where n is the number of structural levels observed in the scattering pattern. The first term describes a large-scale structure of average size R_{gi} composed of small-scale structures of average size $R_{g(i+1)}$. The second term allows for mass-, surface-fractal or diffuse-interfacial power-law regimes for the large structure. Using equation 2 we can parameterize scattering from a system with multiple-size-scale features. It should be noted that, although equation 2 appears cumbersome, no new parameters have been introduced over local fits using Guinier exponentials and power-laws. Equation 2 can be extended to describe an arbitrary number of interrelated structural features at different size-scales. Such extensions, however, can only be justified when data extends over many decades in q.

Figure 1 is a combination of light scattering (9), ultra-low-q SAXS and pinhole SAXS data from a 1% suspension of a conventional fumed silica filler (Hi-Sil 233®) in a rubber matrix. The expanding lines in the structural models of figure 1 indicate magnifications to the next structural-level of the material in decreasing size from left to right. The data has been fit using equation 2 as shown by the dotted line that runs

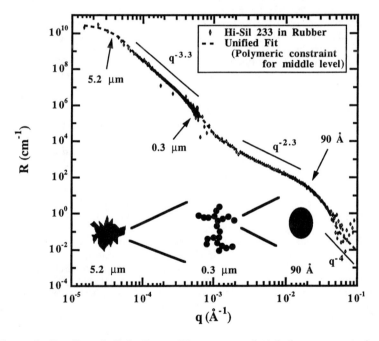

Figure 1. Small-angle light, Bonse-Hart x-ray, and pinhole x-ray scattering for a conventional silica filler (Hi-Sil 233®) in a polymer matrix. The fit is to equation 2 with n=3. The expanding lines in the structural model at the bottom indicate magnifications to the next structural level. These structural levels act as limits to the observed power-laws.

through the data points. The conventional filler displays three levels of structure in the size-range observed. The parameters for the fit using equation 2 are n=3 and $G_1 =$ 3.4×10^{10}, $R_{g1} = 5.2$ μm, $B_1 = 2 \times 10^{-5}$, $P_1 = 3.3$, $G_2 = 9500$, $R_{g2} = 0.28$ μm, $B_2 =$ 0.0023, $P_2 = 2.3$, $G_3 = 69$, $R_{g3} = 90$ Å, $B_3 = 8.5 \times 10^{-7}$, $P_3=4$.

At large sizes, in the light scattering regime (left), 5 μm aggregates of the filler are observed. In the log-log plot of absolute intensity or Rayleigh ratio (10), R, versus q an average structural size is observed as a knee in the scattering pattern. This knee corresponds to a region of exponential decay in scattering where Guinier's Law, or the first term in equations 1 and 2, applies. These 5 μm aggregates can be described as fractally rough particulates over a decade of size. Fractally rough surfaces display power-law scattering with a slope of -3 to -4. At about 0.5 μm this description begins to fail. At this point the radius of gyration of a mass-fractal structure is observed as a second knee in the scattering pattern. Since R_{g2} describes the limit of the low-q power law it is assumed that the large structures are composed of the intermediate-scale mass fractals. The unified approach is the only technique capable of demonstrating this relationship.

Mass fractals display power-law scattering with a slope generally in the range of -1 to -3. The slope of the power law decay is equal to the negative of the mass fractal dimension. An analogy can be made between this branched mass-fractal structure and a branched polymer chain. Both display two size scales, one pertaining to the overall radius of gyration at large sizes and the second pertaining to a substructural or monomeric size at small sizes. Between these two size scales a regime of mass-fractal scattering is observed. In this case the mass-fractal description extends over one and a half decade of size. The power law prefactor, B_2, for this mass fractal regime is calculated from the polymeric mass fractal constraint given above (page 2). At about 100Å the sub-structure of the mass fractal is observed. For the silica filler this sub-structure is analogous to a Kuhn-step or persistence length for a polymer coil. The sub-structure displays close to a power-law of -4 at high-q that describes 100Å glass sub-structures with smooth surfaces. The surface area of the silica filler could be obtained from the high-q power-law regime if data extending further in q had been obtained.

Figure 2 shows a similar scattering pattern from a 2-step *in situ* filled siloxane sample. The data has been fit to equation 2 as shown by the dashed line that runs through the data points. Again, the expanding lines in the graphic inset indicate magnification of the field of view to the next structural level. The parameters for the fit using equation 2 are n=3 and $G_1 = 2 \times 10^8$, $R_{g1} = 0.92$ μm, $B_1 = 6.1 \times 10^{-8}$, $P_1 =$ 4.2, $G_2 = 114$, $R_{g2} = 150$ Å, $B_2 = 0.00043$, $P_2 = 2.67$, $G_3 = 1.53$, $R_{g3} = 12.1$ Å, $B_3 =$ 0.00027, $P_3=4$. In this case the intermediate-size radius of gyration, R_{g2}, was not used to limit the large scale power law since the structures are believed to be independent. Poorer fits result from attempts to force this cutoff.

In the 2-step process crosslinks are first formed in the matrix material. The filler is subsequently produced by hydrolysis of tetraethylorthosilicate (TEOS). The knee in the scattering pattern at large sizes (small-q) indicates micron-sized phases in the material. The power law that follows to the right of this micron-size phase is steeper than -4 indicating that the micron-size structure has a diffuse interface. This is in contrast to the fractally-rough interface of the micron-size fumed-silica structure in Figure 1.

Figure 3 describes the difference between a diffuse interface with a power law steeper than -4 and a fractally rough interface with a power law shallower than -4. In the diffuse case, a plot of composition versus angle is constant at a given distance from the center of a spherically symmetric particle. A plot of composition versus radial distance shows a monotonic decay. For a fractally rough particle the angular composition plot shows random discontinuities near the surface while the radial composition plot shows a single sharp transition from the particle phase to the matrix phase.

The power-law scattering regime, that indicates diffuse-surface scaling, extends over a larger range of q than the fractally-rough, surface-scaling-regime in the conventional fumed silica (figures 1 and 2). At about q = 0.01 Å^{-1} a weak second knee is observed in the scattering pattern that describes a structure whose radius of gyration is about 150 Å. As noted above, R_{g2} is not used to cut off the low-q power law from level 1 in the fit. From kinetic studies it is known that the micron scale structure is present long before nanoscale features become apparent. This indicates that the micron and nanoscale features are independent.

The power-law regime corresponding in size to between R_g = 150 and 12 Å reflects a mass fractal structure for the R_g = 150 Å domains. Because the slope is steeper than the mass-fractal power-law slope observed for fumed silica it can be said that the domains are similar in morphology but more ramified or branched than those observed in fumed silica. The polymeric constraint (page 2) was used to limit the power law prefactor in the mass fractal regime of the fit in figure 2. The *in situ*-formed mass-fractal structure has a much smaller overall radius of gyration when

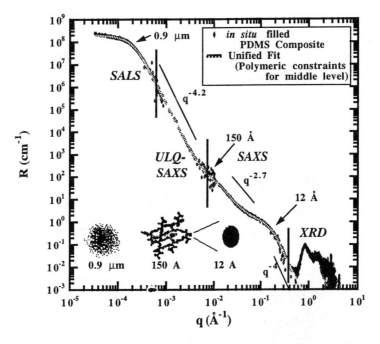

Figure 2. An example of the use of equation 2 in a hybrid silica/siloxane composite material (2-step synthesis with no correlations of nano-phases). The data includes small-angle light (SALS), Bonse-Hart x-ray (ULQSAXS), pinhole x-ray (SAXS) and diffraction data in transmission mode (XRD). The expanding lines in the structural model indicate magnification to the next structural-level in the material. The material displays structural features on three size scales and has fairly well behaved mass scaling behavior that separates these size scale regimes.

compared with Hi-Sil 233. Similarly, the sub-structure of the intermediate mass-fractal aggregate displays a smaller radius of gyration when compared with the conventional filler of figure 1. The mass-fractal scaling regime for the *in situ* filled material of figure 2 is otherwise similar to that of conventional fumed silica. At highest-q the diffraction regime is observed. The humps in this regime correspond to the amorphous halo of PDMS combined with the amorphous halo of the silica filler. The smaller sub-structural elements in the *in situ* filled system are expected to lead to a much higher surface area when compared with the conventional filler.

The high-q limit to the application of equations 1 and 2, that is the limit to the small angle scattering regime, is experimentally reached when the amorphous halo of the diffraction regime begins to dominate the scattering pattern. This is seen in figure 2 at about $q = 0.6$ Å^{-1}. This corresponds to a Bragg size, $2\pi/q$ of about 10 Å. This is the limit to the point-scatterer approximation that is implied in equations 1 and 2.

Correlated Systems

Correlated, nano-phase domains of *in situ* produced silica filler are often observed in scattering patterns from these hybrid-materials, figures 4, 5 and 7. Correlation of these filler domains leads to a correlation peak in the scattering pattern. A description of this correlation peak can be incorporated into the general description of scattering from structurally complex materials given by equations 1 and 2. The scattered intensity for correlated systems is given by (6, *11, 12, 13*),

$$\overline{I(q)} = A\, F(q)^2\, S(q) \tag{3}$$

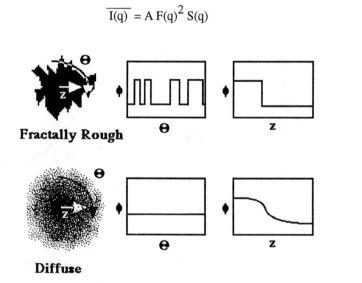

Fractally Rough

Diffuse

Figure 3. Demonstration of the difference between a diffuse and a fractally rough interface. Schematic plots are of composition, ϕ, versus radial angle, θ, at a fixed radial distance close to the surface and of composition versus radial distance from the center of mass, z.

where "A F(q)^2" corresponds to the scattered intensity for non-correlated domains of equation 1 and S(q) is a factor that accounts for weak correlations of the domains. In modifying equation 2 to account for correlations, S(q) is used to modify the two terms of a single structural level. The remaining structural levels are not effected by S(q).

We have previously used a semi-empirical function based on Born-Green theory for S(q) which describes correlations of colloidal particles or domains in terms of a radius of correlation, ζ, and a packing factor, k (*11, 12, 13*),

$$S(q) = \frac{1}{1 + k\,\theta} \tag{4}$$

where θ is the "form factor" for structural correlations occurring at an average radial distance ζ (*13*),

$$\theta = 3\,\frac{\sin(q\zeta) - q\zeta\cos(q\zeta)}{q\zeta^3} \tag{5}$$

and k describes the degree of correlation (0 < k < 5.92). Equation 5 describes the amplitude of scattering from a sphere. Substitution of more complicated amplitude functions are possible under special circumstances such as with very strong correlations (*14*). "k" is equal to 8 times the ratio of the average "hard-core" volume

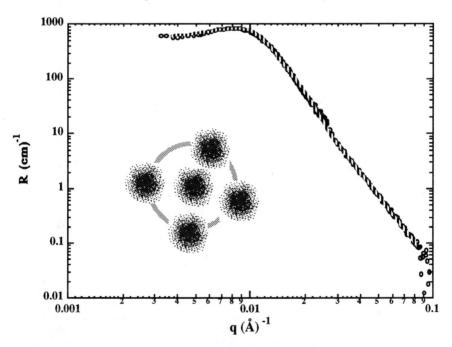

Figure 4. Fit of equation 3 (dashed line) to SANS data (circles) from an *in situ*-filled siloxane rubber with a precursor PDMS molecular weight of 18 kg/mole. Data is from a 1-step synthesis. The structural model (inset) indicates correlated particulates with diffuse interfaces. The shaded circle in the structural model is at a distance ζ from the central particle.

of a domain, v_0, to the average-volume available to the domain. v_1.   The maximum value of k, 5.92, is obtained by calculation of this ratio for hexagonal or cubic close-packed crystal structures. In this model the average-volume available to a domain is the volume not occupied by other similar domains divided by the number of domains in the scattering volume. For application of equations 1 and 2 in equation 3, k must have a low value, k < 4, consistent with weak correlations. Weakly correlated systems are sufficiently random that the ideas discussed above concerning uncorrelated systems are adequate. Weakly correlated particles are expected to display two size scales, ζ, the average Bragg-like spacing between domains and the radius of gyration, R_g, for the domains. Generally, $R_g \leq \zeta$. ζ corresponds well with the Bragg-spacing of $2\pi/q_{max}$ where q_{max} is the value of q for the maximum peak intensity. This approach is applicable to correlated particulate and spinodal morphologies. It can not distinguish between these physically different morphologies.

Equation 3 can be used to determine the correlation distance and the average size of the filler domains as well as the surface characteristics of the filler (11). Figure 4 shows a typical fit (dashed line) to data from an *in situ* filled silica-siloxane

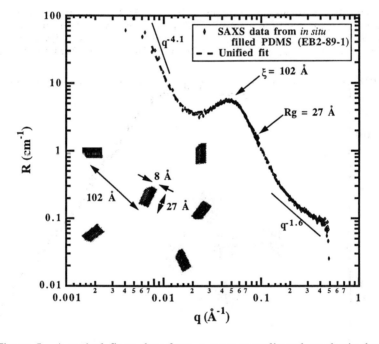

Figure 5. A typical fit to data from a more complicated synthesis that displays low-dimensionality power-law scaling of correlated nano-domains. Data is from a 1-step synthesis. The power-law at high-q is close to -1.6. Since B_2 does not follow the polymeric constraint (page 2) the nano-domains are probably asymmetric. This is shown in the structural model, inset.

composite prepared in a 1-step synthesis. The q-range of this data is much smaller than the data of figures 1 and 2. In the data of figure 4 only the nano-scale features are observed. The schematic representation of this nano-structure, shown in the inset, shows an average correlation distance indicated by the shaded ring of radius ζ and a particulate domain morphology. As mentioned above, equation 3 could also describe a spinodal structure. The correlation distance for the fit is $\zeta=532\pm1$Å, degree of correlation $k=1.99\pm0.01$, $R_g=166.1\pm0.2$ Å, $G=1508\pm4$ cm^{-1}, $P=4.24$, $B=2.23e-6$ cm^{-1}. Since P is greater than 4 the nano-phase/matrix interface is diffuse.

In a 1-step synthesis crosslinking of the matrix and formation of the filler occur simultaneously. The 1-step synthesis characteristically leads to correlation of the nano-scale domains. These correlated domains occur either as diffuse-interfaced structures as evidenced by a power-law steeper than -4 (figure 4) or as low-dimensionality structures with power-laws between -1 and -2.8 as shown in figure 5. The dashed line in figure 5 is a fit to equation 2 using equation 3 to modify structural level 2. Levels 1 and 3 are not modified by equation 3. The parameters for the fit using equations 2 and 3 (for level 2) are $n=3$ and $B_1 = 5.3 \times 10^{-8}$, $P_1 = 4.1$, $G_2 = 8.1$, $R_{g2} = 26.8$ Å, $B_2 = 5.3 \times 10^{-8}$, $P_2 = 1.57$, $k_2 = 2.2$, $\xi_2 = 102$ Å, $G_3 = 0.12$, $R_{g3} = 8.1$ Å. G_1 and B_3 are 0 since the low-q Guinier and high-q power law are not observed. Since B_2 does not follow the polymeric mass-fractal constraint (page 2) the substructure is probably asymmetric. For example a rod structure would yield a power-law slope of -1 while a lamellar or platelet structure a power-law slope of -2. Beaucage (6) has termed these structural or apparent mass fractals. In this case the structure could be distorted lamellae as shown in the inset of figure 5. These asymmetric nano-structures are formed under special synthetic conditions such as when crosslink sites are topologically constrained or when specialized silica precursors are used (*15*). If the distorted lamellae were more closely packed a more specialized correlation amplitude function might be used to replace equation 5. For example a lamellar correlation amplitude that is one dimensional yields a correlation or stacking distance of 92 Å for the data of figure 5. This compares well with a correlation distance of 102 Å for the three-dimensional spherical correlation model used in the fit of figure 5. The fit to the data for these two models is otherwise indistinguishable from that of figure 5. Since the correlation distance is much larger than the radius of gyration for the nano-domains the spherical correlation model seems appropriate here. This assumes that the lamellae are randomly oriented with respect to each other.

The presence of correlated domains does not alter the power law scaling regime. In figure 6 this is demonstrated by varying "k" for the fit of figure 4.

1-STEP VS. 2-STEP SYNTHESIS.

An example of an application of this fitting approach to small-angle scattering data involves a study of the TEOS/hydroxyl-terminated PDMS system (*11*). As is detailed in reference 11, TEOS is hydrolyzed to silica in the formation of these insitu composites. Synthesis of this hybrid material can be carried out in a single-step whereby formation of crosslink sites at the hydroxyl end-groups and formation of silica filler occurs simultaneously. Alternatively, in a 2-step process crosslink formation can occur in a first step using a stoichiometric amount of TEOS. Formation of silica in the crosslinked PDMS matrix is carried out in a separate step by swelling with a TEOS/catalyst mixture. In the 1-step synthesis correlated domains are generally observed. Data are fit using equation 3. The packing factor, "k", is generally close to 2 indicating weak correlations, figures. 4, 5 and 7. In the 2-step synthesis the domains are generally less correlated. In this case, data can be fit with a non-correlated model, equations 1 or 2, figures 2 and 8.

For the 1-step synthesis the radius of gyration for the nano-domains, determined using equation 3, scales with the molecular weight of the precursor hydroxyl-

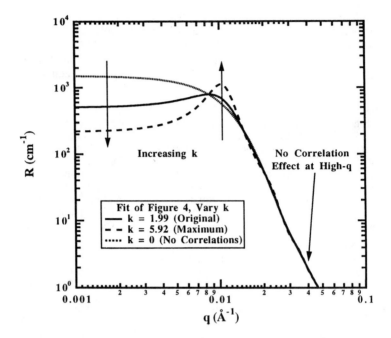

Figure 6. Demonstration of the effect of varying the packing factor "k" on the scattering pattern for the data of figure 4. Packing of the domains does not affect the power-law scaling regime at high-q.

terminated PDMS raised to a power of about 0.3 to 0.4 depending on the amount of catalyst present, figure 9. For the 2-step synthesis, in which crosslinks are first formed followed by swelling with the glass precursor and *in situ* formation of the filler, the radius of gyration follows a much weaker power-law dependence on molecular-weight especially at high loading levels, figure 10.

There are at least two explanations for the observed differences between the 1-step and the 2-step synthesis (*11*). The simplest explanation is that in the 1-step synthesis hydroxyl end-groups on the PDMS chains, serve as nucleation sites for the formation of silica domains. These hydroxyl groups eventually form crosslink sites whether in the precipitated silica phase or in the PDMS phase. Partially hydrolyzed-TEOS and water are expected to be selectively attracted to regions rich in the hydroxyl end-groups where an alcohol-like environment is found. The end-to-end distance for a Gaussian polymer scales with molecular weight raised to a power of 1/2. When filler particle nucleation is limited to chain-end-bound crosslink sites, the amount of silica precursor available for filler growth is proportional to the average volume associated with a crosslink site. The volume of a filler particle would be proportional to the molecular weight to the 3/2 and the radius of the particle proportional to the molecular weight to the 1/2. If a small number of nucleation sites are not associated with chain ends the molecular weight dependence is expected to be weaker. This is especially the case at the low-crosslink density, high-precursor-molecular-weight end of the spectrum. This model is termed the chain-end model.

Depletion of the solution of reacting TEOS enhances the correlation of the filler

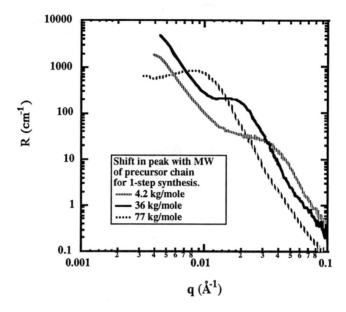

Figure 7. Small-angle neutron scattering data from a 1-step synthesis varying molecular weight of the precursor PDMS. Nano-phases in the 1-step synthesis are typically correlated as indicated by the correlation peak. Correlated data is fit using equation 3 (not shown). The power law at high-q indicates diffuse-interfaced domains.

phase. Where end-groups are close, filler particle growth is limited by the amount of TEOS available. Growth is encouraged where chain-ends are well separated due to ample supply of TEOS. Ionic repulsion may also play a role in the development of correlations.

In the chain-end model, preformed crosslink sites are sterically unfavorable sites for the formation of silica filler particles. This is the case for the 2-step synthesis. In the 2-step synthesis the formation of the filler phase is not intimately related to the network. Because of this only a weak dependence on the precursor molecular weight is expected. Nucleation of silica domains would occur at incidental heterogeneous nucleation sites of which there are fewer especially in the low PDMS molecular-weight-end of the spectrum. Thus one might expect larger particles compared to the 1-step synthesis at high crosslink-densities (figures 9 and 10).

An alternative explanation (*11*) for the relationship between filler size and network molecular weight involves the rubber elasticity of the network. This approach is based on the work of de Gennes concerning phase-separation in a crosslinked polymer/polymer system (*16*). de Gennes and later Bettachy (*17*) introduce a term describing the elasticity of the network into the free energy. This term opposes large-scale phase-separation. The elasticity term leads to the prediction of a correlated phase-separated structure at early stages. The correlation distance is expected to scale with the square-root of the molecular-weight between crosslinks due to the elasticity term. This would lead to a correlation peak in the scattering pattern that is expected to scale with the molecular-weight to the 1/2.

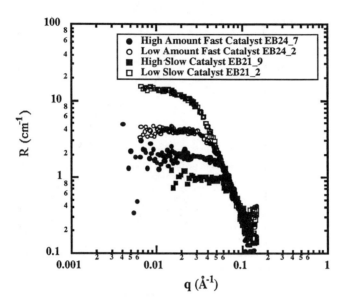

Figure 8. Small-angle neutron scattering from a 2-step synthesis showing the absence of significant phase correlation. Non-correlated data is fit using equations 1 or 2 (not shown). The power law at high-q indicates diffuse-interfaced domains. Data courtesy C. C. Han, B. Bauer to be published.

In the hybrid system the growing glass phase excludes the PDMS network. In this way an analogous condition to that of the de Gennes model exists in which the elasticity of the network might inhibit large-scale phase-separation. Since this explanation for correlations of domains is based on a thermodynamic approach it is termed a pseudo-equilibrium model here. For the pseudo-equilibrium model the association between network elasticity, phase-size and correlation distance would be expected to be most evident in the case of the 2-step synthesis. The approach is somewhat obscure in the case of a 1-step synthesis in which crosslink sites form simultaneously with filler domains. Further, the effect of polymer entanglements on this approach is not clear. At high molecular weights, chain entanglements are expected to act as physical crosslinks such that the elasticity of the network and the phase size should plateau at molecular weights above the entanglement molecular weight. For PDMS the entanglement molecular weight is in the tens of thousands of gm/mole. An entanglement effect is not observed in these systems. An argument can be made that in the 1-step synthesis local crosslinks form rapidly and long before network percolation is reached. These local crosslinks might be sufficient to locally restrict the growth of the silica phase.

In the presence of high catalyst concentrations, where the formation of the silica phase occurs at a more rapid rate, the observed scaling of phase-size with molecular weight is closer to $M^{1/2}$ when compared with lower catalyst concentrations for the 1-step synthesis (figure 9). For the chain-end model there is an inherent competition between crosslink formation and silica formation. If crosslinks form first, silica formation at the cross-link sites is sterically inhibited. Acceleration of the silica-

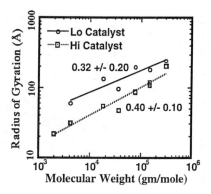

Figure 9. Dependence of the radius of gyration on molecular weight for the 1-step synthesis . Data is from fits using equation 3. Several models predict a power-law of slope 1/2.

domain formation through a higher catalyst concentration is expected to lead to a closer approach to the 1/2 scaling law according to the chain-end model.

For the pseudo-equilibrium model, high catalyst concentrations in the 1-step synthesis leading to faster formation of the silica phase would be expected to lead to less correlated domains and departure from the 1/2 scaling law since the silica formation might out-pace the formation of the network. Data from high and low catalyst concentrations support the chain-end model, figure 9.

The source of correlated phases which scale with molecular weight of the precursor is not resolved. Support for the idea that the glassy phase preferentially forms at chain ends comes from the fact that the 1-step synthesis leads to better correlated phases that more closely follow an $M^{1/2}$ law. Additionally, the observation that higher catalyst concentrations lead to a closer approach to the $M^{1/2}$ law also supports this model. An experiment is suggested whereby the crosslink density is varied through introduction of non-polar crosslinking groups along the chain while retaining hydroxyl terminated endgroups. Through variation in the molecular weight between these non-hydroxyl crosslinking groups the network elasticity could be modified independently of the distance between hydroxyl groups. The end-group model would predict scaling of the particulate size with the molecular weight between hydroxyl groups while the pseudo-equilibrium model would suggest a dependence on the overall crosslink density. We are in the process of carrying out this experiment.

Summary

The unified approach to small-angle scattering (6) is useful in examining multiple-size-scale structures in silica/siloxane composites. It was demonstrated that the ideas developed for random structures are also applicable to weakly correlated systems. The unified approach is flexible enough to extract useful information from data in which micro- and nano-structures overlap. This approach is the first to simultaneously account for Guinier scattering and power-law scattering of arbitrary type, i.e. mass-, surface-fractal, Porod or diffuse-interfacial scattering all of which are observed in the in situ-filled siloxane systems.

A discussion of possible sources of correlated phases in *in situ* produced PDMS composites suggests future experiments in which various models for correlated phases can be tested.

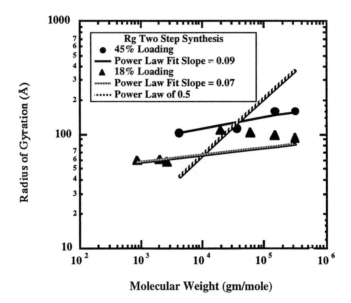

Figure 10. Dependence of the radius of gyration on molecular weight for the 2-step synthesis. In the 2-step synthesis the radius of gyration has a much weaker dependence on precursor molecular weight than in the 1-step synthesis.

Acknowledgments

This work was performed at Sandia National Laboratories supported by the US Department of Energy under contract #DE-AC04-94AL85000.

Literature Cited.

1. See for example the review by Paul Schmidt, Schmidt, P. W. *J. Appl. Cryst.* **1992,** *15*, 567 for a general survey of power-law scattering, seminal work by Porod and Ruland in Porod, G. *Acta Physica Austriaca* **1948**, 2, 133, Porod, G *Kolloid-Z.* **1951**, 124, 83, Porod, G *Kolloid-Z.* **1952**, 125, 51, Porod, G *Kolloid-Z.* **1952**, 125, 108 and Ruland, W. J. Appl. Cryst. **1971**, 4, 70, and references in Korberstein, J. T.; Morra, B.; Stein, R.S., *J. Appl. Cryst.* **1980**, *13* , 34. Also see Debye, P.; Henderson, H. R.; Brumberger, H., *J. Appl. Phys.* **1957**, *28*, 679 and Fischer, M. E.; Burford, R. *J. Phys. Rev.* **1967**, *156*, 583 for references to models of complex systems.
2. National Institute of Standards and Technology beam-line at Brookhaven National Laboratory's National Synchrotron Light Source. G. Long, D. Fischer, G. Beaucage, D. Schaefer.
3. 10-meter small angle x-ray camera at Oak Ridge National Laboratory. J. S. Lin, G. Beaucage, D. W. Schaefer.
4. Sans measurements were performed at Oak Ridge National Laboratory's High Flux Isotope Reactor, National Institute of Standards and Technology's Neutron Scattering Facility and at Los Alamos National Laboratory's LANSCE facility.
5. Lin, J. S. *J. Polym. Sci., Polym. Phys. Ed.* **1990**, *86*, 434.

6. Beaucage G., submitted *J. Appl. Crys.* This approach has been extended to arbitrary polymeric-mass-fractals, Beaucage, G., submitted *J. Appl. Cryst.*

7. For example, "Igor" by WaveMetrics Inc., P.O. Box 2088, Lake Oswego, Oregon 97035. Also the error function is available in some versions of C.

8. For example an empirical function that describes the error function is given in, Press, W. H.; Flannery, B. P.; Teukolsky, S. A.; Vettering, W. T. *Numerical Recipes in C, the Art of Scientific Computing;* Camb. U. Press.: New York, NY, pp. 176.

9. Light scattering data was scaled to overlap with Bonse-Hart x-ray data. This is only possible when close to a half decade of overlap in q is available and when contrast in the light scattering experiment is weak. For purposes of the discussion presented here this is a good approximation.

10. The equivalence of the Rayleigh ratio common in light scattering measurements and the absolute intensity or cross section, $d\Sigma/d\Omega$, measured in neutron and x-ray scattering experiments was detailed by Higgins and Stein, Higgins, J. S.; Stein, R. S. *J. Appl. Cryst.* **1978**, *11*, 346. The Rayleigh ratio is equivalent to the probability that a neutron, for example, will be scattered into solid angle Ω for unit volume of the sample. Absolute intensity in cm^2 can be converted to Rayleigh ratio, which has units of cm^{-1}, by dividing by the irradiated volume. Absolute intensities in cm^{-1} are commonly used at national user facilities since sample and instrumental dimensions are removed by dividing by the irradiated volume. Also see Russell, T. P.; Lin, J. S.; Spooner, S.; Wignall, G. D. *J. Appl. Cryst.* **1988**, *21*, 629.

11. Beaucage, G.; Schaefer, D. W.; Ulibarri, T.; Olivier; B. J., in preparation and short summary of some of this work in MRS Bulletin April 1992 Ulibarri, T.; Beaucage, G.; Schaefer, D. W.

12. Guinier, A; Fournet, G. *Small-Angle Scattering of X-rays*; John Wiley and Sons Inc.: New York, NY, 1955, pp. 42-65. Also see Porod, G. In *Small Angle X-ray Scattering*; Glatter, O.; Kratky, O., Eds.; Academic Press: New York, N. Y., 1982 as well as Feigin, L. A.; Svergun, D. I. *Structural Analysis by SAXS and SANS*; Plenum Press: New York, NY, 1987.

13. Guinier, A.; Fournet, G. *Small-Angle Scattering of X-rays*; John Wiley and Sons Inc.: New York, NY, 1955; pp. 55.

14. We have used a function, θ, based on the form factor for spheres to describe correlations. Since this function is damped by the packing factor, "k", the exact form of this function is not critical when k is small. The spherical function is convenient since it is well behaved and easily interpreted in a physical sense, that is as an average radius of correlation. In highly correlated systems, k > 4, the sphere function may become inappropriate, especially when fits to second and higher order peaks are desired. In these cases more specialized functions may be appropriate. We have also used equations 3, 4 and 5 to describe packing effects in filled systems. When k is small, k < 2, a peak is not observed but a shift in R_g and G is observed which can be described using this approach, figure 6. It is often necessary to obtain data on a dilute suspension of the filler particle, $\approx 1\%$, to used this approach to describe such packing.

15. Ulibarri, T. A.; Black, E.; Loy, D.; Jamieson, G., in preparation.

16. de Gennes, P. G. *J. de Phys. Lett. France* **1979**, *40*, 69.

17. Bettachy, A.; Derouiche, A.; Benhamou, M.; Daoud, M. *J. Phys. I* **1991**, *1*, 153.

RECEIVED September 15, 1994

Chapter 10

Composite Polymer Colloid Nucleated by Functionalized Silica

E. Bourgeat-Lami[1], P. Espiard[1], A. Guyot[1], S. Briat[2], C. Gauthier[2], G. Vigier[2], and J. Perez[2]

[1]Laboratoire de Chimie et Procédés de Polymérisation, Centre National de la Recherche Scientifique, B.P. 24, 69390 Vernaison, France
[2]Group d'Etude de Métallurgie Physique et Physique des Matériaux, Institut National des Sciences Appliquées, Unité de Recherche Associée 421 du Centre National de la Recherche Scientifique, Villeurbanne, France

Emulsion polymerization of ethylacrylate has been carried out in the presence of functionalized and non-functionalized commercial silica. Both polymerization kinetics and particle size were studied. Particular attention was focused on the characterization of grafting onto the silica surface. It is demonstrated that the nature of the interface depends on the type of silica employed. Polyethylacrylate formed is only adsorbed to the surface of non-functionalized silica, while the existence of a chemical bond between polymer molecules and functionalized silica has been demonstrated using different techniques. Because of this bonding to the rigid surface, the molecular mobility of the attached polymer chains is hindered and their molecular conformation is changed. As a consequence, we show that the mechanical strengh of the composites is strongly influenced by the nature of the interfacial interaction between the continuous and dispersed phase.

During the last ten years, a growing interest has been devoted to hybrid materials containing both inorganic and organic parts dispersed at the molecular level or in nanometric domains (1-3). In particular, composite polymers are specially attractive in many fields such as surface coatings, adhesives, synthetic elastomers, latex paints and other speciality products. So far, different approaches have been used to develop novel hybrid materials. More recently, encapsulation of small inorganic silica particles by an emulsion polymerization process has been investigated by some of us (4-6). The main advantage of the emulsion polymerization process is the possibility to produce an ideal polymer - filler system characterized by an inorganic core surrounded by a polymer shell. However, the main difficulties consist of the strong tendency of silica particles to form aggregates in water and of the possibility of some latex particles to nucleate besides silica so that the expected polymeric boundary layer on the inorganic particle surface will not be formed. The conditions required have been extensively studied and presented in some previous papers (5-6). But the fact remains that in such composite materials, it is quite difficult to obtain an homogeneous distribution of the filler without any aggregates throughtout the whole polymeric matrix.

The purpose of this paper is to present new results along this line concerning composite polymer colloid nucleated by functionalized and non-functionalized silica particles of commercial origin. Kinetic curves as well as particle sizes are presented.

We also discuss the morphological aspect of our materials and give some explanations concerning the above observations. However, the main point of our study focuses on the nature of the interface. It is then demonstrated that if we carried out the polymerization reaction in the presence of the functionalized particles, some chemical bonding could be formed on the interface of polymer and filler. As a consequence, the mechanical strength of the composite material is improved because of the interfacial interaction between the continuous and dispersed phase.

Experimental

Materials   Two kinds of nano-sized silica particles (Hoechst France) were used. The hydrophilic Klebosol silica was dispersed in water (32% solid content w/w) and stabilized by ammonium ions. The hydrophobic Highlink OG 100 was functionalized with methacryloyl groups and dispersed in the monomer used for grafting (i.e. the 2-hydroxy-ethylmethacrylate: 33% solid content w/w).

Ethyl acrylate was distilled in vacuo before used, and the water was deionized. The other reagents were used as supplied.

Emulsion Polymerization   The 250 ml jacketed glass reactor was charged with the aqueous dispersion of silica. The nonylphenyl polyoxyethylene surfactant (NP30: 0,09g/l) was then added in an amount corresponding to the partial coverage of silica without the formation of micelles in water. pH was fixed at 9.5 by adding small quantities of ammonium hydroxide. Degassing was carried out overnight under gentle stirring. The following morning, part of the monomer (10%) was introduced and its adsorption at the silica surface was allowed to proceed for one hour at ambient temperature without stirring. The temperature was then raised to 60°C and the initiator (potassium persulfate 0.7 g/l), dissolved in 10 cc water and added at 6ml/h to start the polymerization. Beginning one hour later, the remaining part of the ethyl acrylate was introduced continuously at 1ml/h. Sodium dodecyl sulfate (SDS 1g/l) was also dissolved in 10 cc water and added slowly at 2ml/h. The stirring rate was controlled at 300 rev/min. The reaction was carried out for up to 30 hours, and small samples were taken at various times for the determination of conversion and particle size distribution.

Characterization   Physical properties of the inorganic materials were determined using various techniques. Density measurements were performed with a picnometer and the BET method was used to evaluate the specific area. Particle size and size distribution were characterized using both dynamic light scattering and transmission electron microscopy (Jeol apparatus). Finally, the number of double bonds available at the silica surface was determined by carbon analysis or iodometry titration (7) after washing of physically adsorbed molecules. Intensive washing was performed using a series of four centrifugations and redispersions into water.

The monomer to polymer conversions were determined gravimetrically. The latex particle size was obtained by conventional light scattering for intermediate samples and electron microscopy for the final products. The adsorbed polymer was separated from silica according to Scheme 1. The latex was first centrifuged (20 000 rev/min during one hour) to separate the serum. The non-grafted polymer was extracted by toluene at reflux temperature overnight. The solution was then centrifuged to separate the non-grafted polymer dissolved in toluene from the grafted polymer remaining on the silica particles. The amount of non-grafted polymer was estimated by measuring the solids content in toluene. Molecular weights of adsorbed polymers were determined by light scattering (Zimm plot).

Mechanical Properties   Films were obtained after evaporation and maturation in an aerated oven at 30°C during three weeks. Differential scanning calorimetry (DSC) was performed on a Perkin Elmer apparatus from - 80°C to + 40 °C at a heating rate of 10°C/mn. Low temperature was attained by liquid nitrogen. Dynamic mechanical spectrometry was carried out using a forced oscillation pendulum working at 0.3 Hz in a temperature range from 100 to 450 °K. Static mechanical properties were determined using an Instron machine (1 kN cell) with dumbbell shaped specimens of standard size.

Results and Discussion

Inorganic Materials   The different types of silica used are presented in Table I.

Table I. Principal characteristics of the inorganic materials

Samples	Nature of the surface	Dispersant
Klebosol	SiO_2 —O-NH_4 / —OH	Water
Highlink OG 100	SiO_2 — OCH_2CH_2O-$\overset{O}{\overset{\|}{C}}$ $\overset{H_3C}{\diagdown}C = CH_2$	2- hydroxy-ethylmethacrylate

The klebosol silica particles are non-functionalized commercial silica i.e. silica containing only hydroxyl groups on its surface. The particles are characterized by an average diameter of 13 nm and a surface area of 200 m²/g as reported in Table II. They are stabilized in water by ammonium ions and carry a negative charge since the pH of the 32 weight % dispersion is 9,2.

Table II. Physical properties of the inorganic materials

Samples	Density (g/cm³)	Specific area (m²/g)	Double bond content [a] (μmol/m²)	Diameter [b] (nm) DLS	Diameter [c] (nm) TEM	Dw/Dn [c]
Klebosol	2.1	200	0	16	/	/
Highlink	2	196	0.3	25	12	1.07

[a]As determined after intensive washing of 2- hydroxy-ethylmethacrylate.
[b]As determined by dynamic light scattering (DLS).
[c]As determined by transmission electron microscopy (TEM).

The Highlink silica are functionnalized commercial silica particles carrying methacryloyl groups on its surface. The functionnalization has been performed according to a direct condensation route of 2- hydroxy-ethylmethacrylate with the surface silanol groups. This results in the formation of a Si-O-CH$_2$ type of linkage which is known to be hydrolytically very unstable in aqueous media (8). This instability could account for the very low double bond content at the silica surface after washing and redispersion into water. In the present work, we decided to use the original silica dispersion containing 67 weight % of 2- hydroxy-ethylmethacrylate. Silica particles are dispersed in water before use. The functionalized silica particles consist in polydisperse beads of 13 nm diameter as evidenced by transmission electron microscopy (Figure 1). Using dynamic light scattering, a size of 25 nm is measured suggesting the formation of some aggregates of at least two particles (Table II).

Emulsion Polymerization of Ethyl Acrylate at the Silica Surface

Polymerization Results and Encapsulation State   The initial silica dispersion is diluted in water and a non-ionic emulsifier is added to create an affinity between silica and monomer. Ethyl acrylate and sodium dodecyl sulfate are slowly introduced by a semi continuous feed process to promote the polymerization onto the silica surface. In addition, the surfactant concentration is kept at values lower than the CMC in order to avoid the formation of new latex particles besides the organic phase encapsulating the silica. Encapsulation is demonstrated by measuring the final particle size and after comparison of the number of final latex particles to the number of initial silica particles (R ratio).
    A set of experiments have been performed. In the first series, functionalized silica is engaged. Monomer/silica weight ratios from 9/1 to 48/1 were investigated increasing the monomer/water weight ratio from 2.8 to 14.7%. In all of these experiments, the silica content is constant so that the initial particle number is maintained at the same level. In another experiment, we used non-functionalized silica particles with monomer/silica weight ratio of 16/1 and a solid content of 4.9%. Experimental conditions and some results are summarized in Table III. Polymerizations are compared to polymerization carried out in the absence of silica where all the ingredients are introduced in batch. A final experiment was performed in which the emulsifier is incorporated at once to improve the stability of the system.

Table III. Emulsion polymerization of ethyl acrylate in the presence of functionalized (*) and non functionalized silica (**). Batch process without silica (***). Emulsion polymerization with functionalized silica and SDS introduced in batch (****).

Monomer/Silica weight ratio	SiO$_2$ (%)	Conversion (%)	Solid content (%)	Diameter nm DLS	Np/l silica [a] .10^{-17}	Np/l latex .10^{-17}	R [b]
9/1 *	11	100	2.8	144	1.8	0.15	0.08
16/1 *	6	100	4.9	130	1.8	0.4	0.2
48/1 *	2	99	14.7	120	1.8	1.5	0.8
16/1 **	6	100	4.9	104	1.8	0.7	0.4
/ ***	0	100	4.9	75	0	2	/
16/1 ****	6	100	4.9	71	1.8	2.3	1.3

[a] Silica content = 0.3%, diameter = 25 nm.
[b] R = Np latex / Np silica

The above experimental data indicate that the polymerization reaction takes place in a surface layer region close to the filler since a monodisperse population of latex is obtained as demonstrated by dynamic light scattering. However, the R values, lower than one, demonstrate that the number of latex particles is not as great as expected. In fact, the final particles are larger than the particles obtained in the batch process. This tends to prove that when the emulsifier (SDS) is added continuously, the latex particles form some aggregates because of the surfactant deficiency. As a matter of fact, when the SDS is introduced at once (last line of Table III), a lower particle size is obtained (71 nm) similar to that measured in the batch process. The corresponding R value is then close to one. The reported diameters do not change much with the monomer content. The particle size should be enhanced at higher initial monomer concentration. On the contrary, the increase in the latex particle number with increasing the monomer content suggests that besides the formation of some aggregates, new particles are continuously formed in the course of polymerization. It can be then concluded that the polymerization reaction does not proceed solely around the silica particles.

Observation of Composite Powders by Electron Microscopy   The morphological aspects of our materials have then been examined using transmission electron microscopy in order to determine the final particle size and to draw conclusions on the state of encapsulation (Figure 2). A complete study concerning those morphological characterizations will be soon published (C. Gauthier, G. Thollet, G. Vigier, E. Bourgeat-Lami and A. Guyot, *Polymer for Advanced Technologies*, Wiley interscience eds, **1994.**, in Press).

Microscopic observations give evidence of the presence of latex particles very close to the inorganic materials. It appears that the silica surface is covered with latex particles at the scale of several beads. This could account for the aggregates observed by dynamic light scattering. The latex size is then determined from the micrographs. Counting is done on the polymer beads only. Results are reported in Table IV. The average particle size is found to be around 46 nm for the higher silica content into the polymer and 52 nm for the two others samples with polydispersity index around 1.03. It appears that the latex size slightly depends on the monomer to silica weight ratio. As expected, the lower particle size is obtained for the lower monomer content. However, increasing the monomer to silica weight ratio, the particle size does not significantly change. This is probably the consequence of the formation of new particles, the presence of which leads to a decrease in the average particle size. Our experimental results demonstrate once more that the silica particles are not the exclusive locus of polymerization.

Table IV. Particles size and polydispersity index from TEM observations. Experimental data in the presence of functionalized silica (*) and in the absence of silica (**)

Monomer/Silica weight ratio	SiO$_2$ (%)	Solid content (%)	Diameter nm DLS	Diameter nm TEM	I [a]
9/1 *	11	2.8	144	46	1.03
16/1 *	6	4.9	130	52	1.03
48/1 *	2	14.7	120	53	1.04
/ **	/	4.9	75	45	1.07

a I= Dw/Dn

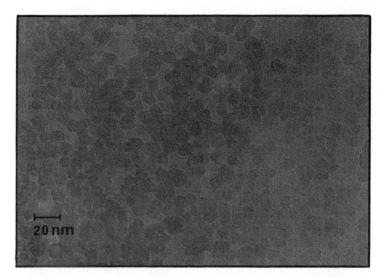

Figure 1.   Transmission electron micrograph of Highlink OG 100 silica particles.

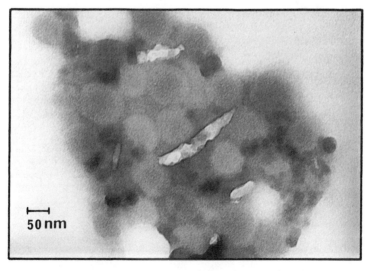

Figure 2.   Transmission electron micrograph of the sample containing 6% of functionalized silica.

Kinetic Study   Typical kinetic curves are presented in Figure 3; the arrow indicates the end of the monomer addition process. Whatever the monomer content, 100% conversions are obtained at the end of the reaction. The instantaneous conversion curve, reported in Figure 4, indicates that the semi-continuous addition rate is quite good. The inhibition period observed at the beginning of the reaction can be attributed to the presence of a stabilizer added to the silica dispersion in order to avoid the polymerization of 2- hydroxyethyl methacrylate. In the batch process, kinetics are much more rapid and 100 % conversions are obtained in only two hours. In any case, the monomer to polymer conversions are not limited as previously observed in this type of reaction (6). It is supposed that the comonomer takes part in the emulsion polymerization process thus modifying the kinetics. Additional experiments have been performed in the absence of comonomer i.e. after washing and redispersion of silica into water as described previously. As expected, we reach an upper limit of 74% conversion.

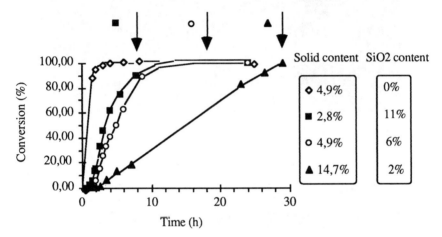

Figure 3.   Global time-conversion curves for the polymerization of ethyl acrylate.

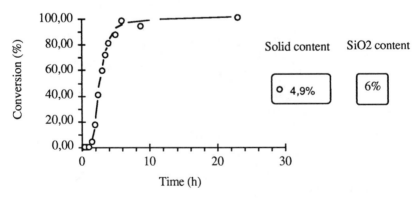

Figure 4.   Instantaneous time-conversion curve for the polymerization of ethyl acrylate.

All the above data seem to demonstrate that even if the experimental conditions have been choosen in order to promote the polymerization all around the inorganic surface (*6*), we did not succeed in the formation of a polymer film covering the silica particles. Obviously, we did not succeed in doing a perfect encapsulation. The main difference existing between this new system and the one described in a previous study (*6*) is first the type of silica particles, which are smaller in this study and have a higher dispersibility in water (E. Bourgeat-Lami, P. Espiard and A. Guyot, Part I - Submitted to *Polymer*, **1994**), and second the presence in this new type of commercial silica of an excess of monomer that could take part into the polymerization reaction.

Because of their lower size and their higher dispersion state in the polymerization medium, it is probable that insufficient amounts of non-ionic emulsifiant have been added to cover the larger surface of the particles. Indeed, calculation shows that only 10% of the overall surface is recovered by NP 30 assuming a molecular surface area of 200 Å2. So, we suppose that without enough stabilising emulsifier the original silica particles tend to gather and form strings, and an ideal core-shell morphology will not be obtained. On the other hand, the presence of the 2-hydroxyethyl methacrylate, highly soluble in water, could promote polymerization in the aqueous phase. Then, the polymerization rate will be increased and new particles will be formed besides the existing silica beads.

Nethertheless, it is expected that some chemical bonds are formed in the presence of functionalized silica so that the polymer which originates close to the silica surface is chemically linked to this surface. Evidence of such bonding will be presented below.

Grafting process

Polymer extraction and molecular weights   A strategy similar to that described previously (P. Espiard and A. Guyot; Part II - Submitted to *Polymer*, **1994**), was used to separate the grafted polymer from the adsorbed polymer. After the reaction, the mixture was separated by means of centrifugation and the adsorbed polymer was extracted with toluene at reflux temperature as presented in Scheme 1.

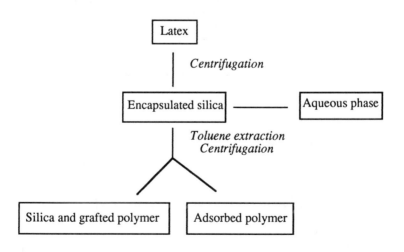

Scheme 1.   Latex treatment for the determination of the adsorbed polymer content.

The data concerning both functionalized and non-functionalized silica are presented in Table V. We show that in the case of non-functionalized silica, all the polymer formed can be dissolved in boiling toluene (fourth line of Table V). It can be concluded, as expected, that polyethylacrylate is only adsorbed onto the silica surface. In contrast, when functionalized silica is used, part of the polymer remains attached to the silica surface after toluene extraction. The amount of grafted polymer depends on the silica content as shown in Table V (first three lines). As a consequence, the grafted/silica weight ratio does not change with increasing monomer content, but rather, depends only on the initial amount of silica.

Table V. Characterization of the extracted polymer. Polymerization is carried out in the presence of functionalized silica (*), non functionalized silica (**), without silica (***).

Monomer/Silica weight ratio	SiO_2 (%)	Solid content (%)	Extracted polymer %	Grafted polymer (%)	Grafted/silica weight ratio (g/g of silica)	Mw of extracted polymer
9/1 *	11	2.8	60	40	3.9	420 000
16/1 *	6	4.9	73	27	3.9	288 000
48/1 *	2	14.7	92	8	3.9	54 000
16/1 **	6	4.9	100	0	0	/
/ ***	0	4.9	100	/	/	315 000

Molecular weights of extracted polymers decrease drastically with increasing monomer content. This may be due to the feed policy used in the polymerization process. When the amount of monomer is larger, most of the monomer is added in starved conditions as shown in Figure 3. In this event, the molecular weight may be limited by the instantaneous concentration of monomer.

Kinetics of grafting   A kinetic study has been performed for the latex containing 6% of functionalized silica. Samples are taken at different times and the amount of grafted polymer is determined as described previously. The fraction of grafted polymer compared to the total amount of polymer formed is plotted against time in Figure 5.

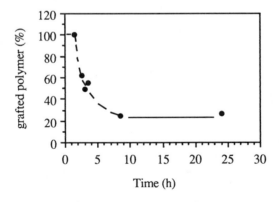

Figure 5.   Kinetics of grafting. 6% silica into the polymer.

From the latter results, it can be concluded that grafting takes place early during the polymerization process. In the first hour, the totality of polymer formed has been grafted to the silica surface. This proportion is slowly decreasing because of the formation of the adsorbed polymer. Finally, the total amount of grafted polymer remains constant for 100% conversion.

Characterization of grafting   That grafting takes place is clearly proved by two techniques.

Differential Scanning Calorimetry   Glass transition temperatures of the different composite materials have been first determined from DSC measurements (Figure 6). Films obtained from extracted polyethylacrylate exhibit a Tg value around -10°C as expected (Figure 6a). The film resulting from the encapsulating latex exhibits a wide range of Tg (Figure 6b). Two slopes are observed. The first one at lower temperature might correspond to the glass transition temperature of the free polymer while the second one, at higher temperature might be due to the existence of the attached polymer to the silica surface. As a matter of fact, the Tg values of functionalized silica surrounded by the grafted polymer as well as functionalized silica alone (Figures 6c and 6d) proved to be higher than the conventional Tg value of polyethylacrylate. It is then demonstrated that the polymeric interface formed at the silica surface has a lower mobility. Indeed, the attached polymer chains exhibit a higher value of Tg because of their close binding to the rigid surface. Thus, the observed increase of Tg can be considered as a direct consequence of grafting.

Dynamic Mechanical Spectrometry   The mechanical relaxation temperature, T α, associated to the glass transition phenomena of our materials has been determined from the Log (Tan δ) peak temperature (Figure 7). Three films have been studied: one is from polyethylacrylate latex without silica and the two others contain 6 and 11% of functionalized silica. In all cases, the modulus G' is first slowly decreasing and then drops suddenly due to the main mechanical relaxation. A rubbery plateau is observed after T α which is rather short for the film containing 6% silica and more extended for the others. The Log (Tan δ) temperature plot of polyethylacrylate shows a prominent peak at -6°C which corresponds, for this particular frequency, to the glass transition temperature measured from DSC.

A very different behaviour is observed for the film containing functionalized silica. Two T α transition temperatures can be determined from the Log (Tan δ) temperature plot. The first one can be associated to the Tg of polyethylacrylate while the second one is shifted towards higher temperatures (+10°C) and could correspond to the relaxation phenomena of the interfacial layer. As a consequence, since a constant amount of polymer is grafted per gramme of silica, the more silica is introduced into the film, the greater is the amount of grafted polymer and the higher should be the intensity of this second peak. Our experimental results support this statement. We can therefore demonstrate that the properties of the polymer which is bound to the filler surface differ from those of the bulk polymer. As before, dynamic mechanical spectrometry measurements give proof of the existence of an interfacial polymer layer with a lower molecular mobility.

Mechanical Properties

The mechanical properties of the composite materials containing 6% of non functionalized and functionalized silica was tested and compared with those obtained for polyethylacrylate as well as for a post synthesis blending of polyethylacrylate and functionalized silica. Figure 8 illustrates the conventional stress-strain curves obtained in elongation experiments.

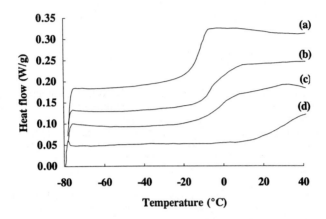

Figure 6.   DSC thermograms of a: extracted polymer, b: encapsulating polymer, c: grafted polymer onto silica, d: functionalized silica.

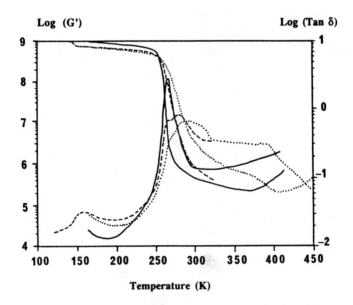

Figure 7.   Storage modulus and Log (Tan δ) versus temperature. Polyethylacrylate (——), polyethylacrylate containing 6% of functionalized silica (- - - -), polyethylacrylate containing 11% of functionalized silica (·····).

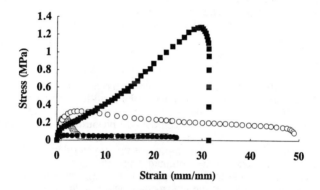

Figure 8. Stress-strain curves. Polyethylacrylate (□), polyethylacrylate containing 6% of non functionalized silica (●), post-synthesis blending of polyethylacrylate and functionalized silica (O), polyethylacrylate containing 6% of functionalized silica (■).

When the film does not contain silica, the curve goes through a yield point for low strain. Rupture then takes place at relatively high elongations. Very different behaviour is obtained in the presence of functionalized silica. The initial part of the curve is reversible and typical of an elastic deformation. At higher elongation, a non-linear plastic deformation is observed; the material behaves like a crosslinked elastomer. Finally, the rupture takes place after extensive deformations. Similar behaviour was described by some of us recently (P. Espiard, A. Guyot, J. Perez, G. Vigier and L. David, Part III - Submitted to *Polymer*, **1994**), and it was demonstrated that in the case of functionalized silica, strong reinforcement effects originate from the polymer forming covalent bonds on it surface. As a matter of fact, when non-functionalized silica is involved, very low stress are obtained up to very high elongations. The same mechanical behaviour is observed in the post synthesis blending of functionalized silica and polyethylacrylate except that it occurs at higher stress.

Regardless of the history of our samples, these experimental data indicate that in the absence of chemical bonds at the filler surface the composite material does not exhibit very strong properties. In the case of the post synthesis blending, the improvement we observed may be due to the presence of 2- hydroxyethylmethacrylate in the material that could polymerize during the formation of the film. Nevertheless, it is then clear that grafting during polymerization is primordial to obtain those remarkable mechanical properties.

Conclusion

With materials of different types in contact with each other, the control of the interface appeared to be a matter of great significance in the resulting composite materials. We show that the existence of a chemical bond between polymer molecules and the filler surface promotes strong adhesion between the two phases. The use of functionalized silica particles from commercial origin proved to be of great interest in such a purpose. So, a new encapsulating polymerization process by ethylacrylate has been performed resulting in stable latexes. Films obtained from the latexes display remarkable mechanical properties similar to those of crosslinked elastomers reinforced with solid particles. It is demonstrated that the intermediate layer which interacts strongly with both the filler surface and the matrix polymer would provide the observed improvement.

Literature Cited

(1)   E.L. Warrick and O.L. Pierce; *Rubber Chem. Technol.* 52 (**1979**) 437.
(2)   G.L. Wilkes, H.A. Huang and R.H. Glaser; in "Silicon based Polymer Science. A comprehensive ressource", J.M. Ziegler and F.G. Feazon (Eds.), Advances in Chem. Ser. 224 (**1990**) 207-236.
(3)   J.E. Mark; *J. Appl. Polym. Sci. Appl. Polym. Symp.* 50 (**1992**) 273.
(4)   P. Espiard, A. Revillon, A. Guyot and J.E. Mark; A.C.S. Preprints, *Polymer Material Science and Engineering.* 64 (**1991**).
(5)   P. Espiard, A. Revillon and A. Guyot; *Double liaison.* 38 (**1991**) 285-298.
(6)   P. Espiard, A. Revillon, A. Guyot and J.E. Mark in "Polymer Latexes", E.S. Daniels, E.D. Sudol and M. El Aasser (Eds.); Am. Chem. Soc. Symp. Series 492, Chap 24 (**1992**) 387.
(7)   G.M. Brauer and E. Horowitz, in "*Analytical Chemistry of Polymer*"; Vol XII, Part III G. M. Kline Ed.; Interscience, New York, **1962**, Chap 1; p 85-91.
(8)   V. Bazant, V. Chvalovsky and J. Rathousky in "*Organosilicon Compounds*"; Academician Oto Wichterle Ed; Academic Press, New York, **1965**.

RECEIVED October 20, 1994

Chapter 11

Vinyl-Polymer-Modified Hybrid Materials and Photoacid-Catalyzed Sol–Gel Reactions

Yen Wei, Wei Wang, Jui-Ming Yeh, Bin Wang, Dachuan Yang, James K. Murray, Jr., Danliang Jin, and Gu Wei

Department of Chemistry, Drexel University, Philadelphia, PA 19104

We have developed a general method for the preparation of vinyl polymer-modified sol-gel hybrid materials. The polymer precursors are obtained by chain copolymerization of 3-(trialkoxysilyl)propyl methacrylate or other alkoxysilyl vinyl monomers with conventional vinyl monomers such as methyl methacrylates, acrylonitrile and styrene. The alkoxysilyl groups in the polymer precursors undergo the sol-gel reactions with inorganic precursors such as the alkoxides of silicon, titanium, and/or aluminum to afford monolithic, transparent hybrid materials without macroscopic phase separation. This approach makes it possible to incorporate virtually any conventional vinyl polymers into the inorganic matrices. We have also demonstrated that the vinyl polymer components in the inorganic matrices could be further chemically modified to give new hybrid materials or nanocomposites. The first photochemical synthesis of vinyl polymer-inorganic hybrid materials has been achieved via photoacid-catalyzed sol-gel reactions. The exploratory applications are also discussed.

In the past two decades, the field of materials science has witnessed the emergence of sol-gel process for the preparation of ceramics or organic modified hybrids materials from precursor compounds in the presence of an acid or base catalyst at relatively low temperatures.[1-3] For example, acid-catalyzed hydrolysis and polycondensation of tetraethyl orthosilicate (TEOS) yield a SiO_2 glass at temperatures much lower than the conventional fusion process.[1] Many researchers have demonstrated the successful combinations of various polymers and copolymers with inorganic structures like the SiO_2 at the molecular level to form a class of organic-inorganic sol-gel materials. There are two major families of these organic-modified sol-gel materials: (1) organic-inorganic composites in which the organic/polymer components are trapped in the inorganic networks and (2) organic-inorganic hybrid materials in which the organic/polymer components are covalently bonded to the inorganic networks. Though all these terms have been loosely used, one significant feature of these materials is that the organic/polymer components are uniformly distributed, with or without formation of covalent bonding, in the inorganic networks and therefore the macroscopic interfaces, as existing in the conventional composite materials, between the polymers and the inorganic components are eliminated. Many physical properties of these sol-gel materials could be designed and controlled by varying the nature and composition of both the polymer and inorganic components. A number of polymers have been incorporated into SiO_2 and/or TiO_2 glass networks, including polydimethylsiloxane,[4] poly(tetramethylene oxide),[5] polyimides,[6] poly(arylene ether ketone) and poly(arylene ether sulfone),[7] polymethacrylates,[8,9] polyoxazolines.[10] The structure-property relationship in the new materials was studied to certain extent. Numerous commercial applications have been explored.

We have been interested in the synthesis of new polymer-inorganic hybrid sol-gel materials and their applications in dental restorative materials, coatings, membranes, catalysts or catalyst supports, reaction templates, and lithography.[11-19] Recently, we have developed a general method for the preparation of polymer sol-gel precursors containing the reactive alkoxysilyl groups via chain copolymerization of 3-(trimethoxysilyl)propyl methacrylate or other alkoxysilyl vinyl monomers with conventional vinyl monomers (e.g. methyl methacrylates, acrylonitrile and styrene). The polymer precursors were then hydrolyzed and co-condensed with inorganic precursors such as the alkoxides of silicon, titanium, aluminum and/or zirconium to afford monolithic, transparent hybrid materials without macroscopic phase separation. This approach makes it possible, in principle, to incorporate any conventional vinyl polymers into the inorganic matrices. We have also demonstrated that the vinyl polymer components in the inorganic matrices could be further chemically modified to give new hybrid materials or nano-composites. The first photochemical synthesis of vinyl polymer-inorganic hybrid materials has been achieved via photoacid-catalyzed sol-gel reactions. In this article, these developments and our latest results in this area will be described systematically and some of the potential applications of these new materials will be discussed.

Results and Discussion

Vinyl Polymer-Modified Hybrid Materials

The general equations for the preparation of new vinyl polymer precursors containing alkoxysilyl groups and their sol-gel reactions with inorganic precursors are illustrated in Scheme I.   Thus, the vinyl polymer precursors are prepared by copolymerization of an alkoxysilyl-containing vinyl monomer with a conventional vinyl monomer at various compositions. The content of the alkoxysilyl-containing monomer is usually low to afford the copolymers whose properties are more close to the conventional vinyl polymers. These polymer precursors are then hydrolyzed and co-condensed (sol-gel reactions) with one or more inorganic precursors at ambient temperature to afford the vinyl polymer-modified hybrid sol-gel materials. In these new materials, the vinyl polymer chains are uniformly distributed in and covalently bonded to the inorganic matrices at molecular level. There is no macroscopic phase separation between the organic polymers and inorganic oxide matrices as evidenced by the facts that the hybrid materials are transparent and therefore the scale of phase separation, if any, should be below the wavelengths of visible light (i.e. <400 nm) and that the glass transition temperature (T_g) of the polymer components is either unmeasurable or significantly altered in comparison with the bulk polymers.

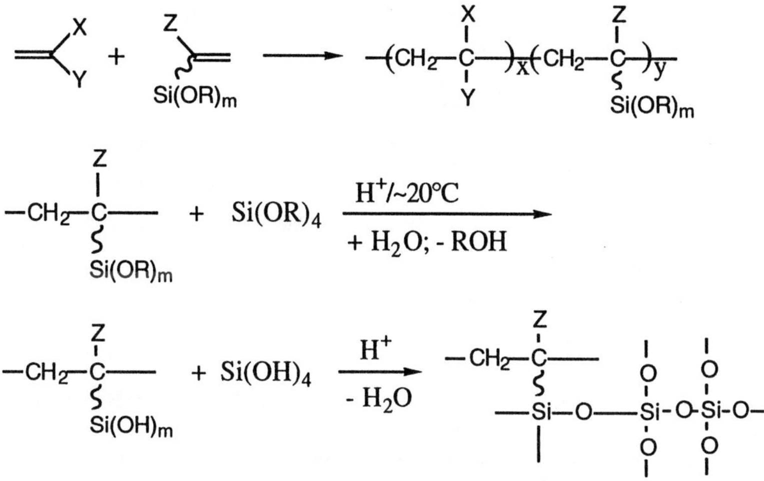

Scheme I. Preparation of the sol-gel reactive polymer precursors and subsequent sol-gel reactions with inorganic precursors. X, Y and Z are various functional groups; R: alkyl groups; Si can be replaced by other metals such as Al, Ti, Zr, etc.

The alkoxysilyl-containing monomers that have been employed in our studies include 3-(trimethoxysilyl)propyl methacrylate (MSMA), 3-(triethoxysilyl)propyl methacrylate (ESMA) and styrylethyltrimethoxysilane (STMS). The conventional vinyl monomers include methyl methacrylate (MMA), ethyl methacrylate (EMA), sec-butyl methacrylate (SBMA), styrene (St), acrylonitrile (An), etc. The inorganic precursors include tetraethyl orthosilicate (TEOS), titanium tetraisopropoxide (TIPO), aluminum tri-sec-butoxide (ASBO), etc. Generally, the sol-gel reactions are carried out under an acid catalyst (e.g. aqueous HCl) in an organic solvent, e.g. tetrahydrofuran (THF) and N-methylpyrrilidinone (NMP), at room temperature. By optimizing the reaction conditions, the hybrid materials in a form of disc or thin films can be made transparent, monolithic and free of cracks.

Several methods were used to synthesize the polymer precursors. For example, poly[methyl methacrylate-co-3-(triethoxysilyl)propyl methacrylate] [P(MMA-ESMA)] was prepared through the group-transfer polymerization route.[11-13] P(MMA-ESMA) precursors with designed number of trialkoxysilyl groups were prepared by group-transfer copolymerization of allyl methacrylate with various amount of methyl methacrylate with trimethylsilyl ketene acetal as initiator and tris(dimethylamino)-sulfonium bifluoride (TASHF$_2$) as catalyst, followed by hydrosilylation of the ester allyl groups in the polymer in the presence of the Speier's catalyst (H$_2$PtCl$_6$), as shown in Scheme II. This method offers excellent control of molecular weight of the polymers.

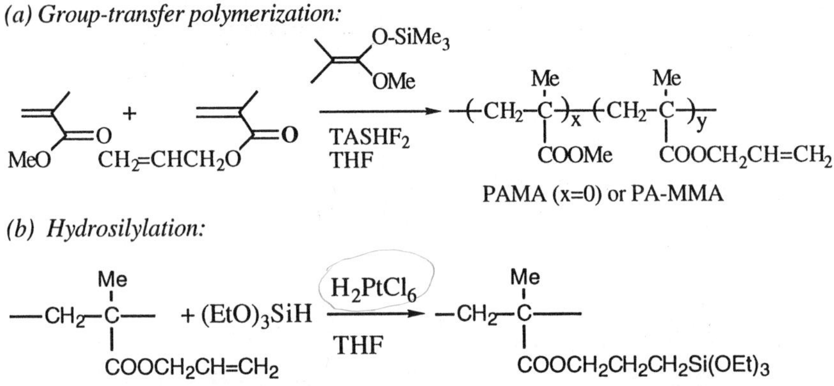

(a) Group-transfer polymerization:

PAMA (x=0) or PA-MMA

(b) Hydrosilylation:

Scheme II. Synthesis of polyacrylate precursors via group-transfer polymerization.

A more general method is free-radical copolymerization of the commercially available 3-(trimethoxysilyl)propyl methacrylate or other alkoxysilyl-containing monomers (e.g. styrylethyltrimethoxysilane) with methyl methacrylate[13-14], styrene[15], or acrylonitrile[16], as shown in Scheme III. The synthesis by this method has only one step and the starting materials are all commercially available. The number of reactive trialkoxysilyl groups (i.e. the value of y) can also be designed and controlled. Therefore, this method would significantly lower the cost of the potential large scale production of the new hybrid materials.

These polymers (x=0) and copolymers (x≠0) with trialkoxysilyl groups were then hydrolyzed and co-condensed with tetraethyl orthosilicate (TEOS) and/or other inorganic precursors via an acid-catalyzed sol-gel route to yield the new hybrid materials as exemplified in Scheme IV. We have studied the effect of various reaction conditions, such as type and amount of catalysts and solvents, amounts of the polymers, inorganic precursors and water, reaction temperature, and rate of evaporation of the volatiles, to assure the optical transparency and the integrity (i.e. monolithic and crack free) of the sol-gel materials of various organic-inorganic compositions.[11-16] One of the general process with TEOS as the inorganic precursor is shown in Scheme V.

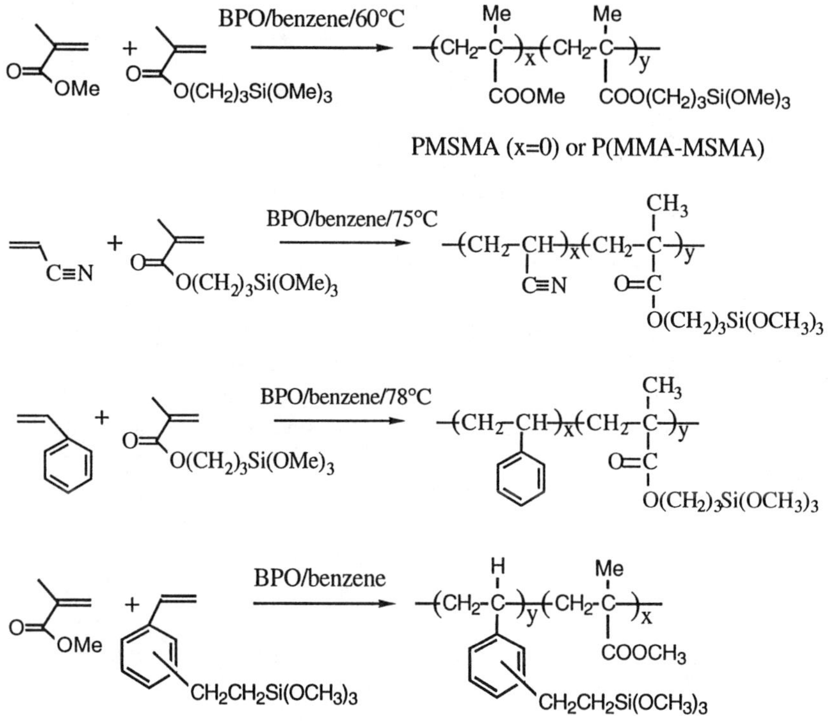

PMSMA (x=0) or P(MMA-MSMA)

Scheme III. Preparation of alkoxysilyl-containing polymer precursors by free-radical copolymerization method.

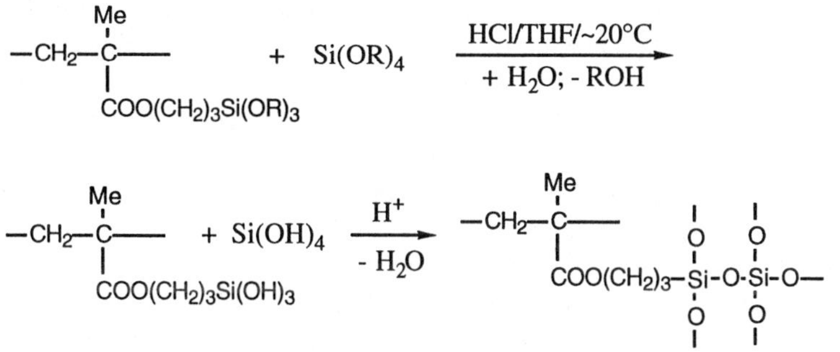

Scheme IV. Example of sol-gel process for the preparation of vinyl polymer-modified hybrid materials.

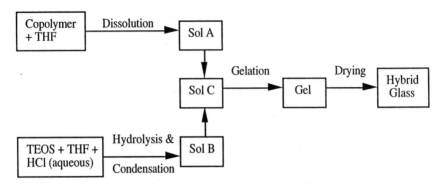

Scheme V. A general hybrid sol-gel process diagram.

All the hybrid materials were characterized by FTIR spectroscopy in which all the characteristic bands of the polymer components were identified. The differential scanning calorimetry (DSC) results showed a general lack of well-defined glass transition temperatures for the polymers in the silica matrices, indicating that the polymer chains are uniformly distributed in the materials. Thermogravimetric analysis (TGA) technique was used to determine the silica contents in the hybrid materials by decomposing all the organic contents at high temperatures (e.g. >700°C). The SiO_2 contents in the hybrids obtained from TGA generally agree well with the calculated values based on the stoichiometry of the starting materials. The TGA measurements also reveal that the thermal stability of some polymer components (e.g. polyacrylates[11]) is enhanced since their thermal decomposition temperatures are higher than the bulk polymers. In addition to the thermal stability, many other physical properties, such as bulk density, refractive index, and hardness, exhibit strong correlation with and therefore can be tailored by variation of the organic-inorganic compositions of the hybrid materials. All the spectroscopic and thermal analytical results are consistent with the picture that in the hybrid materials the vinyl polymer chains are incorporated into the inorganic matrices at molecular level without macroscopic phase separation.[11-17] Transparent, monolithic hybrid materials were also derived from TEOS and the copolymers of styrylethyltrimethoxysilane, ethyl methacrylate, sec-butyl methacrylate, etc. As a noteworthy example, under carefully controlled conditions, we were able to produce monolithic poly(methyl methacrylate)-SiO_2 hybrid materials having excellent optical transparency, which has been considered as a strong evidence for the high uniformity of the polymer chain distribution in the SiO_2 matrix.[11-14] However, it has been pointed out recently[2] that the apparent good transparency of PMMA-SiO_2 materials may also be attributed to the fact that the refractive index of PMMA ($n = 1.4920$)[20] is very close to that of SiO_2 glass (1.4589)[21]. To test this possibility, we decided to use functionalized polystyrene chains as the polymer component in the hybrid silica materials because the refractive index of polystyrene (1.59-1.60)[20] is significantly greater than that of SiO_2 glass. We have found that the polystyrene-silica hybrid materials can also be obtained with excellent optical transparency over the entire composition range and with the refractive index changing continuously between the limits of polystyrene and silica glass.[15] This observation further supports our conclusion that the polymer chains are uniformly distributed in the inorganic matrices.

There are numerous potential applications for these vinyl polymer-inorganic hybrid materials. We are particularly interested in the dental applications of the polymethacrylate-silica hybrid materials because both the polymethacrylate and silica are widely used in the conventional dental restorative composite materials. Our preliminary results show that the mechanical properties of the polymethacrylate-silica hybrids were improved over the conventional polymethacrylate-silica composite materials at similar organic-inorganic compositions.[22]

Reactions of the Polymer Components in the Hybrid Materials

We have found that the vinyl polymer components in the inorganic matrices can undergo a variety of reactions. This allows us to prepare new polymer-inorganic hybrid materials via chemoselective reactions of the polymer components or to modified the properties of the hybrid materials. For example, the polyacrylonitrile component in the hybrid silica materials was found to undergo pyrolysis in an inert atmosphere towards the formation of carbon fibers of micro- or nano-structures in the silica matrix.[16] Recently, we have also chemoselectively hydrolyzed the nitrile groups of polyacrylonitrile component to yield new water-soluble polyacrylics modified silica hybrid materials as illustrated in Scheme VI.[17]

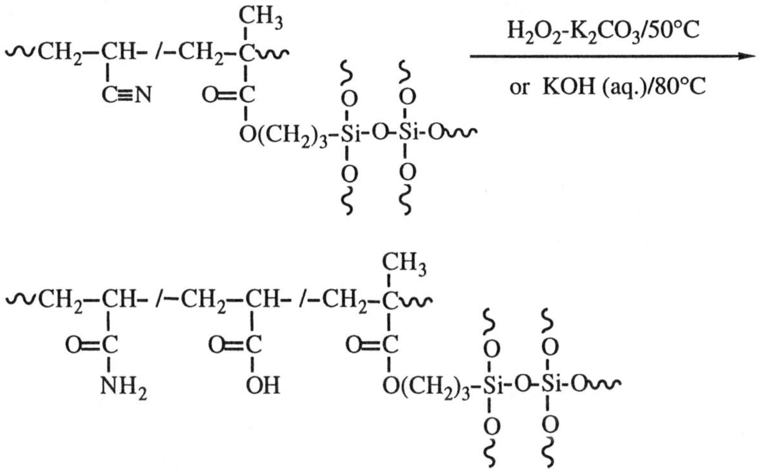

Scheme VI. Synthesis of water-soluble polyacrylics-modified silica sol-gel materials via Radziszewski hydrolysis of polyacrylonitrile component.

It is important to note that after the chemoselective hydrolysis with hydrogen peroxide in the presence of a base catalyst (Radziszewski reaction[23]), the nitrile groups are converted to the water-soluble acrylic amides or acids while the ester groups remained intact and the polymer chains are still *covalently bonded to* the silica matrix. Sulfonation of the polystyrene component in the polystyrene-silica hybrid materials was also performed as shown in Scheme VII.[24] The resultant new materials contain sulfonic acid groups which can be readily converted to various salts as well. The cationic exchange capacity of the sulfonated hybrid materials were found to increase as the polystyrene content in the original hybrids or the extent of sulfonation is increased. These materials may also be used as solid polymer electrolyte materials.

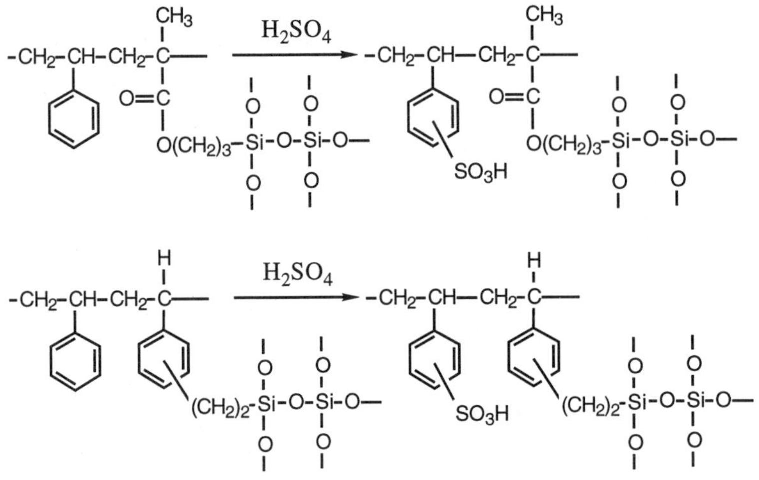

Scheme VII. Sulfonation of polystyrene-silica hybrid materials.

Another very interesting aspect of the vinyl polymer-modified hybrid sol-gel materials is that the organic content in these hybrid materials could be removed thermally without destroy the integrity of the inorganic matrices. For example, a poly(methyl methacrylate)-SiO_2 hybrid sol-gel material synthesized at room temperature was found to have a surface area of ~ 0.1 m^2/g (measured with BET method with krypton as analysis gas on an AccuSorb BET Multiple-Points Porosimeter). After thermal treatment at ~ 400 °C in nitrogen, the polymer content in the material was completely removed leaving a monolithic silica glass with a drastically increased surface area of 629 m^2/g and an average pore diameter of ~ 28 Å.[25] By tailoring the nature and the composition of the polymers, it is possible to *control* the size and volume of the pores and/or micro-channels in the porous inorganic glasses. These glasses could also be made to contain other metal oxides such as Ti, V, Zr, etc., via multicomponent sol-gel process.   There are many potential applications under exploration for these porous glass materials including catalysts/catalyst supports, reaction templates, host matrices for fabricating capacitors, and as size-selective membranes for construction of biosensor devices.

Photochemical Synthesis of Polymer-Modified Sol-Gel Materials

Recently, we have achieved for the first time a successful photochemical synthesis of polyacrylate-inorganic hybrid sol-gel materials using the photoacids, such as diphenyliodonium chloride (DIC) and triphenylsulfonium hexafluoroantimonate (TSHF), as catalysts under UV radiation.[18,19]   Poly[methyl methacrylate-*co*-3-(trimethoxysilyl)propyl methacrylate] [P(MMA-MSMA)] of various copolymer compositions and TEOS were selected for detailed study as the inorganic and organic sol-gel precursors, respectively.   Both the photoacids, i.e. DIC and TSHF, yield protonic acid upon photolysis.[26-28]   For example, the main photochemical processes for DIC are depicted in Scheme VIII, in which RH represents a proton source, e.g., the solvent or reactants.   The protonic acid, HCl, is produced as a byproduct of secondary reaction and, therefore, will function as the catalyst in the sol-gel reactions.[26-27]

$$Ar_2I^+\,Cl^- \xrightarrow{\ h\nu\ } ArI^{\bullet +}\,Cl^- + Ar\bullet$$

$$ArI^{\bullet +}\,Cl^- + R\text{-}H \longrightarrow ArI + HCl + R\bullet$$

Scheme VIII. Photolysis of DIC to generate HCl.

To prepare the polyacrylate-silica hybrid materials, the polymer precursors were hydrolyzed and co-condensed with various amounts of TEOS in the presence of the photoacid (DIC or TSHF) under UV radiation.   The reaction system was housed in a light-proof box as illustrated in Figure 1.   The rate of sol-gel reactions could be qualitatively represented (a) by the radiation time that is required for the organic/water phase separation to disappear and is mainly indicative of the rate of hydrolysis or (b) by the gelation time that is required for the reaction system to lose its fluidity and is mainly indicative of the rate of polycondensation.   Longer radiation or gelation time indicates slower reactions.   It should be noted that the reaction system in the absence of the photoacid neither became homogeneous nor showed any gel formation after exposed to the UV radiation.   On the other hand, no sol-gel reaction was observed when the system in the presence of the photoacid was kept in dark for over 24 h.   The radiation time decreases as the photoacid concentration was increased.   Furthermore, the rate of reactions increased when the light intensity is increased by shortening the distance between the light source and the reaction mixture or when a UV lamp with short wavelength only was used as anticipated.[26-28]

Amount of the polymer precursor in the system was found to influence the reaction rate significantly. The higher the polymer content the shorter the gelation time. This trend as shown in Figure 2 might be explained based on the fact that the polymer chain bears more reactive alkoxysilyl functional groups and therefore reaches gelation (i.e., formation of crosslinked network) sooner than the inorganic precursor TEOS. For the same reason, when the polymer precursors containing 49 mol-% or more MSMA units were employed with TEOS, the polymer gelated and precipitated before the gelation of the entire reaction system occurred. Therefore, for these polymer precursors, the synthesis procedure was modified slightly by allowing TEOS to undergo the photoacid-catalyzed sol-gel reactions first to form a homogeneous solution, to which the polymer solution was then added. By this procedure, TEOS may hydrolyze and condense to afford an inorganic oligomeric or polymeric precursor having a functionality comparable with the organic polymer precursors. The thermal, optical and mechanical properties of all the hybrid samples are close to those reported for the hybrid materials of the similar compositions prepared with conventional acids (e.g. HCl) as catalysts.[11-14]

Similar results were obtained with TSHF as the photoacid. The TSHF-catalyzed sol-gel reactions appeared to proceed faster than the DIC-catalyzed reactions. As shown in Table I, the systems with DIC catalyst required a longer UV radiation time for the phase separation to disappear than those with TSHF catalyst.

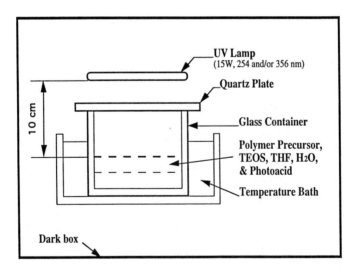

Figure 1. The experimental setup for the photoacid-catalyzed sol-gel reactions for the preparation of the polymer-modified hybrid materials.

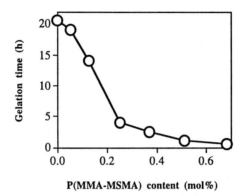

P(MMA-MSMA) content (mol%)

Figure 2 . Effect of the polymer precursor content (mol-%) on the rate of sol-gel reactions as represented by the gelation time.

Table I. Effect of photoacids, diphenyliodonium chloride (DIC) and triphenylsulfonium hexafluoroantimonate (TSHF) on sol-gel reactions [a]

Sample code	$\dfrac{[\text{Photoacid}]}{[\text{TEOS}]}$	Reactants composition		Radiation time for homogeneous sol (min)
		Polymer(g)	TEOS(g)	
PG38[b]	0.01	0	11.4	130
PG25[b]	0.01	0.934	11.4	130
PG64A[c]	0.01	0	11.4	23
PG64B[c]	0.01	0.934	11.4	25

(a) Amount of photoacid DIC: 0.1732 g (0.547 mmol), TSHF (used as 50% solution in propylene carbonate): 0.546 g (0.547 mmol); H_2O: 3.942 g (219 mmol); THF (solvent): 7.675 g. Reaction temperature: 25°C. (b) Photoacid: DIC. (c) Photoacid: TSHF.

We have also prepared TiO_2 and Al_2O_3 materials and their binary and/or ternary hybrids with P(MMA-MSMA) and/or SiO_2 by the photoacid-catalyzed sol-gel reactions of titanium(IV) isopropoxide (TIPO), aluminum(III) sec-butoxide (ASBO), TEOS and/or P(MMA-MSMA) precursors. Different alkoxides of different elements are known to have a wide range of reactivities toward hydrolysis. The titanium and aluminum precursors, TIPO and ASBO react with water much faster than TEOS. If TIPO and TEOS or ASBO and TEOS were allowed to react with water under the same conditions as used for TEOS alone, the hydrolysis and condensation of TIPO or ASBO occur rapidly before the reactions of TEOS, and the particles of TiO_2 or Al_2O_3 precipitate from the system. This makes it difficult to prepare homogeneous multicomponent sol-gel systems. To overcome this problem, some researchers[29] introduced a chelating organic ligand into the solution to reduce the hydrolysis rate of the highly reactive metal alkoxides. We have obtained the homogeneous multicomponent gels by partially prehydrolyzing the slow-reacting component, i.e. TEOS, first under photoacid catalysis. TIPO or ASBO was then added to the solution to form a clear sol. Finally the polymer precursor was added to the system to afford the multicomponent hybrid materials. The whole process is illustrated in Scheme IX and some of the representative stoichiometric data for the synthesis are given in Table II. The compositions of the hybrid products differ slightly from the those of the starting materials which might be attributed to the incomplete condensation and/or evaporation of solvents and byproducts. We are currently studying the detailed kinetics of the photochemical reaction system and its dental and lithographic applications.

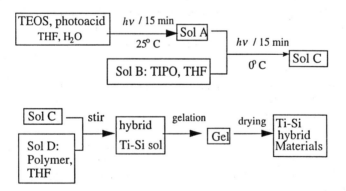

Scheme IX. A sol-gel process diagram for the synthesis of ternary hybrid materials (e.g. polymer-Ti-Si hybrids) with compensation for the reactivity differences.

Table II. Synthesis and composition of the hybrid materials derived from the sol-gel reactions of poly[methyl methacrylate-co-3-(trimethoxysilyl)propyl methacrylate], tetraethyl orthosilicate (TEOS), titanium isopropoxide (TIPO), and/or aluminum tri-sec-butoxide (ASBO) in the presence of triphenylsulfonium hexafluoroantimonate (TSHF) catalyst under UV radiation at 25 °C [a]

Sample code	MSMA unit in the polymer (mol-%)	Reactants composition		[Polymer] [TEOS]	Inorganic (wt-%)[a]	
		Polymer (g)	TEOS (g)		calc.	anal.
PG64A[b]	—	0	11.4	0	100	98
PG64B[b]	12	0.930	11.4	0.144	83	77
PT5C[c]	—	0	2.85	0	100	92
PT7[c]	20	0.45	2.85	0.216	82	77
PA3C[d]	—	0	2.85	0	100	93
PA4[d]	20	0.45	2.85	0.216	84	79

(a) The SiO_2, SiO_2-TiO_2 and SiO_2-Al_2O_3 contents in the sol-gel materials were calculated from compositions of the reactants (calc.) and were determined by TGA at 750 °C (anal.). (b) Amount of TSHF: 0.546 g (0.547 mmol); H_2O: 3.942 g (219 mmol); THF (solvent): 7.675 g. (c) Amount of TSHF: 0.137 g (0.137 mmol); TIPO: 3.89 g (14.0 mmol), H_2O: 0.247 g (13.7 mmol); THF (solvent): 3.84 g. (d) Amount of TSHF: 0.137 g (0.137 mmol); ASBO: 3.37 g (13.7 mmol); H_2O: 0.247 g (13.7 mmol); THF (solvent): 3.84 g

Conclusions

In summary, we have demonstrated, with examples, that the vinyl polymer-modified hybrid materials can be prepared by a general method that involves the sol-gel reactions of polymer precursors with inorganic precursors. The polymer precursors that contain the reactive alkoxysilyl groups can be synthesized via chain copolymerization of the alkoxysilyl vinyl monomers, e.g. 3-(trimethoxysilyl)propyl methacrylate and styrylethyltrimethoxysilane, with conventional vinyl monomers, e.g. methyl methacrylates, acrylonitrile and styrene. The polymer precursors are then hydrolyzed and co-condensed with inorganic precursors, such as tetraethyl orthosilicate to afford monolithic, transparent hybrid materials without macroscopic phase separation. Many physicochemical properties, such as refractive index, hardness, thermal stability and bulk density, of the hybrid materials can be controlled by choosing appropriate vinyl polymers and their compositions in the inorganic matrices. We have also described that the vinyl polymer components in the inorganic matrices could be further modified to give new hybrid materials or nano-composites. For examples, the polyacrylonitrile component in the hybrid silica materials undergoes pyrolysis towards the formation of carbon fibers of micro- or nano-structures in the silica matrix; the nitrile groups of polyacrylonitrile could also be chemoselectively hydrolyzed to yield new water-soluble polyacrylics-modified silica hybrid materials.

The first photochemical synthesis of vinyl polymer-inorganic hybrid materials has been achieved via the photoacid-catalyzed sol-gel reactions. Polymethacrylate-inorganic two-component or multicomponent hybrid materials can be prepared photochemically by using photoacids such as diphenyliodonium chloride (DIC) and triphenylsulfonium hexafluoroantimonate (TSHF) as catalysts in the sol-gel reactions of poly[methyl methacrylate-co-3-(trimethoxysilyl)propyl methacrylate] with tetraethyl orthosilicate, titanium(IV) isopropoxide, and/or aluminum(III) sec-butoxide at various compositions. The preliminary kinetic studies show that TSHF is a more effective photoacid than DIC and that the gelation of the polymer precursors occurs sooner than that of the inorganic precursors.

Experimental Section

Tetraethyl orthosilicate (TEOS), titanium tetraisopropoxide(TIPO), aluminum tri-sec-butoxide (ASBO) (all Aldrich), DIC (Janssen), and TSHF (50% solution in propylene carbonate, P&B) were used as received. Methyl methacrylate (MMA), ethyl methacrylate (EMA), sec-butyl methacrylate (SBMA), styrene (St), and acrylonitrile (An) (all Adrich) were purified based on the standard monomer purification procedures. 3-(Trimethoxysilyl)propyl methacrylate (MSMA, Aldrich) was purified by distillation under a reduced pressure. Styrylethyltrimethoxysilane (STMS, United Chemical Technologies) was purified by column chromatography on silica gel. The polymer precursor, P(MMA-MSMA), was prepared by either group-transfer polymerization followed by hydrosilation[11-13] or by free-radical copolymerization of MMA with MSMA.[13,14] The polystyrene[15] and polyacrylonitrile[16,17] precursors were synthesized by free-radical copolymerization. All the polymer precursors were characterized by IR, ^{1}H NMR, elemental analysis, DSC, and TGA.

The detailed synthesis and characterization of polymethacrylates[11-14], polystyrene[15] and polyacrylonitrile[16]-modified hybrid sol-gel materials were given in our previous reports. As an example for the sol-gel process, the preparation of polystyrene-silica hybrid materials was achieved by the following procedure[15]. 0.390 g (2.74 mmol based on repeating units) of the polymer precursor containing 73 mol-% of styrene units and 27 mol-% of MSMA units was dissolved in 1.560 g of tetrahydrofuran (THF) to give a solution denoted as Sol A. To a solution of 0.865 g of THF and 0.3 g of HCl (0.2 M, 0.06 mmol), 1.259 g TEOS (6.0 mmol) was added under vigorous stirring for 30 min to yield a clear solution denoted as Sol B. Combination of Sol A and Sol B in a 50-mL beaker under vigorous stirring for 10 min resulted in a clear homogeneous solution. This solution was allowed to stand open to ambient without stirring to evaporate the solvent (i.e. THF) and the low molecular-weight products of hydrolysis and condensation (i.e. MeOH, EtOH and H_2O) until a clear gel formed. To obtain a crack-free sol-gel material, the evaporation was then slowed down by covering the beaker with a paraffin film having a number of holes made with a syringe needle. After drying for about 20 days at room temperature, a transparent, monolithic disc was obtained. The product was further heated in an oven at 200 °C for 2-4 hours until the weight became essentially constant (~ 0.7 g). Based on the TGA, this polystyrene-SiO_2 hybrid glass contained 54 wt-% of SiO_2 component.

As an example for the reactions of the vinyl polymer components in the hybrid sol-gel materials, the chemoselective hydrolysis of poly(acrylonitrile-MSMA) was achieved by the following procedure[17].   0.05 g of finely ground powder of the polyacrylonitrile-SiO_2 hybrid material[16] containing 72 wt-% SiO_2 and 0.17 mmol of nitrile groups was placed in 0.5 mL mixture solvents of acetone and water (50/50, v/v) in a 25-mL round-bottom flask equipped with a condenser and a magnetic stirring bar. Then 0.15 mL of 30 wt-% aqueous H_2O_2 (1.33 mmol) and 0.05 g (0.36 mmol) K_2CO_3 were added to the system with stirring.   The reaction mixture was heated to and kept at 50 °C for 4 h.   After cooling to room temperature, the reaction mixture was neutralized with 1 N HCl followed by filtration.   The white solid product was thoroughly washed with distilled water and methanol followed by drying in vacuo at 90 °C overnight.   Spectroscopic characterization showed that the nitrile groups in acrylonitrile units were selectively hydrolyzed to water-soluble amide groups while the ester groups in MSMA units remained intact.

As a typical procedure for the photoacid catalyzed sol-gel reactions[18,19], a solution of 0.173 g (0.547 mmol) diphenyliodonium chloride, 11.4 g (54.7 mmol) TEOS and 0.93 g (7.87 mmol based on the repeating unit) of P(MMA-MSMA) containing 12 mol-% of MSMA units in 7.67 g tetrahydrofuran (THF) was prepared in a 100-mL beaker.   To this solution was added 3.942 g (219 mmol) H_2O to give a two phase mixture.   The beaker was then covered with a quartz plate and placed in a double-wall jacketed cell through which water was circulated from a Polytemp thermostat to maintain a constant temperature of 25±0.5 °C.   The reaction mixture was stirred magnetically and irradiated with a UV light housed in a light-proof hood (see Figure 1 in the text).   The light source employed was a 15-watts Rayonet Photochemical Reactor Lamp comprising both the short (254 nm) and long (356 nm) wavelengths at a distance of 10 cm between the lamp and the surface of the reaction mixture. After the organic-water phase separation disappeared and the reaction mixture became homogeneous and transparent (~130 min), the UV radiation was stopped.   The solution was transferred to a 50-mL beaker and kept open to ambient in dark for gelation. After the gelation (~14 h), the beaker was covered with a paraffin film with 4 to 5 needle-punched holes to allow a slow drying at room temperature for ~2 months to afford a monolithic, transparent hybrid sol-gel glass disc having a diameter of ~2.5 cm and a thickness of ~2 mm.   The glass was then heated at 200 °C for approximately 4 h till the sample weight became constant.

Acknowledgments

This work was supported by the National Institutes of Health (Grant No. RO1-DE09848).   YW is grateful to E.I. Du Pont de Nemours & Company, Inc. for a Young Faculty Award.   We thank Dr. R. Bakthavatchalam (DuPont-Merck) and Mr. C.K. Whitecar (ISI) for their earlier contribution to this work; Dr. M.K. Hutchins (LNP Co.) for some spectroscopic characterization of the hybrid materials.   Professor G.R. Baran (Temple University), Dr. G.F. Cowperthwaite (EssChem Co.) and Professor A.W. Addison (Drexel University) for many valuable discussions.

References

(1) For reviews, see: (a) Brinker, C. J.; Scherer, G. W. *Sol-Gel Science, the Physics and Chemistry of Sol-Gel Processing*; Academic Press: San Diego, 1990. (b) Ulrich, D. R. *J. Non-Cryst. Solids* **1990**, *121*, 465. (c) Schmidt, H.; Sieferling, B.; Phillip, G.; Deichmann, K. *Ultrastructure Processing of Advanced Ceramics*; Mackenzie, J. D.; Ulrich, D. R. Eds.; Wiley: New York, 1988, p. 651. (d) Schmidt, H. *J. Non-Cryst. Solids* **1989**, *112*, 419. (e) Wilkes, G. L.; Huang, H.; Glaser, R. H. *Silicon-Based Polymer Science (Advances in Chemistry Series 224)*, Ziegler, J. M.; Fearon, F. W. Eds.; Am. Chem. Soc.: Washington, DC, 1990, pp. 207-226; (f) Dagani, R., *Chemical & Engineering News* **1991**, *69(21)*, 30. Also see Refs. 2 and 3.

(2) Calvert, P. *Nature* **1991**, *353*, 501.

(3) Novak, B.M. *Adv. Mater.* **1993**, *5*, 422.

(4) (a) Wilkes, G. L.; Orler, B.; Huang, H. *Polym. Prepr. (Am. Chem. Soc., Div. Polym. Chem.)* **1985**, *26*, 300. (b) Huang, H.; Orler, B.; Wilkes, G. L. *Macromolecules* **1987**, *20*, 1322. (c) Mark, J. E.; Jiang, C. Y.; Tang, M. Y. *Macromolecules* **1984**, *17*, 2613. (d) Mark, J. E.; Ning, Y. P.; Tang, M. Y.; Roth, W. C. *Polymer* **1985**, *26*, 2069. (e) Wang, S. B.; Mark, J. E. *Macromol. Reports* **1991**, *A28*, 185. (f) Haruvy, Y.; Webber, S. E. *Chem. Mater.* **1991**, *3*, 501.

(5) (a) Huang, H.; Wilkes, G. L. *Polym. Prepr. (Am. Chem. Soc., Div. Polym. Chem.)* **1987**, *28*, 244. (b) Glaser, R. H.; Wilkes, G. L. *Polym. Bull.* **1988**, *19*, 51.

(6) Nandi, M.; Conklin, J.A.; Salvati Jr., L.; Sen, A. Chem. Mater. 1991, 3, 201.

(7) (a) Wang, B.; Wilkes, G. L.; Hedrick, J. C.; Liptak, S. C.; McGrath, J. E. *Macromolecules* **1991**, *24*, 3449. (b) Morikawa, A.; Iyoku, Y.; Kakimoto, M.; Imai, Y. *Polym. J.* **1992**, *24*, 107.

(8) (a) Pope, E. J. A.; Asami, M.; Mackenzie, J. D. *J. Mater. Res.* **1989**, *4*, 1018. (b) Landry, C.J.T.; Coltrain, B.K. *Polym. Prepr. (Am. Chem. Soc., Div. Polym. Chem.)* **1991**, *32*, 514.

(9) (a) Novak, B. M.; Davies, C. *Macromolecules* **1991**, *24*, 5481. (b) Ellsworth, M.W.; Novak, B.M. *J. Am. Chem. Soc.* **1991**, *113*, 2756.

(10) (a) Chujo, Y.; Ihara, E.; Ihara, H.; Saegusa, T. *Macromolecules* **1989**, *22*, 2040. (b) Chujo, Y.; Ihara, E.; Kure, S.; Suzuki, K.; Saegusa, T. *Polym. Prepr. (Am. Chem. Soc., Div. Polym. Chem.)* **1990**, *31*, 59. (c) Chujo, Y.; Ihara, E.; Kure, S.; Suzuki, K.; Saegusa, T. *Makromol. Chem., Macromol. Symp.* **1991**, *42/43*, 303.

(11) Wei, Y.; Bakthavatchalam, R.; Whitecar, C. K. *Chem. Mater.* **1990**, *2*, 337.

(12) Whitecar, C.K. *M.S. Dissertation*, Drexel University, 1990.

(13) Wei, Y.; Bakthavatchalam, R.; Yang, D. C.; Whitecar, C. K. *Polym. Prepr. (Am. Chem. Soc., Div. Polym. Chem.)* **1991**, *32(3)*, 503.

(14) Wei, Y.; Yang, D. C.; Bakthavatchalam, R. *Mater. Lett.* **1992**, *13*, 261.

(15) Wei, Y.; Yang, D. C.; Tang, L. G.; Hutchins, M. K. *J. Mater. Res.*, **1993**, *8*, 1143.

(16) Wei, Y.; Yang, D.C.; Tang, L.G. *Makromol. Chem., Rapid Commun.* **1993**, *14*, 273.

(17) Wei, Y.; Wang, W.; Yang, D.; Tang, L. *Chem. Mater.* **1994**, in press.
(18) Wei, Y.; Wang, W.; Yeh, J.-M.; Wang, B.; Yang, D.; Murray, J.K., Jr. *Adv. Mater.* **1994**, *6*, 372.
(19) Wei, Y.; Wang, W.; Yeh, J.-M.; Wang, B.; Yang, D.; Murray, J.K., Jr. *Polym. Mater. Sci. Eng.* **1994**, *70*, 272.
(20) *Polymer Handbook*, Brandrup, I.; Immergut, E. H. Eds.; Wiley: New York, 1989, p. V-79 and p. V-82.
(21) Schulze, H. *Glass- Nature, Structure and Properties*, translated by Lakin, M.J.; Springer-Verlag: New York, 1991, p. 220.
(22) Wei, Y.; Jin, D.; Wei, G.; Wang, W.; Yang, D. Unpublished results.
(23) (a) Radziszewski, B. *Ber. Dtsch. Chem. Ges.* **1885**, *18*, 355. (b) Katritzky, A. R.; Pilarski, B.; Urogdi, L. *Synthesis* **1989**, 949. (c) Kabalka, G. W.; Deshpande, S. M.; Wadgaonkar, P. P.; Chatla, N. *Synth. Commun.* **1990**, *20*, 1445.
(24) Wei, Y.; Wang, W. Unpublished results.
(25) Wei, Y.; Yang, D.; Jin, D.; Wei, G. Unpublished results.
(26) For leading reviews, see (a) Crivello, J.V. *CHEMTECH* **1980**, *10*, 624. (b) Crivello, J.V. *Adv. Polym. Sci.* **1984**, *62*, 1. (c) Crivello, J. V. *Polym. Mater. Sci. Eng.* **1989**, *61*, 62. (d) Crivello, J.V. *J. Coat. Tech.* **1991**, *63*, 35. (e) Crivello, J.V. *Polym. Eng. Sci.* **1992**, *32*, 1462. (f) Lohse, F.; Zweifel, H. *Adv. Polym. Sci.* **1986**, *78*, 62. (g) Pappas, S.P. *J. Imaging Tech.* **1985**, *11*, 146. (h) Willson, C.G.; Bowden, M.J. *CHEMTECH* **1989**, *19*, 182.
(27) (a) Crivello, J.V.; Lam, J.H.W. *J. Polym. Sci., Symp.* **1976**, *56*, 383. (b) Crivello, J.V.; Lam, J.H.W. *Macromolecules* **1977**, *10*, 1307. (c) Dektar, J.L.; Hacker, N.P. *J. Org. Chem.* **1990**, *55*, 639.
(28) (a) Crivello, J.V.; Lam, J.H.W. *J. Polym. Sci.: Polym. Chem. Ed.* **1980**, *18*, 2677 and 2697. (b) Dektar, J.L.; Hacker, N.P. *J. Am. Chem. Soc.* **1990**, *112*, 6004.
(29) (a) Wang, B.; Brennan, A. B.; Huang, H.; Wilkes, G. L. *J. Macromol. Sci.-Chem.* **1990**, *A27(12)*, 1447. (b) Yi, G.; Sayer, M. *Ceramic Bulletin* **1991**, *70(7)*, 1173.

RECEIVED October 24, 1994

Chapter 12

Hybrid Organic–Inorganic Interpenetrating Networks

A. B. Brennan, T. M. Miller, and R. B. Vinocur

Department of Materials Science and Engineering, University of Florida, Gainesville, FL 32611–6400

A new class of interpenetrating networks (IPNs) has been synthesized by swelling a sol-gel-derived, poly(tetramethylene oxide)/silica hybrid composite with either methacrylic acid, cyclohexyl methacrylate, or N-vinylpyrrolidone. The monomer-swollen network is then gamma-polymerized *in-situ*. The resulting composites are optically transparent and exhibit significantly different mechanical properties than the original organic-inorganic hybrid. In particular, a 3000% increase in the elastic modulus for the methacrylic acid derived IPN is observed. Similarly, hydrogel-like behavior is also observable, including preferential pH absorption.

Numerous examples exist in recent literature describing the synthesis of organic-inorganic hybrid composites using the sol-gel process.(*1-6*) One such approach involves the *in-situ* precipitation of the inorganic phase, e.g., silica or titania, within an existing polymeric membrane.(*3,4*) This is accomplished by swelling the membrane with the desired metal alkoxide and then converting the absorbed species to an oxygen bridging network by utilizing the sol-gel process. Similarly, these hybrid materials have also been produced by co-reacting the inorganic monomer and the organic polymer, which may or may not be functionalized with alkoxy endcaps, during the initial stages of the sol-gel reaction.(*5,6*) In either case, the resulting material is a mixture of both organic and inorganic phases. The morphology and hence the degree of interaction between these two phases is governed by the traditional sol-gel processing variables such as the water to alkoxy and acid to alkoxy ratios, as well as the type of solvents and catalysts used.(*7-10*) Additionally, variables such as the molecular weight of the organic species, the presence and extent of polymer functionality, and the solubility of the polymer in the sol solution also influence the morphology of these hybrid materials.(*11,12,2*)

0097–6156/95/0585–0142$12.25/0

Therefore, depending upon the processing conditions, a wide range of morphologies and hence material properties can be engineered.

For example, dynamic mechanical spectrometry (DMS) and small angle X-ray scattering (SAXS) experiments indicate that the morphology of the benchmark acid catalyzed hybrid used in this study is composed of both organic and inorganic rich domains, as well as mixed regions of both.(*13,14*) Most likely, this mixed region consists of an entangled network of both physically and chemically cross-linked poly(tetramethylene oxide) (PTMO) and silicon metal alkoxide derived polysilicate chains. Depending upon the processing conditions, the polysilicate domains may be lightly cross-linked inorganic networks composed of predominantly linear chains, or more highly cross-linked, ramified polymeric networks.(*8,9*) For the benchmark hybrids described in this work, the acid catalysis results in an incomplete hydrolysis and condensation of the metal alkoxides.(*15*) Consequently, a predominantly linear, high surface area polysilicate phase is developed that interacts strongly with the elastomeric chains. However, we have recently demonstrated that the degree of interaction between the PTMO and polysilicate phases can be controlled *in-situ* by the chemically induced syneresis and ripening of the polysilicate phase.(*16*) This was accomplished by aging the hybrid in a strongly basic solution of 70% ethylamine in water. The result of such a treatment is the apparent modification of the interface and pore structures of both phases as the polysilicate domains phase separate from the PTMO and densify. This process allows tailoring of the structure-property relations of these hybrid composites. This technique may also allow the production of composite materials with controllable pore size and distribution for use as semi-permeable membranes. In particular, by filling these pores with either moisture sensitive or pH sensitive polymer, some very interesting hydrogel-like materials could be developed.

Therefore, our research is ultimately directed towards the synthesis of semi-permeable membranes of controllable morphology. Presented in this manuscript is the first known synthesis and characterization of such hybrid composite IPNs. These materials were generated by swelling the sol-gel-derived PTMO/silica hybrid in various monomers. These swollen gels were then exposed to a ^{60}Co gamma radiation source and the absorbed monomer was gamma polymerized *in-situ.* The resulting hydrogel materials were transparent, polymer films exhibiting significantly different physical properties than the original composite.

Experimental

A nomenclature detailing the conditions under which this class of organic-inorganic hybrids was processed has been previously developed.(*17*) In keeping with this system, the base composite used throughout this work is designated as a TEOS(40)-PTMO(2k)-100-0.014 HCl gel. Procedures detailing the synthesis of the isocyanatopropyltriethoxysilane end-capped PTMO oligomers have been previously described.(*13*) The initial sol-gel processing of the tetraethoxysilane (TEOS) and end-capped PTMO oligomers was carried out in a co-solvent system of isopropyl alcohol and tetrahydrofuran (THF) (Fisher Scientific - HPLC grade). Both were used as received.

Three monomers were used in the production of the IPNs herein described: cyclohexyl methacrylate (CHMA), N-vinylpyrrlidone (NVP) and methacrylic acid (MAA). All monomers were purified by vacuum distillation to remove inhibitors, prior to the synthesis of the IPNs. Upon conversion of the monomer to polymer and consequently, the conversion of the benchmark hybrid to an IPN, the terms PCHMA IPN, PVP IPN and PMAA IPN will designate the IPNs generated using cyclohexyl methacrylate, N-vinylpyrrolidone, and methacrylic acid, respectively.

The procedures used to produce the sol-gel-derived TEOS(40)-PTMO(2k)-100-0.014 HCl hybrids used in this experiment were previously described in detail.(16) Recapping briefly, 9 g of PTMO, 12 ml of isopropyl alcohol and 3 ml of THF were stirred in a parafilm covered 50 ml Erlenmeyer flask for 20 min. After 20 minutes, 6 g of TEOS was added to the PTMO/solvent mixture via a hypodermic needle. This resulted in a solution composed of 40 Wt% TEOS and 60 Wt% PTMO with respect to one another. The notation (40) following the term TEOS in the sample designation indicates this ratio. Continuing, this new mixture was then covered with parafilm and stirred an additional 7 minutes. After this 7 minutes had elapsed, a mixture of 0.225 g of 10N HCl and 2.475 g of deionized water was added to the PTMO/solvent/TEOS solution. These amounts resulted in a molar ratio of acid:ethoxy of 0.014:1 and a molar ratio of water:ethoxy of 1:1, i.e., 100% of the water required for complete hydrolysis had been added. These ratios also appear in the sample designation. The resulting sol was again covered and stirred vigorously for an additional minute. The solution was subsequently poured into polystyrene petri dishes which are then covered to reduce solvent evaporation. The covered petri dishes were then left at room conditions (23±1°C, ≈50%RH) for 4 days to gel. The dishes were then left uncovered for 2 days to exhume any residual alcohol or solvent. These materials were optically transparent indicating no macro-phase separation had occurred.

Each of these cast gels was then extracted and dried prior to their use in the formation of IPNs to ensure that all samples were equivalent. This was accomplished by soaking the TEOS(40)-PTMO(2k)-100-0.014 HCl gels in THF for 22 hrs. This process swelled the cross-linked network and should remove any residual by-products of the sol-gel process.(10) The gels were removed from the THF and dried in a vacuum oven at 40°C (28 in Hg) for 15 hrs. The gels were then soaked in deionized water for 24 hrs. Again the gels were dried in a vacuum oven at 40°C (28 in Hg) for 15 hrs. Samples treated in this manner were still optically transparent, and are referred to as *standard* gels.

The standard gels were cut into 5 or 6 rectangular pieces approximately 1.2 cm wide by 5 to 7 cm long. Typical sample thicknesses were 0.28 ± 0.05 mm. Each group of 5 or 6 samples was placed in a Pyrex petri dish containing approximately 60 ml of the respective distilled monomer. The petri dishes were then covered and the samples allowed to absorb monomer for 4 hrs. With the exception of the MAA samples, weights of each sample were recorded prior to immersion in monomer. After 4 hrs had elapsed, the samples were removed from their respective petri dishes and the surfaces were blotted dry. Upon completion of the surface blotting each sample was immediately placed in a prefilled test tube containing either pure deionized water or a 10% (Wt/Wt) distilled monomer in water solution.

Water was chosen as the suspending fluid because of its poor ability to swell these hybrid gels.*(16)* However, both the NVP and MAA monomers are miscible in water. Therefore, a 10% (Wt/Wt) monomer in water solution was used to reduce the concentration gradient and hence the driving force for monomer desorbtion. Fortunately, the CHMA monomer is immiscible in water and no such dilute solution was needed. Hence, only deionized water was used.

The test tubes containing the samples were then evacuated under mild vacuum, back-filled with argon and sealed with polyethylene caps. These test tubes were then placed 4 inches from a ^{60}Co gamma radiation source for 2 hrs. The dose rate was approximately 580 rads/min. Thus, the total dose was approximately 0.07 Mrads. Upon removal from the gamma source chamber, the solutions in the test tubes were viscous, indicating polymerization had occurred. Each sample was then removed from its test tube and the gelled residue was lightly wiped off the surface. These IPN samples were then immersed in open beakers containing 500 ml of distilled water for 24 hrs. The samples were removed periodically from the water and wiped clean. This process swelled any partially gelled monomer still clinging to the surface of the samples. The samples were then dried in a vacuum at 40°C (28 in Hg) for 12 hrs (MAA IPNs) or 24 hrs (CHMA and NVP IPNs). The IPNs were then removed from the vacuum and placed in Pyrex petri dishes containing approximately 60 ml of THF for 12 hrs (MAA IPNs) or 24 hrs (CHMA and NVP IPNs). THF is a good solvent for the PTMO/silica hybrid and this swelling should have flushed any monomer and/or low molecular weight polymer from the IPNs. The samples were removed from the THF and were once again vacuum dried at 40°C (28 in Hg) for 12 hrs (MAA IPNs) or 24 hrs (CHMA and NVP IPNs). Upon removal from the vacuum the PCHMA and PVP samples were weighed and all samples were placed in covered polystyrene petri dishes for storage at ambient conditions prior to characterization.

Characterization

Tensile strength and elongation were measured using an Instron Model 1122 equipped with a 200 lb load cell at ambient conditions. The strain rate was 2.54 mm/min. Dog-bone shape specimens were cut using the Type M-III ASTM die described in ASTM test D638M-84. The initial width of the test section was 2.5 mm with a grip-to-grip distance of 25 mm. A minimum of 5 samples were tested for each IPN composition.

Dynamic mechanical spectrometry was performed using a Seiko DMS 200(FT) interfaced with a Seiko Rheostation Model SDM/5600H. Testing of all IPNs was carried out from -150 to 275°C at a heating rate of 0.75°C/min in a dry nitrogen atmosphere maintained by a minimum flow rate of approximately 200 ml/min. Experiments performed to determine the effect of gamma irradiation, i.e., standard and gamma irradiated standard gels, were carried out from -150 to 260°C at a heating rate of 1.5°C/min in the same dry nitrogen atmosphere also maintained at a flow rate of 200 ml/min. All spectra presented are those obtained at 1 Hz.

A Seiko TG/DTA 320 interfaced with the same Rheostation was used to conduct thermogravimetric analysis. All tests were performed over the range of 25 to 1000°C with a heating rate of 10°C/min in a dry air atmosphere maintained at a flow rate of approximately 100 ml/min.

Swelling studies were performed using discs punched from the desired gel using a No. 6 cork borer (9.5 mm in diameter). Typically, three such punched samples were used in the determination of a single datum. Averages and standard deviations are given throughout the work and indicate this 3 sample technique.

Results and Discussion

Effect of Gamma-Radiation. When polymeric materials are gamma irradiated, typically two competing processes are possible. First, chain scission can occur which results in a degradation of the mechanical properties of the polymer. Secondly, systems which are capable of additional cross-linking may undergo such a process. Manifestations of such an increase in network cross-link density would include an increase in the elastic modulus and a decrease in the swelling coefficient after irradiation.(18) Hence, both of these properties were evaluated before and after irradiation by the same ^{60}Co gamma radiation source.

The effect of gamma radiation upon the mechanical properties of the standard TEOS(40)-PTMO(2k)-100-0.014 HCl hybrid composite is shown in Figure 1. The results of the irradiation are an increase in the elastic modulus and strength at break, and a reduction in the elongation at break. These values are listed in Table I.

Table I. Effect of Gamma-Irradiation Upon the Mechanical Properties of a Standard TEOS(40)-PTMO(2k)-100-0.014 HCl Gel

Material	Elastic Modulus (MPa)	Stress at Break (MPa)	% Elongation at Break
Standard Gel Before Irradiation	13.9 ± 0.14	5.16 ± 0.43	35.6 ± 3.0
Standard Gel After Gamma Irradiation	34.7 ± 0.98	8.84 ± 1.6	23.9 ± 3.1

Similar increases in the dynamic mechanical storage modulus, E', of the PTMO/silica hybrid are presented in Figure 2. In Figure 2, the storage modulus is plotted as a function of temperature for the hybrid both before and after exposure to the gamma radiation source. The basic profile of each spectra is the same in that each possesses approximately the same modulus in the low temperature, glassy regime and both exhibit the dramatic decrease in the storage modulus associated with the glass to rubber transition (T_g). However, the gamma radiation treatment does result in an increase in E' in the rubbery plateau, which begins at approximately 275K and ends at the upturn induced by synersis of the polysilicate phase.(16) In particular, the storage modulus increases from 17.7 MPa to 46.8 MPa at 298K for the standard and gamma irradiated standard gels, respectively.

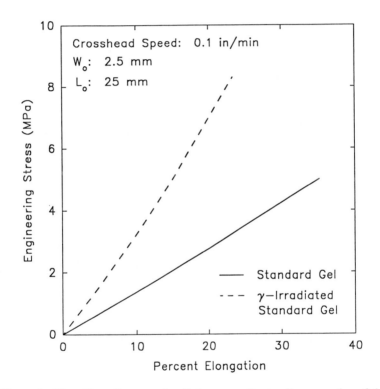

Figure 1. The effect of gamma-irradiation upon the tensile properties of the standard TEOS(40)-PTMO(2k)-100-0.014 HCl hybrid.

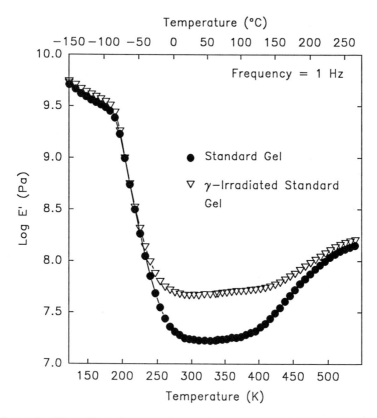

Figure 2. The effect of gamma-irradiation upon the dynamic mechanical storage modulus of the standard TEOS(40)-PTMO(2k)-100-0.014 HCl hybrid.

It can be seen in Figure 3, where the tan δ is plotted as a function of temperature, that again the same basic response can be seen for both materials. The sub-T_g relaxations (ca. 125K), which have been attributed to a crank-shaft type rotation of the tetramethylene units along the PTMO backbone, appear unaffected.(*19*) Similarly, only one relatively broad relaxation is observable when the composite passes through its T_g. However, exposure to gamma radiation appears to result in a tightening-up of the network. This is manifested in the decrease in intensity of the T_g relaxations from 0.33 (centered at 240K) to 0.25 (centered at 225K) for the standard and gamma irradiated standard hybrids, respectively. Similarly, the breadth of this relaxation is also reduced.

The effect of this radiation treatment upon the equilibrium mass uptake of THF is shown in Figure 4. A measure of the equilibrium uptake of a solvent is the swelling coefficient, Q, which is defined as:

$$Q = \frac{Mass_{Equilibrium} - Mass_{Initial}}{Mass_{Initial}} * \frac{1}{\rho_{Solvent}}.$$

Therefore, three samples were punched using the above described procedure (Characterization) from both a standard gel and a gamma irradiated standard gel. Each of the six samples were immersed in Pyrex petri dishes containing approximately 60 mls of THF. Within 9 minutes, both groups of gels had swollen to equilibrium. The standard gel then remains swollen at this equilibrium uptake value. However, the gamma irradiated gel only remains at its initial pseudo-equilibrium for a few minutes before it continues to absorb solvent. Using 16 minutes as a reference point, the gel swelling coefficient of the standard gel was 0.71 ± 0.01 (ml/g), compared to the lesser value of 0.41 ± 0.004 (ml/g) exhibited by the gamma irradiated gel of the same composition. Nevertheless, it is evident that indeed the uptake of solvent by the gamma irradiated gel is greatly reduced.

Based upon the changes in mechanical properties and solvent absorption, it is believed that exposure to the gamma radiation source does not result in a degradation of the mechanical integrity of these materials. Rather, these experiments suggest an increase in the network cross-link density of these standard hybrids with exposure to the ^{60}Co radiation source for 2 hrs at a dose rate of approximately 580 rads/min.

Properties of the IPNs. After determining the effect of gamma radiation upon the standard hybrid it was necessary to determine the maximum uptake of the various monomers so that a quantitative estimate could be made regarding the extent of hydrogel formation expected. This was accomplished by swelling three punched samples from a standard gel in each of the monomers. The monomer uptake was monitored as a function of time for 60 hrs. Typical absorption profiles, expressed as both the percent mass uptake and the gel swelling coefficient of each monomer, are shown in Figure 5. The initial absorption of all three of the monomers appears to exhibit Fickian type diffusion when plotted as a function of the square root of time. Similarly, all of the monomers also possess linear plateaus between 30 min and approximately 12 hrs. It was decided that 4 hrs would provide more than sufficient time for the standard gels used in the IPN synthesis to reach an

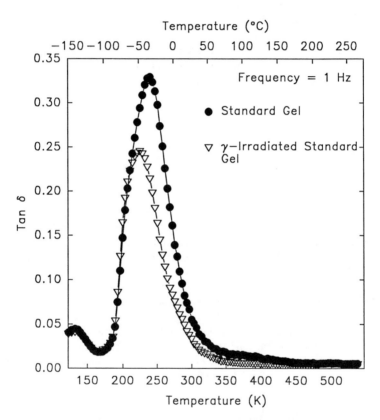

Figure 3. The effect of gamma-irradiation upon the dynamic mechanical tan δ response of the standard TEOS(40)-PTMO(2k)-100-0.014 HCl hybrid.

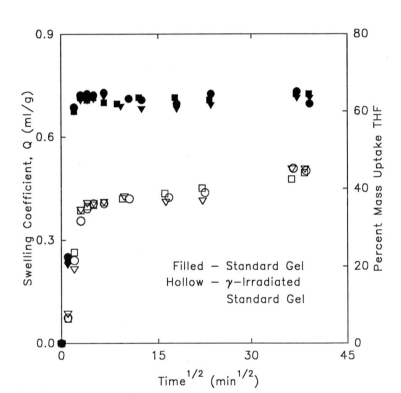

Figure 4. The effect of gamma-irradiation upon the equilibrium mass uptake of THF (3 standard (filled) and 3 gamma-irradiated standard (hollow) TEOS(40)-PTMO(2k)-100-0.014 hybrid samples tested; each sample shown).

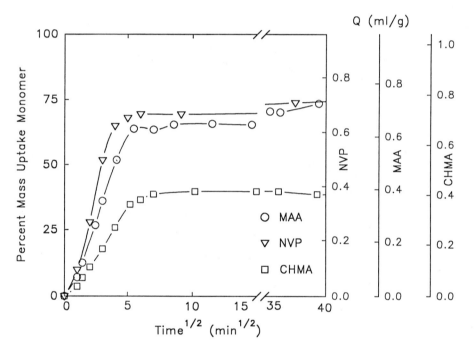

Figure 5. Monomer uptake profiles of a standard TEOS(40)-PTMO(2k)-100-0.014 HCl hybrid (typical results for each monomer).

equilibrium monomer uptake. Values of the swelling coefficient and percent mass uptake of monomer for these smaller punched samples after 4 hrs of swelling are given in Table II.

Table II. Swelling response of monomer exposed TEOS(40)-PTMO(2k)-100-0.014 HCl hybrids (uptake and swelling coefficient values recorded after 4 hrs)

Monomer	Q (ml/g)	% Mass Uptake
NVP	0.662 ± 0.013	69.1 ± 1.4
MAA	0.659 ± 0.009	66.9 ± 0.92
CHMA	0.425 ± 0.012	41.0 ± 1.2

These results suggested that upon completion of the irradiation process the IPNs synthesized from the NVP and MAA monomers would contain a greater mass fraction of the newly formed polymer than that of the CHMA swollen gels. However, this was not observed. Apparently, the miscibility of the NVP and MAA in water, despite the dilute solution employed to lessen the diffusion gradient, allowed a considerable amount of monomer to diffuse out of the hybrid. This diffusion most certainly began to occur immediately upon placing the swollen hybrid into the aqueous solution filled test tubes.

The standard hybrids used in the synthesis of the NVP and CHMA IPNs were weighed prior to exposure to monomer and subsequent gamma-polymerization. Similarly, the IPN samples resulting from the NVP and CHMA monomers were weighed immediately upon removal from the vacuum oven after synthesis and subsequent cleaning (Experimental). TGA was also performed on the newly synthesized IPNs and the percent residue on ignition (%ROI) at 1000°C in air recorded. It is important to note that the standard hybrids have 21.2 ± 0.06 %ROI, and this residue is assumed to be silicon dioxide. Additionally, the lower the %ROI the greater the polymer to silica ratio in the IPN. As can be seen in Table III, both the experimental and calculated percent polymer formed (based on TGA results) confirm significant but less than anticipated hydrogel formation.

Table III. Quantitative assessment of the weight percent polymer formed as a result of gamma irradiation\polymerization

Polymer Formed	Wt% Polymer$_{Exp.}$	%ROI	Wt% Polymer$_{TGA\ Based}$
PMAA	N.A.	11.8 ± 0.40	45.6
PVP	19.2 ± 1.4	18.5 ± 1.31	14.9
PCHMA	24.7 ± 1.6	16.3 ± 0.40	25.2

The mechanical properties of IPNs depend strongly upon the degree of mixing between the phases, i.e., macro-scale mixing or molecular mixing. The solubility parameter, δ, provides a good indicator of how compatible two polymers are. In general, the closer their respective δ-values the higher the degree of mixing expected. In our case the two components are the standard hybrid and the gamma polymerized monomers. Therefore, the first step in predicting and understanding the mechanical behavior of these materials was to determine the solubility parameters of the various polymers present in the IPNs. The δ-values of the PMAA, PCHMA and PVP were available in the literature.(20-22) The range of 26-30 $MPa^{1/2}$ was reported for Rohm and Haas PMAA in strongly hydrogen bonding liquids by Grulke.(20) Due to the significant number of unreacted hydroxy groups present on these acid catalyzed, sol-gel-derived organic-inorganic systems, and the demonstrated ability of these hybrids to hydrogen bond, this value should be representative of the PMAA generated within the benchmark gel.(15,23) Similarly, the value of 25.8 $MPa^{1/2}$ cited for PVP was determined by swelling experiments in solvents of known δ-values on radiation polymerized NVP.(21) Therefore, this value is also believed to be representative of the PVP produced in this experiment. The value of 18.2 $MPa^{1/2}$ for PCHMA was calculated by Frank and Gashgari using the methods of Small and Hoy.(22,24,25) However, the organic-inorganic hybrids used in this study are indeed hybrid composites and it was necessary to experimentally determine the solubility parameter.

The easiest way to determine the δ-value for the polymer network was to swell punched discs in "non-reactive" solvents possessing known solubility parameters.(20) A range from 14.9 to 29.7 $MPa^{1/2}$ was examined, which included hexane, p-xylene, tetrahydrofuran, chloroform, acetone, isopropanol and methanol. These solvents were chosen for two reasons. First, each was easily acquired. More importantly, these solvents would not create conditions of high solubility (pH>10.5) for the polysilicate phase once absorbed.(26) After immersing three punched samples of the benchmark hybrid in approximately 60 mls of each of these liquids of known δ-value, the mass uptake was recorded as a function of time until a constant mass uptake was reached. Usually, equilibrium was reached within 15 minutes. However, the samples remained immersed for 2 hrs to ensure that this was indeed the equilibrium mass uptake value. The results appear in Figure 6 where a maximum in the swelling coefficient is observed at 19.1 $MPa^{1/2}$. Hence, we estimate δ to be equal to approximately 19.1 $MPa^{1/2}$ for the hybrid composites used in the synthesis of all IPNs described in this work. Solubility parameters for all of the polymers generated in this study appear in Table IV, along with the reported glass transition temperature of each as well. It is important to realize that the actual T_g of any of these materials may be affected by moisture content. In particular, the T_g of PVP has been shown to decrease from 448K to 323K when the water content changes from 0 to 15%.(27)

Based on the experimental value of 19.1 it is expected that the highest degree of mixing would occur for the PCHMA containing IPN while the PVP and PMAA systems would exhibit much less mixing. As will be discussed shortly, DMS experiments confirm this hierarchy of mixing.

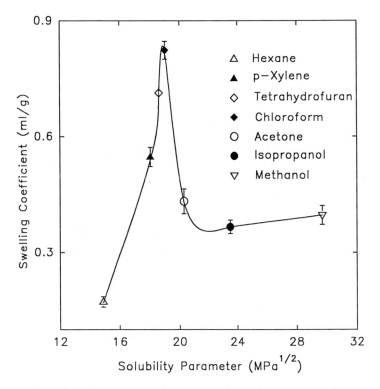

Figure 6. Solubility parameter, δ, determination using solvents of known δ-values.

Table IV. Solubility parameters and glass transition temperatures of the polymers known or expected to be generated in this study

Polymer	δ (MPa$^{1/2}$)	T_g (K)[d]
Standard Hybrid	19.1	197
PMAA	26-30[a]	501
PVP	25.8[b]	423
PCHMA	18.2[c]	356 (atactic)

[a]Data from reference 20.
[b]Data from reference 21.
[c]Data from reference 22.
[d]Data from reference 28.

The tensile strength as a function of elongation for the three IPNs is plotted in Figure 7. It is immediately evident that a significant increase in the mechanical properties of these organic-inorganic hybrids can be induced by the transformation to IPNs. Of particular interest is the much greater elastic modulus and elongation at break for the PMAA containing IPNs. To a lesser extent the same trends are observable for the PVP containing IPNs. However, the mechanical response of the PCHMA IPN possesses an elastic modulus slightly less than the gamma irradiated hybrid, but the percent elongation at break and ultimate strength are significantly increased. The mechanical properties for the IPNs are given in Table V. Similar, data is available in Table I for the standard and gamma irradiated standard hybrid.

Table V. Mechanical properties of the TEOS(40)-PTMO(2k)-100-0.014 HCl derived IPNs

IPN System	Elastic Modulus (MPa)	Stress at Break (MPa)	% Elongation at Break
MAA	1060 ± 63	36.9 ± 4.4	129 ± 34
CHMA	19.1 ± 1.1	20.8 ± 4.0	60.0 ± 9.7
NVP	110 ± 10	25.5 ± 1.3	67.8 ± 6.0

The storage modulus is plotted as a function of temperature in Figure 8 for each of the IPN systems. The dynamic mechanical response of the standard and gamma irradiated hybrid systems are available in Figure 3. The spectrum of the PCHMA gel most closely resembles that of the standard hybrid. In particular, only one significant decrease in the storage modulus is observed and occurs in the range of 180K to 280K. Absent is any indication of a decrease in the storage modulus around 350K, the expected T_g for PCHMA (Table IV). An increase in the storage modulus does begin at ca. 375K and is attributable to the syneresis process mentioned earlier. However, the two spectra of the PVP and PMAA containing

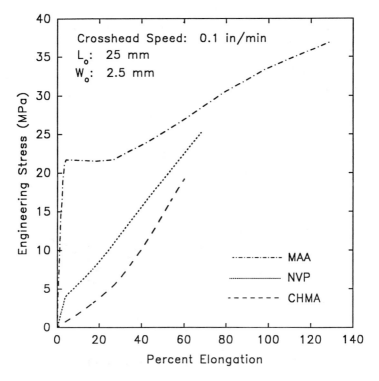

Figure 7. Engineering stress as a function of elongation for the three IPNs (labeled according to the original monomer employed), as well as the gamma-irradiated standard hybrid.

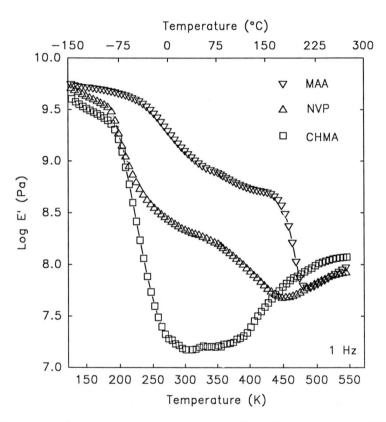

Figure 8. Dynamic mechanical storage modulus, E', for the three IPNs labeled according to the original monomer employed for IPN synthesis.

IPNs are significantly different than that of the standard hybrid (Figure 3). For the PVP sample, a high modulus is observed in the low temperature glassy regime (123K to 175K).   Similarly, the onset of the glass transition of the PTMO\polysilicate hybrid is observable at approximately 180K.   However, the magnitude of this decrease is lessened. Consequently, the modulus in the rubbery region of the original hybrid is significantly increased due to the presence of the higher T_g PVP phase (Tg≈423K) within the IPN. Again, a decrease in the storage modulus begins at approximately 325K and reaches a minimum at ca. 450K. This decrease in the storage modulus is associated with the T_g of the PVP phase. Beyond 450K the same syneresis process is observable as the storage modulus again increases. Similar to the mechanical tensile testing data, the most significant changes are observed for the PMAA containing IPN. Initially, the glassy modulus is comparable to that of the PVP sample. However, as the temperature increases, no observable decrease associated with the T_g of the PTMO\polysilicate hybrid appears until approximately 245K. Even then, only a slight, broad decrease is observable which slowly decays until ca. 450K. At 450K the sharp, downturn in the storage modulus associated with the T_g of the PMAA is observable and a minimum value at 460K, beyond which point the syneresis increase dominates.

The related dynamic mechanical tan δ response of the IPNs as a function of temperature is presented in Figure 9.   Returning to the PCHMA IPN, the presence of a single, relatively narrow relaxation centered at 200K strongly suggests that indeed a high degree of molecular mixing exists between the original, cross-linked PTMO\polysilicate phases and PCHMA phase. Furthermore, there is no evidence of any T_g relaxations for the PCHMA (Table IV). In fact, if not for the difference in the tensile properties of the CHMA IPNs, than the presence of the gamma polymerized phase would be suspect. Continuing, multi-modal relaxations are seen in the spectrum of the PVP at both 205K and 410K, which are associated with the original hybrid and PVP T_g, respectively. Similar relaxations are seen in the spectrum of the PMAA IPN at 270K and 455K. In particular, the PMAA relaxation at 455K is more intense than the T_g relaxation of the PVP IPN. Recall that the magnitude of the decrease in E' associated with the T_g of the PVP gel was much broader and more gradual than that of the PMAA gel also. These two trends suggest that a lesser degree of mixing exists between the PMAA and PTMO/polysilicate phases than between the PVP and PTMO/polysilicate phases. The shifting of the Tg of the original hybrid in the PMAA IPN spectrum to a temperature greater than in the PVP spectrum is undoubtedly due to the higher T_g of PMAA versus that of PVP. Nevertheless, it appears that all of the polymers generated by this *in-situ* polymerization mix on a fairly intimate level with the original PTMO/polysilicate hybrid.   Consequently, the response of these successfully synthesized IPNs to moisture and pH was evaluated.

In an attempt to determine the moisture absorbing ability of these potentially hydrogel-like IPNs, three samples of each IPN were placed in a vacuum oven at 40°C (28 in Hg) for 72 hrs under continuous evacuation. They were then removed, immediately weighed and then immersed in approximately 25 ml of deionized water for 24 hrs. As mentioned earlier, water is a very poor swelling agent for the benchmark hybrid. Therefore, any uptake of water can be attributed solely to the presence of the gamma-polymerized polymer. Additionally, It was also suspected

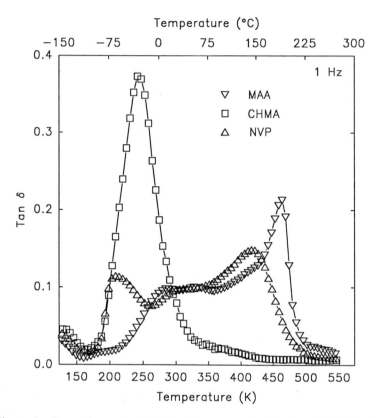

Figure 9. Dynamic mechanical tan δ response for the three IPNs labeled according to the original monomer employed for IPN synthesis.

that these IPNs would exhibit a preferential swelling as a function of pH. In particular, this was expected due to the presence of the acid species on the PMAA. Therefore, two samples were punched from each of the IPNs and each was immersed in approximately 50 ml of standard buffer solutions of pH 2, 4, 6, 8 and 10 for 2 hrs. Table VI presents the results of both swelling studies.

The buffer solutions of strength 8 and 10 contain potassium hydroxide. This base is capable of hydrolyzing and condensing any unreacted alkoxy and hydroxy groups in the original hybrid. Similarly, it can also induce syneresis in the inorganic phase and eventual dissolution of the silica.(*16,27*) To verify that the inorganic portion of the IPN was quantitatively unaffected, TGA was performed on discs punched from the MAA derived IPN exposed to the buffer solutions employed. The results also appear in Table VI, and confirm that indeed, no silica has been dissolved and expelled from the IPN after the 2 hr exposure time.

Conclusions

Gamma radiation does not induce degradation of the benchmark organic-inorganic hybrid system used in this study. Rather, it appears to result in an increase in

Table VI. Moisture and pH sensitivity of the TEOS(40)-PTMO(2k)-100-0.014 HCl derived IPNs

Property	PMAA IPN	PVP IPN	PCHMA IPN
% Mass Uptake Water	7.79 ± 1.96	17.2 ± 8.23	2.46 ± 0.463
% Mass Uptake Buffer			
pH:			
2	-2.03, -2.54	4.98, 10.2	1.11, 3.42
4	1.03, 4.81	9.60, 13.4	0.40, 1.68
6	N.A.	4.71, 14.3	-0.38, -0.87
8	13.5, 14.6	11.1, 21.8	-0.80, 0.78
10	37.9, 60.0	16.7, 17.6	-0.77, 1.56
%ROI			
pH:			
2	13.5, 13.6		
4	13.1, 13.8		
8	17.6, 24.8		
10	15.4, 17.8		

network cross-link density. Novel, transparent interpenetrating networks were successfully synthesized by swelling a sol-gel-derived, organic-inorganic composite system with various monomers and subsequently gamma polymerizing them. Indeed, significant changes in the mechanical and dynamic mechanical properties of these IPNs were observed, especially in the case of the PVP and PMAA samples. In particular, the modulus of the PMAA IPN showed a 3000% increase to a value in excess of 1 GPa. Also, these materials did exhibit hydrogel-like behavior in that they did absorb significant amounts of water. Similarly, the PMAA IPNs demonstrated a preferential pH absorption.

These results are very promising. Of importance now is further characterization to understand the degree of hydrogel behavior these materials exhibit, i.e., the effect of moisture upon the physical properties. These results will be presented as they are obtained.

Literature Cited

1. Schmidt, H. *J. Non-Cryst. Solids* **1985**, *73*, 691.
2. Novak, B. M.; Ellsworth, M.W. *Mater. Sci. and Eng.* **1993**, *A162*, 257.
3. Sur, G. S.; Mark, J. E. *Eur. Polym. J.* **1985**, *21*, 1051.
4. Mauritz, K. A.; Storey, R. F.; Jones, C. K. *Multiphase Polymers: Blends and Ionomers*; Ultracki, L. A.; Weiss, R. A., Eds.; *ACS Symp. Series* **1989**, 401.
5. Landry, C. J. T.; Coltrain, B. K.; Brady, B. K. *Polymer* **1992**, *33*, 1486.
6. Huang, H. H.; Orler, B.; Wilkes, G. L. *Macromolecules* **1987**, *20*, 1322.
7. Sakka, S.; Kamiya, K. *J. Non-Cryst. Solids* **1982**, *48*, 31.

8. Brinker, C. J.; Keefer, K. D.; Schaefer, D. W.; Assink, R. A.; Kay, B. D.; Ashley, C. S. *J. Non-Cryst. Solids* **1984**, *63*, 45.

9. Brinker, C.J.; Scherer, G. W. *J. Non-Cryst. Solids* **1985**, *70*, 301.

10. Brennan, A. B.; Wilkes, G. L. *Polymer* **1990**, *32*, 733.

11. Huang, H. H.; Orler, B.; Wilkes, G. L. *Macromolecules* **1987**, *20*, 1322.

12. Saegusa, T. *J. Macromol. Sci.-Chem.* **1991**, *A28*, 817.

13. Huang, H. H.; Glaser, R. H.; Wilkes, G. L. *ACS Symp. Ser.* **1987**, *360*, 354.

14. Wilkes, G. L.; Brennan, A. B.; Huang, H. H.; Rodrigues, D. E.; Wang, B. W. *Mater. Res. Soc. Symp. Proc.* **1990**, *15*, 171.

15. Glaser, R. H.; Wilkes, G. L.; Bronniman, C. E. *J. Non-Cryst. Solids* **1989**, *113*, 73.

16. Brennan, A. B.; Miller, T. M. *Chem. Mater.* **1994**, *6*, 262.

17. Huang, H. H.; Orler, B.; Wilkes, G. L. *Polym. Bull.* **1985**, *14*, 557.

18. Reichmanis, E.; Frank, C. W.; O'Donnell, J. H.; Hill, D. J. T. *Irradiation of Polymeric Materials: Processes, Mechanisms, and Applications*; ACS Symp. Ser. 1993, *527*, 1.

19. McCrum, N. G.; Read B. E.; Williams, G. *Anelastic and Dielectric Effects in Polymeric Solids*; Dover Publications, Inc.: New York, NY, 1967, pp. 564-565.

20. Grulke, E. A. In *Polymer Handbook*; Brandrup, J.; Immergut, E. H., Eds., Wiley-Interscience: New York, NY, 1989, VII-519.

21. Tashmukhamedow, S. A.; Sagdullaev, B. U.; Darabaev, A. Sh.; Tillaev, R. S. *Uzb. Khim. Zh.* **1972**, *16*, 47.

22. Frank, C. W.; Gashgari, M. A. *Macromolecules* **1979**, *12*, 163.

23. Landry, C. J. T.; Coltrain, B. K.; Wesson, J. A.; Zumbulyadis, N; Lippert, J. L. *Polymer* **1992**, *33*, 1496.

24. Small, P. A. *J. Appl. Chem.* **1953**, *3*, 71.

25. Hoy, K. L. *J. Paint Technol.* **1970**, *42(541)*, 76.

26. Iler, R. K. *The Chemistry of Silica*; John Wiley & Sons: New York, NY, 1979, pp. 40-54.

27. Tan, Y; Challa, G. *Polymer* **1976**, *17*, 739.

28. Peyser, P. In *Polymer Handbook*; Brandrup, J.; Immergut, E. H., Eds.; Wiley-Interscience: New York, NY, 1989, VI-209.

RECEIVED October 11, 1994

Chapter 13

A New Route to Polymer-Filled Glass

Hybrid Interpenetrating Networks with Appreciable Toughness

Kenneth G. Sharp

Central Research, DuPont, Wilmington, DE 19880–0323

Hybrid inorganic/organic materials can be generated in non-aqueous systems by rapid formation of polysilicate networks in the presence of dissolved organic polymers. The enabling chemistry is the direct reaction of formic acid with simple tetraalkoxysilanes. The acid acts as solvent, water source and catalyst for both hydrolytic and condensation reactions. Gelation times can be two to three orders of magnitude faster than for the conventional aqueous sol-gel counterpart. Transparent composites can be fabricated from a variety of polymers soluble in the acid. The polysilicate phase can become continuous at levels as low as 5 wt%, causing substantial modification of system modulus and polymer viscoelastic behavior. Continuity was established from careful pyrolysis of the polymer at 650°. The hybrids were formulated as interpenetrating networks based on NMR evidence against covalent bonding between the phases. Small angle X-ray scans show no scattering intensity maxima with scattering angle and a linear power law (Porod) slope of ca. 2.4, indicating mass fractal scattering centers. TEM analysis suggests polymer domains a few nm in size. For the flexible polymer poly(vinyl formate), hybrids showed high levels of toughness and yield behavior. Data from dynamic mechanical studies and some applications-related behavior are also discussed.

The inorganic constituent of hybrid materials is most commonly generated through the sol-gel-based conversion of alkoxysilanes into polysilicate structures. When the process involves dissolved organic polymers, conventional sol-gel chemistry has its shortcomings. First, the requirement of stoichiometric quantities of water for the reaction limits the range of solvating characteristics of the liquid medium. Secondly, the reaction can be quite slow at the low pH's associated with lightly branched network precursors most suitable for interpenetrating network (IPN) preparation. Finally, reactions rates can be further diminished by the presence of cosolvents such as ethanol or THF.

We report here the use of formic acid as an alternative non-aqueous medium for sol-gel chemistry. The acid serves as solvent, water source and catalyst for both hydrolysis and condensation reactions. The system allows rapid formation of

0097–6156/95/0585–0163$12.00/0

polysilicates in the presence of a wide range of dissolved organic polymers and can lead to continuous inorganic phases at levels of about 3 vol%. Flexible polymers can show qualitatively different behavior when confined in such networks; modified behavior includes toughness, yield under stress and suppression of viscous flow. Topologically, such materials can be classified as semi- or full IPNs, depending upon the continuity of the polymeric phase.

Numerous workers have generated inorganic/polymer hybrid materials in which inorganic networks were developed in the presence of dissolved, fluid or swollen organic polymers. In the majority of instances, the two components are covalently bonded by design (1-4) or circumstance (5). Hybrid materials in which the constituent phases may form simple IPNs without *necessarily* being covalently bonded (6) are less well known. Nevertheless, several distinct synthetic pathways to hybrid IPNs have been established. Dried silica gels with open porosity have been infused with monomers such as methyl methacrylate which are then polymerized *in situ* (7,8). Polymers dissolved in the aqueous media normally used for sol-gel reactions can be captured by polysilicate networks grown in their presence (9-11). Inorganic phases have also been generated in the interior of swollen polymers such as poly(dimethylsiloxane) (PDMS) (12), and Nafion (E. I. du Pont de Nemours & Company) membranes (13). The PDMS hybrids contain particulate (14) inorganic species. The Nafion-based materials include polysilicate clusters which are capable of phase continuity under some conditions (15) and apparently contain a stratified pure inorganic layer near the surface (16). Bicontinuous networks can be generated from simultaneous end-linking and silicate formation in hydroxy-terminated PDMS (17). Finally, a novel series of "non-shrinking" hybrids has been formulated by concurrently polymerizing the polysilicate and the alcohol liberated from the starting tetraalkoxysilane (5,18). Because of the susceptibility of silanol bonds to re-esterification by hydroxyl-bearing polymers, it is likely that the components are covalently bonded.

For the synthetic pathway in which organic polymers are dissolved, the requirement for stoichiometric amounts of the hydrolytic agent water necessitates water-soluble polymers or the use of co-solvents such as THF to solubilize the polymer.

Non-aqueous Sol-gel Chemistry

The infinite network polysilicate (19) components are synthesized by newly developed (20) two component non-aqueous sol-gel chemistry: the reaction of tetraalkoxysilanes with formic or other strong carboxylic acids. Water is not required as an initial reagent. Hydrolytic processes are enabled via the *in situ* generation of water from the reaction of the acid with liberated alcohols. Silica gels generated from this reaction show unusually fine pore structure; in some cases nitrogen uptake is not observed in a four hour adsorption isotherm at 77°K. Such materials do possess open porosity, as evidenced by significant uptake of CO_2 at 195°K and substantial differences in bulk and skeletal densities.

As a synthetic platform for hybrid synthesis, this reactive system possesses several potential advantages relative to its aqueous counterpart. The first is that network formation is remarkably rapid, tending to avoid or minimize phase separation of the organic polymer. A poly(methyl methacrylate) (PMMA)/ polysilicate hybrid system described by Landry, Coltrain, et al. (21) required 200 hours for gelation in a closed system at 60°. At a comparable tetraethoxysilane (TEOS)/PMMA ratio, the formic acid-based counterpart gels in 15 *minutes* at that temperature. The second advantage is that formic acid is an excellent solvent for several important families of polymers -- including polyamides, methacrylates, and cellulosics -- which are not water soluble. Finally, the extremely small pore sizes associated with the silica gel generated in the absence of the polymer may lead to a greater level of efficiency of entraining polymer chains in the "pores" of the inorganic network. Another important feature of the solvent system is that hydroxyl-containing polymers will be extensively esterified by the solvent.

Material Expectations

Modulus.   The compositional dependence of tensile or flex modulus in polymer/polymer IPNs has been successfully modeled by a variety of approaches. The simple mechanical models of Takayanagi (*22*) have been modified (*23,24*) for isotropic bicontinuous phase systems. The modified versions accommodated the rapid increase in system modulus with increases in the higher modulus minority component in poly(ethyl acrylate)/poly(styrene-co-methyl methacrylate) IPNs (*25*).

   In hybrid materials, an increase in system modulus (as determined from stress/strain curves) with increasing inorganic content has been observed in systems both with (*1,26*) and without (*13*) bonded phases. Qualitatively similar behavior based on a subjective scale of brittleness has also been observed for semi-IPNs (*10*). The ability to program moduli over a tremendous range has been demonstrated by Wolter (*27*) in systems containing both multiple acrylate and alkoxysilane functionality in the same molecular precursors.

Stress/strain Behavior.   On the scale of phase separation observed in many polymer/polymer IPNs, reducing the domain size in simultaneous IPNs (by closely matching network formation rates) can lead to a *reduction* in tensile strength (*28*).

   Although organic IPNs having appreciable toughness are well known (*29*), little modeling of IPN behavior exists relative to such properties as fracture toughness or phenomena such as yielding. Termonia (*30*) has calculated two-dimensional stress-strain behavior for homopolymer IPNs which comprise long and short-chain components. The molecular weight between entanglements and cross-linking efficiency of the short chains were varied, and some chain slippage during strain was allowed. Simultaneous IPNs demonstrated poor toughness behavior due to the poor connectivity of the long chain component. Sequential IPNs showed higher modulus and strength behavior, but limited toughness due to poor extensibility. Much enhanced toughness was, however, observed if the long chains were allowed to form macrocycles which were *threaded* by the short chains, as in rotaxanes. Although the topology differs from those normally encountered in hybrid materials, the clear advantage of a threaded system leads to some intriguing questions. Would a linear polymer permanently entangled in the extremely small pores of a silica gel show such behavior? What magnitude of chain slippage would occur? What modifications of mobility and viscoelastic behavior of the polymer near or above its T_g could one expect?

   For hybrid materials, covalently bonded systems with appreciable toughness have been synthesized by markedly different synthetic routes and correspondingly different microstructural natures (see Discussion section).

   Other hybrid materials synthesized from dissolved polymers and silicate precursors have proved to be very brittle and less tough than their organic polymeric constituent (*9,10,31*). In these systems, the components may be only hydrogen bonded (*9,31*) or, in the case of poly(vinyl alcohol) PVOH (*10*) or poly(vinyl acetate) (PVAc) (*11*) possibly linked through transesterification reactions as well. In many other studies of hybrids, stress/strain or other data relating to toughness have not been reported.

Results and Discussion

New Route to Polymer/Polysilicate IPNs. The recent discovery that polysilicate networks can be grown rapidly in non-aqueous systems (*20*) enables the facile growth of such networks around dissolved polymers which are not water soluble. The formation of the silicate network involves hydrolytic (*32*) cleavage of Si-OR bonds followed by formation of condensed polysilicate species, as in conventional aqueous-based sol-gel chemistry. However, in the present system, the initial presence of water is not required; it is generated *in situ* during the reaction. Strong mineral acid catalysts normally used in hybrid syntheses are likewise unnecessary. Despite this situation, the reaction is some 2-3 orders of magnitude faster than the corresponding (at comparable measured pH) aqueous-based sol-gel reactions. For hybrid systems in which prevention of gross phase separation is only possible under kinetic control, the enhanced rate of formation of the inorganic network may be a significant factor in generating materials with minimal heterogeneity.

The addition of simple tetraalkoxysilanes (more commonly, a solution of the silane in formic acid) to a solution of polymers such as PVOH, PMMA, poly(vinyl pyrollidone) (PVP) or cellulose acetate in 96% formic acid leads to formation of clear, monolithic gels in approximately 30-60 minutes at room temperature. The coating solutions and materials derived from them appear clear and homogenous over the entire compositional range of the systems. The hybrid gel times resemble those of silane/formic acid systems in the absence of the polymer, although they are typically somewhat shorter. As mentioned above, polymers containing hydroxyl groups are extensively esterified by the solvent. Accordingly, PVOH is essentially completely converted into the amorphous, flexible polymer poly(vinyl formate) (PVF) during dissolution in formic acid. The latter system is the most extensively investigated to date, and will be the focus of the discussion.

Air drying of gelled material can lead to monolithic structures; however, drying times for sizable bulk structures are extensive because of the enormous capillary pressure which can develop at higher drying rates (*33*). Such pressures can easily exceed the material strength. Fortunately, films from drawdown or dip coating operations can be rapidly dried without material failure, and free standing films can be obtained from coating on polypropylene substrates. Films comprising flexible polymers such as PVF and up to about 30% polysilicate by weight are quite flexible after air drying (*34*) (Figure 1) despite the presence of a continuous inorganic network (see below).

The most closely related materials to the PVF hybrids previously described are the PVAc/polysilicate hybrids with the polymer dissolved during the silicate formation. The Kodak group led by Landry and Coltrain has provided a number of important publications on physical properties of this (*9*) and related systems. The authors cited infrared and other evidence of substantial hydrogen bonding between the inorganic and organic components.

An ostensibly related PVAc/polysilicate system has been described by Pascault, et al. (*11*) The two groups differ, however, as to the nature of the bonds between the phases (see below).

Modulus. Dynamic mechanical analysis (DMA) was used to examine the storage and loss moduli of thin films at a frequency of 1 Hz. Figure 2 is an overlay of storage moduli for dried films of hybrids containing from 5 to 20 wt% polysilicate and pure PVF generated from the solution of PVOH in formic acid. Data for a reheated 20% sample are also shown. The glass transition temperature increases by about 10-25° in the series and the behavior at temperatures above the T_g is profoundly altered. The pure polymer enters viscous flow near 100°. The composites show a minimum in modulus near or above 100°, then a steadily increasing modulus up to 250°, the limit of the experiment. The modulus of the

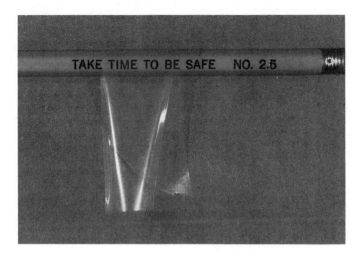

Figure 1. Air dried polysilicate (20%)/PVF film.

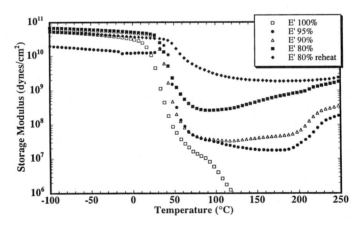

Figure 2. Storage modulus (E') from dynamic mechanical spectra for polysilicate/PVF films vs. pure polymer.

reheated (20%) sample remains above the high temperature limiting modulus of the first scan. Material integrity can be maintained above T_g even with a 5% silica content; this loading corresponds to about 3 vol%!

The development of a plateau modulus which exhibits a minimum with temperature has been reported for both covalently bonded (1) and non-bonded systems (9,11). However, in the Kodak PVAc system at a silicate content of 10%, the material succumbs to viscous flow without any sign of a plateau modulus (35).

Although the effect of extensive drying during the scan could contribute to increased modulus, the principal effect is believed to be the formation of a more substantial inorganic network through additional silicon-based condensation reactions.

From the loss portion of the scan (Figure 3), it is evident that the glass transition becomes attenuated and asymmetric to higher temperature at the higher silicate contents and with reheating. Relaxation behavior in these materials contrasts with that of graft-bonded hybrids (1) in which the T_g is profoundly shifted to higher temperature, and high polysilicate non-bonded ones (10) in which a glass transition could not be observed in dynamic scanning calorimetry. For the more closely related PVAc/polysilicate systems, Pascault (11) has observed a minimum in T_g as silicate content increases. This surprising result could be related to plasticizing effects of retained liquids (the samples were dried at 40°). The Kodak group (35) observed a constant onset of the transition with composition. Both groups also observed an attenuation and broadening of the transition in the hybrid relative to the parent polymer.

Stress-strain Behavior and Toughness. The effect of the confining silica network on stress-strain behavior is illustrated in Figure 4. The room temperature modulus of the hybrid can be nearly an order of magnitude higher than that of the pure polymer (36). Some samples exhibited excellent strain-to-break behavior despite their relatively high moduli. Tensile strengths of >60 MPa and energy to break values as high as 1400 N-mm have been observed. The latter lies in the range of "supertough" polyamide values (37). The hybrids accordingly can be tougher than either pure constituent.

Toughness appears to be maximized in the region of 10-20 wt% polysilicate. At this level, the inorganic network is substantial enough to lead to high modulus materials, but the confined polymer can still contribute to flexibility and energy dissipation without undergoing viscous flow. In this range, the hybrids show prominent yield behavior.

Tensile strength in film samples showing brittle fracture (generally, above 25% polysilicate) was often compromised by the presence of flaws such as small gel particles or edge defects induced during sample preparation.

The materials show sharply contrasted behavior when compared to several other apparently non-bonded hybrid IPNs. With PVAc — whose Tg is very similar to that of PVF — the Kodak group reports (9) strain to break values of 3% or less with tensile moduli quite similar to ours (ca. 3.2 GPa). Their tensile moduli increased only modestly with polysilicate content, and the hybrids show less energy to break than the parent polymer. Stress/strain data were not reported by Pascault (11). The poly(ethyl oxazoline) (PEOX)/polysilicate hybrids claimed to be single phase semi-IPNs are also quite brittle (10).

Polysilicate-based hybrids with appreciable toughness have been reported by the Wilkes, Mark, and Novak groups. Wilkes' (1) involved a trialkoxysilane-grafted poly(tetramethylene oxide) which was co-condensed with TEOS; both the number of grafts and the amount of TEOS were varied in this comprehensive study. Mark's earlier work (12) featured the in situ generation of polysilicates or -titanates inside a swollen PDMS elastomer. A more recent approach (38) from

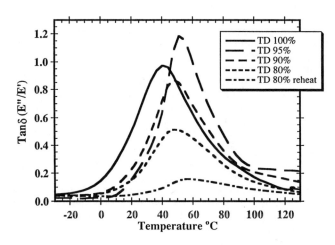

Figure 3. Loss behavior (tan δ) from dynamic mechanical spectra for polysilicate/PVF films vs. pure polymer.

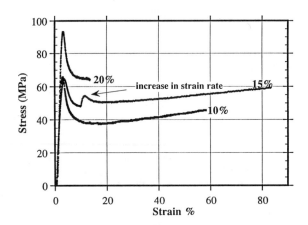

Figure 4. Stress/strain behavior for polysilicate/PVF films.

that laboratory features poly(phenylene terephthalimide) end-capped with trialkoxysilane and co-reacted with TMOS. Novak's non-shrinking systems (39) are apparently simultaneous IPNs, probably with adventitious covalent bonding between the phases. Mechanical data on these materials were generated on monolithic samples measured in compression, and are accordingly more difficult to compare directly with those of other workers.

Yield behavior is an intriguing phenomenon for the PVF/polysilicate materials. Unlike polymer/polymer IPNs (40), the stress does not decline significantly after the yield point. When samples are subjected to repeated strain past the yield point, they retain a significant portion of the initial modulus and continue to show yield behavior for several cycles. The behavior suggests that the inorganic network is only partially destroyed during yielding and perhaps that partial regeneration of network bonds can occur even at room temperature. Several of the tough hybrids from other laboratories have also shown yield behavior.

It is difficult to discern a pattern which would specify general criteria for hybrid toughness. It does seem at this point that the prospects are improved by the use of flexible polymers, acid catalysis, and covalent interphase bonds.

Softening Behavior and Creep. As expected from DMA scans, thermomechanical analysis reveals vastly different softening behavior for the hybrids and pure polymer. Figure 5 shows mechanical deformation vs. temperature for a hybrid containing 75% PVF vs. the pure polymer. The materials were fabricated in the form of cylindrical disks. Only the faintest remnant of the glass transition persists in the hybrid.

Polymer creep can also be significantly suppressed by the silica confinement. A PVF hybrid containing only 5 wt% silica showed a dimensional change of only 0.8% in 16 hrs under a 0.67 N. load. A comparable pure polymer sample rapidly extended to 300% of its original length under the same load.

Morphology and Structure. Two of the more salient questions about the nature of these composites relate to morphology. Under what circumstances is the inorganic network continuous? Are the two components covalently bonded to each other?

Polysilicate phase continuity. Several independent observations argue persuasively for a continuous, load-bearing inorganic network above roughly 3 wt% polysilicate. First, the development of a plateau modulus above the polymer T_g; secondly, the increase in tensile modulus with incorporation of polysilicate (greater than can be realized in systems containing particulate fillers); finally, the suppression of viscous flow and creep of the polymer at very low polysilicate contents.

A more direct piece of evidence was also obtained. A small piece of film containing 20% polysilicate was coiled into a cylinder and heated in air in a thermogravimetric sample holder until the polymeric phase was completely pyrolyzed. The residue was a beautifully transparent shrunken replica of the original film (see Figure 6). A similar experiment with a 5% inorganic sample also led to a continuous, although very fragile piece of "char". The temperature at which the polymer was removed (650°) is well below the viscous sintering temperature for silica (33).

Interphase bonding. Carbon NMR was used to explore the issue of bonding between the two components. First the rate of reaction of PVOH (containing 15% unhydrolyzed acetate groups) with TEOS in the absence of formic acid was established to be negligible on the reaction time scale. Next, the reaction of PVOH with formic acid was shown to result in nearly complete esterification of the hydroxyl groups (and resultant loss of tacticity) during the

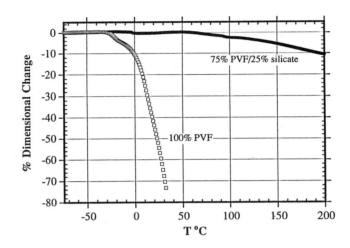

Figure 5. Thermomechanical behavior for polysilicate/PVF vs. pure polymer disks. The vertical displacement of a probe under a force of 0.1 N was determined as a function of temperature.

Figure 6. Glassy "char" of 20% polysilicate/PVF film after polymer pyrolysis in air at 650°. The film was formed into a right cylinder prior to the experiment. 80x magnification.

dissolution process. Finally, the solution of PVF in formic acid was combined with TEOS and spectra acquired prior and subsequent to gelation. No change in the resonance of the polymer carbon atoms could be detected. C-O-Si species are significantly shifted from $\underline{C}$OOCR groups. See Figure 7 for the relevant spectra.

The extent of interphase bonding is a subject of active controversy — even for the ostensibly same PVAc/polysilicate system. The Kodak group (9) used a combination of mechanical, relaxation and spectroscopic analyses to conclude that only hydrogen bonding exists between the phases. However, Pascault, et al., (11) used solvent extraction behavior (including molecular weight of the extracted material) to conclude that the two phase are covalently bonded. In the present work, our current conclusion from the continuity of the inorganic network, plateau modulus and creep suppression and especially the carbon NMR studies is that the phases are not *substantially* covalently bonded and that the confining effect of the network on the polymer chains is sufficient to rationalize the observed behavior.

Extent of Polysilicate Network Development. Silicon NMR provides a means for assessing the extent of network development. The chemical shift differentiation of the number of network-forming bonds per silicon atom is well established. Figure 8 shows the solid state spectrum of a silica gel generated from TEOS and formic acid both with and without polymer present. The greater intensity of the Q^4 peak in the hybrid relative to the pure silica gel indicates a greater extent of network formation in the presence of the polymer. This phenomenon has been observed by Wilkes (41) and is attributed to a lengthening of time between gelation and vitrification. It is a common observation in thermosetting systems (42).

SAXS. Small angle X-ray spectra have been used both in sol-gel systems and hybrids to characterize the fractal nature of the inorganic sol or network species (33) and to determine correlated spatial separations of scattering domains (43). Identification of fractal species is enabled by log-log (Porod) plots of scattering intensity vs the scattering wave vector K (a.k.a. h or q) $(4\pi \sin(\theta)/\lambda)$ or s (2 $\sin(\theta)/\lambda$), where 2θ is the scattering angle and λ is the wavelength of the radiation. (We prefer the use of s so that the length scale of the scattering centers is associated with 1/s.) For mass fractal objects, mass scales with size according to a fractional exponent D_f. Slopes of linear regions in a Porod plot can be directly associated with this exponent of the scattering entity (44), per

$$I \propto K^{-D_f}, \text{ where I is the scattering intensity}$$

Polysilicate species generated from reaction of TEOS and formic acid show an increase in the Porod slope as the sol ages, much as in conventional acid-catalyzed aqueous systems. When the gel is dried, the scattering centers become non-fractal (Euclidean), showing a power law dependence of slope -4 with an onset of ca. 1/s = 20 nm.

When the polysilicate network forms in the presence of polymer, the Porod plot of the gel prior to drying resembles that of the pure wet silica gel. However, a marked change occurs on drying: the fractal structure (with a D_f of 2.4) is apparently retained (Figure 9). The collapse into non-fractal scattering centers is prevented by the entrained polymer chains. The Kodak PVAc hybrids also show mass fractal behavior which is nearly independent of composition at the highest s values (9). The slope falls more quickly with decreasing s than does that of the corresponding PVF hybrids.

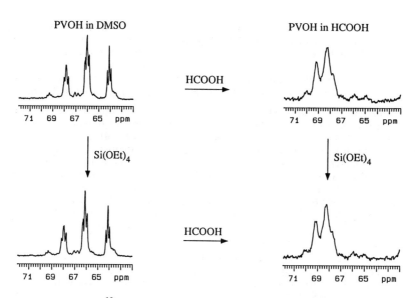

Figure 7. ¹³C NMR spectra related to interphase covalent bonding.

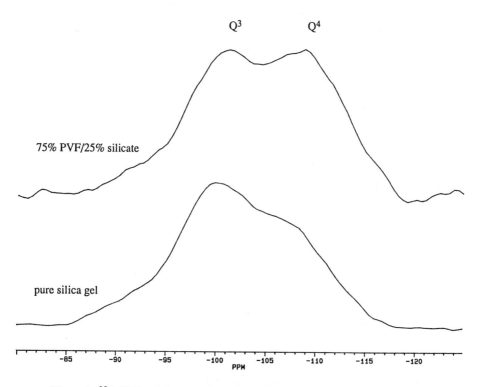

Figure 8. ²⁹Si NMR solid state spectra of 75% PVF hybrid and corresponding pure inorganic phase. $Q^3 = \underline{Si}(OSi)_3(OR, OH)$; $Q^4 = \underline{Si}(OSi)_4$.

Figure 10 shows plots of scattering intensity vs. the scattering wave vector for a representative hybrid (85% PVF). No intensity peak indicative of spatially correlated domains is apparent. The only significant feature is a knee in the curve at s = .01. The power law slope of the curve at smallest s is close to -4, suggesting scattering from particles larger than 100 nm. This behavior differs from the hybrids of Wilkes (43) which clearly show intensity peaks near a length scale of 80 nm.

Transmission Electron Microscopy. Micrographs were obtained from cross-sectional film slices in which the organic polymeric component had been stained with phosphotungstic acid (the acid does not stain silica gel). Figure 11 shows micrographs for such a sample of 75%PVF/25% polysilicate along with its unstained counterpart. Polymer domains (dark areas) on the order of 3-5 nm are apparent. The sample is much thicker (ca. 90 nm) than the domain size, but the stain may not have permeated the entire sample. Since, on a volumetric basis, the polymer constitutes more than 80% of the sample, it seems likely that much of the polymer is intermingled with the polysilicate at a scale unresolvable by TEM. This observation is believed consistent with the presence of mass fractal inorganic structures over a substantial range in size.

Porosity. Nitrogen adsorption at 77°K was used to assess porosity of a cryoground monolithic sample. B.E.T. analysis showed essentially no internal surface area. The corresponding silica gel displays open porosity with surfaces areas of several hundred m^2/g.

Abrasion resistance. Hybrid coatings containing polysilicate in combination with either PVF or poly(ethylene/vinyl formate) were prepared on PMMA slabs by dip coating. Several coatings containing between 60 and 80 wt% polysilicate showed promise as abrasion resistant coatings. Figure 12 shows a dark field micrograph of the coated/uncoated transition with very little abrasion apparent on the coated portion after being rubbed with steel wool.

Oxygen permeability. Free standing films of pure PVF and a hybrid containing 25% polysilicate were examined for oxygen permeability at 81% relative humidity. The hybrid film showed an Oxygen Permeation Value (45) (OPV) of 8.5; the pure polymer was too soft for measurement (see Experimental section). One would expect PVF to closely resemble PVAc, whose OPV is 150. This reduction is much greater than would be realized by the inclusion of comparable loadings of fillers: even plate-like materials such as mica or talc would confer only a 2-3x reduction in permeation at this loading. Although perhaps not sufficiently low to be of commercial interest, the OPV value is within a factor of two of the widely used poly(vinylidene chloride), and remarkably low given the nature of the polymer which is its majority constitutent.

Reaction Mechanism. The following features and reactive scheme appear consistent with NMR observations of the formic acid/TEOS and various model systems (20).

1. The acid catalyzes the hydrolysis of Si-OR bonds by any water present.
2. The acid generates water via esterification of any alcohol present (the latter is generated by reactions (1) and (3) below).
3. The acid greatly increases the rate of silicate condensation via formation of silyl formate groups, which can rapidly react with silanols to regenerate the acid.
4. The acid esterifies the hydroxyls on the polymer.

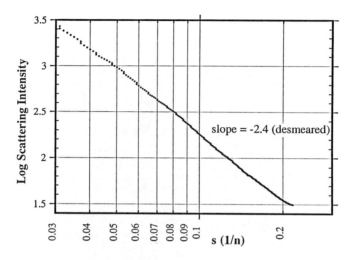

Figure 9. Porod plot from SAXS data. The linear power law relationship is consistent with mass fractal scattering centers.

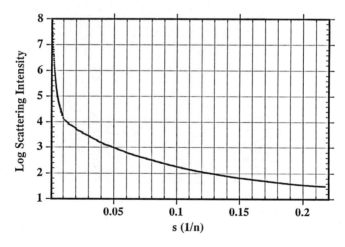

Figure 10. Intensity/wave vector plot from small angle scattering data. 15% polysilicate/PVF film. Slit-smeared data are shown so as to extend the range of s.

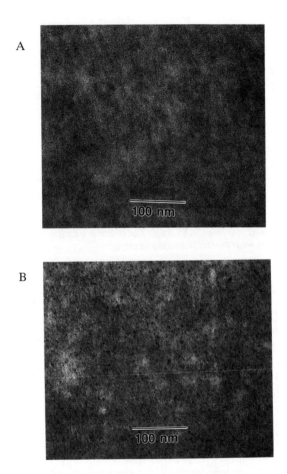

Figure 11. Transmission electron micrographs of 25% polysilicate/PVF hybrid with (A) and without (B) polymer stain (phosphotunstic acid).

Figure 12. Abrasion resistant coating on PMMA. The hybrid coating is 75% polysilicate/25% poly(ethylene-co-vinyl formate). Scratches made with steel wool. 80x magnification.

5. The acid provides extensive hydrogen bonding to both components — the polymer and the developing polysilicate network — enabling system homogeneity.

metathesis

$$\text{HCOOH} + \text{SiOEt} \rightleftharpoons \text{SiOOCH} + \text{EtOH} \quad (1)$$

$$\text{HCOOH} + \text{SiOH} \rightleftharpoons \text{SiOOCH} + \text{H}_2\text{O} \quad (1A)$$

esterification

$$\text{EtOH} + \text{HCOOH} \rightleftharpoons \text{EtOOCH} + \text{H}_2\text{O} \quad (2)$$

hydrolysis

$$\text{SiOEt} + \text{H}_2\text{O} \rightleftharpoons \text{SiOH} + \text{EtOH} \quad (3)$$

condensation

slow $\quad 2\,\text{SiOH} \longrightarrow \text{SiOSi} + \text{H}_2\text{O} \quad (4)$

slow $\quad \text{SiOH} + \text{SiOEt} \longrightarrow \text{SiOSi} + \text{EtOH} \quad (5)$

fast $\quad \text{SiOH} + \text{SiOOCH} \longrightarrow \text{SiOSi} + \text{HCOOH} \quad (6)$

very slow $\quad \text{SiOOCH} + \text{SiOEt} \longrightarrow \text{SiOSi} + \text{EtOOCH} \quad (7)$

Experimental

Reagents. Formic acid is unstable as a strictly anhydrous material (46), decomposing into CO and water, and is available commercially with 4 wt% (or more) water as a stabilizing agent. The 96% grade was purchased from Kodak and generally used as received. TEOS was purchased from Kodak and shown by both gas chromatography and ^{29}Si NMR to be essentially free from impurities. Polyvinyl alcohol was used DuPont Elvanol 90-50 (99% hydrolyzed PVAc) or 51-05 (85% hydrolyzed). The two polymers were determined to have number average molecular weights of 32,000 and 15,000, respectively. The lower molecular weight polymer showed lower modulus and tensile strength values when cast pure from solution, presumably because of fewer chain entanglements. No substantial difference in hybrid behavior was observed.

Simple alkoxysilanes such as TEOS and TMOS are completely miscible (or very rapidly become so) with formic acid. Most commonly, a fresh solution of the silane in the acid is added to a corresponding polymer solution.

Film or Monolith Generation. The resultant solution is homogeneous and can be used to prepare dip or drawdown coatings at once, or poured into containers to generate shaped monoliths after gelation and drying.

Abrasion resistance. The ability of hybrid films to withstand exposure to abrasive media was determined on 25x75 mm pieces of PMMA which were dip coated with the hybrid solutions at a withdrawal rate of approximately 40 mm/min. Samples were then air cured at 100°, with durations between 1 and 16 hours. The specimens were rubbed with 0000 steel wool until the uncoated polymer was profoundly abraded; the coated portion was then examined microscopically (80x) for scratch development relative to the uncoated portion.

Oxygen permeability. Gas permeability was assessed on a 4" circular piece of 25% polysilicate/75% PVF film 42 microns in thickness. The sample was mounted in a Modern Controls unit, equilibrated with 81% relative humidity, and the oxygen transmission rate monitored for nine days. The Oxygen Permeation Value (OPV) (45) during the final three days of the trial was approximately 8.5. The corresponding pure PVF film was so soft that it migrated into its confining screen and confounded the experiment.

Small-Angle X-ray. Scattering data were acquired with a Kratky camera over the range $2\theta = 0.6$ to 4.0 degrees. Intensities were normalized to unit sample thickness. Very small angle data (2θ down to .003°) were obtained on a Bonse-Hart instrument. The considerable overlap between the two instruments allows normalization of the respective intensity curves, as in Figure 10. The 1.54 Å wavelength of Cu K$_\alpha$ radiation was used. Data for Figure 9 correspond to slit desmeared conditions, but slit smeared values were used for Figure 10 so as to extend the range of observation.

NMR Spectrometry. All spectra were run with a 90° pulse and inverse gated decoupling. Silicon spectra were obtained at 79.46 MHz with Cr(acac)$_3$ as a relaxation agent using a 10 second recycle delay. Carbon spectra were run at 100.58 MHz. Solid state ^{29}Si spectra were run at 39.73 MHz on samples spun at 3.5 kHz. Spectra were taken with a 1000 second recycle delay.

Transmission Electron Microscopy. Micrographs were obtained on a cryo-microtomed section of a free standing film. The surface of the film was floated on a methanol solution of phosphotungstic acid, which was independently shown to stain the polymer but not the polysilicate.

Dynamic Mechanical Spectra. DMA scans were taken on a Rheometrics RSA II instrument at a scan rate of 10°/min, a strain of 0.1%, and a frequency of 1 Hz.

Conclusions

Organic/inorganic hybrid materials are rapidly formed when poly(vinyl alcohol) and TEOS are co-dissolved in formic acid; the polymer is converted into the flexible, amorphous poly(vinyl formate). Inorganic contents as low as ca. 3 vol% were found to be capable of forming continuous, load bearing networks which prevent viscous flow of the polymer. Direct evidence of polysilicate phase continuity was obtained from the glassy replica of films after polymer pyrolysis. The hybrids show glass transitions and polymer domains less than 5 nm. A surprisingly high level of toughness accompanied by yield behavior was observed. Scattering studies indicate mass fractal inorganic constituents which collapse during drying in the absence of the polymer. NMR studies have not provided evidence for covalent bonds between the phases.

Acknowledgments

The author is indebted to John Coburn, Laurine Galya, Mimi Keating, Frank Wilson, and Barbara Wood for analytical data and interpretation and to George Scherer for numerous enlightening discussions.

References

1. Huang, H. H.; Wilkes, G. L.; Carlson, J. G. *Polymer* **1989**, *30*, 2001-12.
2. Chujo, Y.; Ihara, E.; Kure, S.; Saegusa, T. *Macromolecules* **1993**, *26*, 5681-6.
3. Schmidt, H. *J. Non Cryst. Solids* **1985**, *73*, 681-91.
4. Wei, Y.; Bakthavatchalam, R.; Whitecar, C. K. *Chem. Mater.* **1990**, *2*, 337-9.
5. Novak, B. M.; Davies, C. *Macromolecules* **1991**, *24*, 5481-3.
6. For convenience, we will use the term "non-bonded" for hybrid sytems in which cross-linking, end-linking or grafting chemistry was not intentionally provided. Adventitious bonding between components is certainly possible in some of the non-bonded systems.
7. Pope, E. J. A.; Asami, M.; Mackenzie, J. D. *J. Mater. Res.* **1989**, *4*, 1018-26.
8. Abramoff, B.; Klein, L. C. *Proc. Spie Int. Soc. Opt. Eng.* **1990**, *1328 (Sol-Gel Opt.)*, 241-8.
9. Landry, C. J. T.; Coltrain, B. K.; Landry, M. R.; Fitzgerald, J. J.; Long, V. K. *Macromolecules* **1993**, *26*, 3702-12.
10. David, I. A.; Scherer, G. W. *Polym. Prepr. (Am. Chem. Soc., Div. Polym. Chem.)* **1991**, *32*, 530-1.
11. Girard, R. E.; Lam, T. M.; Pascault, J. P. *Macromol. Chem. Phys.* **1994**, *195*, 149-58.
12. Ning, Y. P.; Mark, J. E. *J. Appl. Polym. Sci.* **1985**, *30*, 3519-22.
13. Mauritz, K. A.; Storey, R. F.; Jones, C. K. In *ACS Symp. Ser (Multiphase Polym.: Blends Ionomers);* American Chemical Society: Washington, D. C., **1989**; Vol. 395 pp 401-17.
14. Mark, J. E.; Ning, Y. P.; Jiang, C. Y.; Tang, M. Y.; Roth, W. C. *Polymer* **1985**, *26*, 2069-72.
15. Davis, S. V.; Mauritz, K. A. *Polym. Prepr. (Am. Chem. Soc., Div. Polym. Chem.)* **1993**, *34*, 608-9.
16. Mauritz, K. A.; Scheetz, R. W.; Pope, R. K.; Stefanithis, I. D.; Wilkes, G. L.; Huang, H. H. *Polym. Prepr. (Am. Chem. Soc., Div. Polym. Chem.)* **1991**, *32*, 528-9.
17. Schaefer, D. W.; Jian, L.; Sun, C. C.; McCarthy, D.; Jiang, C. Y.; Ning, Y. P.; Mark, J. E. *Polym. Prepr. (Am. Chem. Soc., Div. Polym. Chem.)* **1989**, *30*, 102-4.

18. Ellsworth, M. W.; Novak, B. M. *Chem. Mater.* **1993**, *5*, 839-44.
19. We prefer the term "polysilicate" to "silica" in describing the inorganic component in that it more accurately reflects the presence of unreacted (Si-OR) and uncondensed (Si-OH) bonds which prevent attainment of SiO_2 stoichiometry and skeletal density (2.2 for amorphous silica) under the usual conditions of hybrid synthesis.
20. Sharp, K. G. *J. Sol-gel Sci. Tech.* **1994**, *2*, 35.
21. Landry, C. J. T.; Coltrain, B. K.; Wesson, J. A.; Zumbulyadis, N.; Lippert, J. L. *Polymer* **1992**, *33*, 1496-506.
22. Takayanagi, M.; Harima, H.; Iwata, Y. *Mem Fac. Eng. Kyushu Univ* **1963**, *23*, 1.
23. Kraus, G.; Rollman, K. In *ACS Symposium Series Multicomponent Polymer Systems*; J. Platzer, Ed.; American Chemical Society: Washington, D.C., 1971; Vol. 99.
24. Bauer, P.; Henning, J.; Schreyer, G. *Angew. Makromol. Chem.* **1970**, *11*, 145.
25. Huelck, V.; Thomas, D.; Sperling, L. *Macromolecules* **1972**, *5*, 340, 348.
26. Zhao, M. X.; Ning, Y. P.; Mark, J. E. *Ceram. Trans. (Adv. Compos. Mater.)* **1991**, *19*, 891-8.
27. H. Wolter and W. Storch, presented at the Spring, 1994 Materials Research Society meeting, San Francisco, CA
28. Sperling, L. H. *Interpenetrating Networks*; Plenum Press: New York, 1981.
29. Donatelli, A.; Sperling, L.; Thomas, D. *Macromolecules* **1976**, *9*, 676.
30. Termonia, Y. *Macromolecules* **1991**, *24*, 1392-6.
31. Landry, C. J. T.; Coltrain, B. K.; Brady, B. K. *Polymer* **1992**, *33*, 1486-95.
32. Ethanol is observed as a product even in systems which are initially anhydrous.
33. Brinker, C. J.; Scherer, G. W. *Sol-gel Science*; Academic Press: San Diego, CA, 1990.
34. The films are to some extent plasticized by the retained pore fluids unless vacuum dried at elevated temperatures.
35. Fitzgerald, J. J.; Landry, C. J. T.; Schillace, R. V.; Pochan, J. M. *Polym. Prepr. (Am. Chem. Soc., Div. Polym. Chem.)* **1991**, *32*, 532-3.
36. As generated from solution and dried at moderate temperatures, the lower molecular weight PVF used appears to posses few chain entanglements.
37. Epstein, B. N., U. S. Patent 4,174,358
38. J. E. Mark, presented at the Spring, 1994 Materials Research Society meeting, San Francisco, CA.
39. Novak, B. M. *Adv. Mater.* **1993**, *5*, 422-33.
40. Beecher, J. F.; Marker, L.; Bradford, R. D.; Aggarwal, S. L. *J. Polym. Sci.* **1969**, *26C*, 117.
41. Glaser, R. H.; Wilkes, G. L.; Bronnimann, C. E. *J. Non Cryst. Solids* **1989**, *113*, 73-87.
42. Gillham, J. K. *Polym. Eng. Sci.* **1979**, *19*, 676.
43. Rodrigues, D. E.; Brennan, A. B.; Betrabet, C.; Wang, B.; Wilkes, G. L. *Chem. Mater.* **1992**, *4*, 1437-46.
44. Schaefer, D. W.; Keefer, K.D. in *Better Ceramics through Chemistry*, Brinker, C. J.; Clark, D.E.; Ulrich, D. R., eds. North Holland, New York, **1984**, *Materials Research Society,* Vol. *32*, pp 1-14.
45. OPV has units of cc·mil/100·in²atm·24 hrs. Larger numbers indicate higher permeability.
46. Barnham, H.; Clark, L. *J. Am. Chem. Soc.* **1951**, *73*, 4638.

RECEIVED October 11, 1994

Chapter 14

Solidification of Colloidal Crystals of Silica

Hari Babu Sunkara, Jagdish M. Jethmalani, and Warren T. Ford[1]

Department of Chemistry, Oklahoma State University,
Stillwater, OK 74078

Photopolymerization of methyl acrylate, methyl methacrylate, and a mixture of methyl methacrylate and 2-hydroxyethyl methacrylate, each containing ordered 3-(trimethoxysilyl)propyl-coated monodisperse 152 nm diameter silica particles, produces polymer composites in which the ordering of the particles is maintained. The silica particles formed face centered cubic colloidal crystals in both the monomers and the polymers with (111) lattice planes parallel to the film plane. Both the monomer and the polymer dispersions are selective optical filters of a narrow bandwidth of visible light by Bragg diffraction.

Organic and inorganic polymer hybrid composites are of current research interest because their mechanical and optical properties are often better than those of either component. Silica has been widely incorporated in the organic polymer matrix to reinforce the polymer, and the degree of mixing of the two phases influences the strength of the materials. Optically clear composites of silica- or glass fiber-filled poly(methyl methacrylate) (PMMA) have been prepared either by using silica spheres of diameters much smaller than the wavelength of visible light or by carefully matching the refractive index of the glass fiber with the PMMA (*1,2*). However, in some composites of silica-PMMA, segregation of the particles causes poor optical quality (*3*).

Monodisperse colloidal polymer particles in aqueous and nonaqueous dispersions form crystal-like arrays that diffract light of wavelength corresponding to the interparticle spacings and Bragg's law (*4-18*). The ordered arrays of particles in dispersions are known as colloidal crystals. The composites described here constitute a new class of materials in which the compatible colloidal silica particles are ordered in the polymer matrix, and the silica particles are covalently attached to the polymer. The 264-μm thick composite films transmit more than 70 % of incident visible light normal to the film plane at wavelengths where the light is not diffracted, even though they contain 20 volume percent of 152 nm diameter silica particles. The films are iridescent because the crystalline arrays of silica particles in the polymer matrix diffract visible light. Liquid colloidal optical rejection filters having a narrower bandwidth of rejected light than commercial filters have been developed by exploiting the diffraction of colloidal crystals of polystyrene latexes in water (*8-10*). In these filters, the colloidal crystals of latexes orient with d_{hkl} planes parallel to the plane of

---

[1]Corresponding author

0097–6156/95/0585–0181$12.00/0

the quartz cell. A major drawback with these filters is that weak shear, gravitational, electrical, and thermal forces disturb the order, due to the low elastic modulus of the colloidal crystals, on the order of 10^{-2} to 10^{-3} Pa (7,11,12). Sturdier colloidal crystals in a highly resistant medium may have wider use as selective filters and in other optical devices. Our approach to stable colloidal crystals is solidification of colloidal crystallites of silica in a polymer matrix. Two patents have appeared recently on colloidal crystals in polymer matrices (13,14). One reports no examples (13), and the other describes colloidal crystals of polystyrene latexes in hydrogel matrices such as polyacrylamide (14). These filters are not as rigid as the ones we have prepared with PMMA (15) because the hydrogels retain a large amount of water.

Theory

Diffraction from colloidal crystals of monodisperse charged polystyrene particles in water has been successfully explained by dynamical diffraction theory, which was originally proposed for atomic crystals, by taking into account the interaction between incident and diffracted beams in the medium (16,17):

$$\lambda_{corr} = \lambda \left(1 - \frac{\psi'_o}{2\sin^2 \theta} \right)$$
(1)

where $\qquad \lambda = 2 n_s d_{hkl} \sin \theta \quad$ and $\quad \psi'_o = 3\phi \frac{(m^2 - 1)}{(m^2 + 2)}$

ψ'_o is the real part of the crystal polarizability, λ is the Bragg diffracted wavelength, d_{hkl} is the interplanar spacing, θ is the Bragg angle, m is the ratio of the refractive index of the particles to that of surrounding medium, ϕ is the particle volume fraction, and n_s is the refractive index of the suspension calculated by

$$n_s = n_m(1 - \phi) + n_p\phi$$
(2)

where n_m and n_p are the refractive indexes of the medium and the particles (16). The d spacings of the most common structures of colloidal crystals can be calculated theoretically from the volume fraction of the particles.
For fcc (111)

$$\phi = \frac{2\pi}{3} \left(\frac{D_o}{a} \right)^3$$
(3)

$$d = a / \sqrt{3}$$
(4)

where a is the lattice constant and D_o is the particle diameter.

The bandwidths $(\Delta\lambda_o)'$ of the Bragg diffracted peaks from the dispersions containing colloidal crystals can be calculated from the equation 5 given by Kosan and Spry (18).

$$(\Delta\lambda_o)' = \frac{w_y |K\Psi_H| \lambda_o}{\sqrt{b} \sin \theta}$$
(5)

$$\text{where} \quad \Psi_H = \frac{2}{\pi^2(3)^{3/2}} (m^2+1) \left(\frac{3}{m^2+2}\right) (\sin u - u \cos u)$$

$$u = \pi\sqrt{3/2} \left(\frac{D_o}{D}\right) \quad \text{and} \quad D = \frac{a}{\sqrt{2}}$$

w_y is a numerical factor = 1.155 for Ewald theory, K is the polarization factor, equal to unity for σ polarization and lcos 2θl for π polarization, b is the ratio of direction cosines of the light rays, which is unity for Bragg diffraction, m is the ratio of the refractive index of the particles to that of the medium, and D is the nearest neighbor spacing.

Experimental

Materials. Water was deionized, treated with active carbon, deionized, and distilled in glass. Tetraethyl orthosilicate (TEOS, Aldrich), absolute ethanol (Aaper), and 3-(trimethoxysilyl)propyl methacrylate (TPM, 97%, Aldrich) were distilled prior to use. Ammonium hydroxide (Baker), methyl methacrylate (MMA), methyl acrylate (MA), 2-hydroxyethyl methacrylate (HEMA, Aldrich), 2,2,2-trifluoroethyl acrylate (TFEA, Aldrich) and 2,2-dimethoxy-2-phenylacetophenone (DMPA, Aldrich) were used as received.

Colloidal silica dispersions in aqueous ammoniacal ethanol were prepared by the Stöber method (*19,20*). 3-(Trimethoxysilyl)propyl methacrylate coated silica particles were prepared by the procedure of Philipse and Vrij (*21*). The density of TPM silica particles was measured using a specific gravity bottle at 25 °C. The dispersions of TPM silica in ethanol were dialyzed against methanol using regenerated cellulose dialysis tubing with molecular weight cutoff of 50,000 (Spectra/Por 7, Spectrum) to remove any unreacted silane, water and ammonia. The particle concentration in methanol was increased to 56 wt % by distilling methanol under reduced pressure. Small portions of about 5-7 mL of a concentrated dispersion of TPM silica particles in methanol were dialyzed against MMA and MA. After replacing the monomer 4-5 times during a period of 24 h, the ^{1}H-NMR spectra of the dispersions in monomers showed no methanol peaks. We prepared initially a high particle concentration about 50 wt % in monomers, and subsequently diluted to 35-40 wt %.

Dispersions containing 1 wt % DMPA initiator were transferred by syringe into glass sandwich cells made from 1" x 3" x 1 mm microscope slides and 264-μm Teflon spacers. After colloidal crystals formed, the dispersions were polymerized at ambient temperature for 4-5 h using a medium pressure 450 W mercury vapor lamp 2 cm from the cell with the film plane horizontal. The lamp was cooled by circulating water in a quartz immersion jacket.

The particle diameters were measured using transmission electron microscopy (TEM) and dynamic light scattering (DLS) as described before (*22*). The structural and orientational details of the colloidal crystallites of TPM silica particles both in colloidal dispersion and in thin polymeric films were observed by the orthoscopic images of the crystals between crossed polarizers. The transmission spectra of the samples were recorded using a single beam Hewlett Packard 8452A UV-vis diode-array spectrophotometer.

Results

Colloidal TPM-Silica Particles. We chose methacryloxypropyl functionalized silica (**3**) because grafting of the polymer to the surface of the particles may enhance the tensile strength of the polymer film. As shown in the Scheme, monodisperse silica particles (**2**) were produced from TEOS (**1**) by the seed growth technique. Table I lists the particle diameters from TEM and DLS, and their polydispersity indexes (standard deviation of D_n/mean). The number average diameters of silica were the same from the two different batches (3HB-18 and 3HB-76), which differed slightly in the polydispersity indexes. We did not notice any increase in the particle diameter from TEM after coating the particles with silane coupling agent (3HB-76 and 3HB-77). The particle diameters from DLS measurements of diluted dispersions of particles in ethanol were always greater than those of the dried particles. We measured the density of the TPM silica at 25 °C to be 1.79 g/mL.

Table I. Particle Diameters (nm) of TPM Silica

Sample	TEM			Polydispersity Index	DLS
	D_n	D_w	D_z		D_z
3HB-18[a]	153	157	159	0.08	166
3HB-76[a]	152	152	153	0.04	159
3HB-77[b]	151	153	155	0.04	---

[a]Parent silica.
[b]TPM coated silica.

In contrast to parent silica, which is hydrophilic and charge stabilized, the TPM silica particles are electrosterically stabilized and are colloidally stable in ethanol and in ethanol-toluene mixtures. We transferred the TPM silica particles from ethanol to the desired monomer by dialysis (*23*). The TPM silica particles were stable in a variety of acrylic and methacrylic ester monomers, and the colloidal dispersions were iridescent.

Morphology of the Crystallites. The transmission polarizing microscope was used to visualize directly the colloidal crystallites of silica dispersed in various monomers. The dispersions between crossed polarizers showed a mosaic of crystalline domains of various colors ranging 100-500-μm in size. Crystal growth was slower in the center than at the edges where the dispersion was in contact with the cured epoxy resin used to seal the cells. The crystallites at the edges appeared blue and green, pillar shaped, and larger than the crystallites in the center. Over 4-7 days at 25 °C with the film plane horizontal the entire cell filled with crystallites, and the sample turned from cloudy to iridescent. The crystals in the bulk appeared black, and some of the colored crystallites at the edges turned black over this period. Upon rotation of the cell 40-55° to the incident plane polarized light, the crystals in bulk appeared colored, and the color changed with the angle of incident light.

Bragg Diffraction. We analyzed the transmitted light from the colloidal dispersions using a UV-vis spectrophotometer. Figure 1 shows the visible spectra of 264-μm thick dispersions containing 152 nm TPM silica particles in MMA and in MMA/HEMA (65/35 vol %). The dispersion in the cell appeared cloudy initially and then turned to iridescent with time. The diffracted wavelength decreased as the particle concentration was increased from 35 to 40 wt % of silica in MMA. The visible spectra of MMA/TFEA (65/35 vol %) and MA dispersions are shown in Figure 2. The spectrophotometer detects the light transmitted through an ~8 mm circular cross-section of the cell, which contains approximately 10^2 crystallites. The spectra were reproducible as long as the light focused on the center of the cell containing crystals with lattice planes parallel to the film plane. The crystals at the edges which have different orientation gave irreproducible results.

Table II reports the volume fraction (ϕ) of particles, the approximate refractive indexes (n_D) of the monomers and monomer mixtures, the time at which the spectral data was collected after filling the cells, the diffracted wavelengths (λ_{max}) from the dispersions, and the lattice spacings (d_{111}) of the particles in 264-μm thick cells. The refractive index of the TPM silica particles is 1.449 (6). The lattice spacings of the crystallites calculated from equation 1 and the bandwidths calculated from equation 5 are also reported in Table II. The spectral bandwidths from all the dispersions were in the range of 3.4 to 6.0 nm.

Table II. Properties of TPM Silica-Monomer Dispersions[a]

ϕ	Monomer	n_D	time (h)	λ_{max} (nm)	d_{111} (nm)	bandwidth (nm) obsd	calcd
0.227[b]	MMA	1.4142	8	532	188.1	3.6	3.1
0.195	MMA	1.4142	72	552	195.2	3.4	3.1
0.195	MMA/HEMA	1.4267	3	562	197.0	5.0	2.0
0.195	MMA/TFEA	1.3917	43	548	196.9	5.3	5.1
0.193	MA	1.4040	18	560	199.4	6.0	4.1

[a]Particle diameter 152 nm.
[b]Particle diameter 153 nm.

When the dispersions turned from cloudy to iridescent, indicating formation of colloidal crystals throughout the sample, they were photopolymerized. The maximum extinction coefficient of the initiator (at 336 nm) in ethanol was 274 cm^{-1} mole^{-1} L. The TPM groups on the silica copolymerize with monomer or comonomers as shown in the Scheme. Figures 3-5 show the visible spectra from the colloidal dispersions of 152 nm TPM silica particles in MMA/HEMA, MMA and in MA before and after polymerization. The polymer films gave broader peaks at shorter wavelengths than the corresponding monomer dispersions as reported in Table III.

The polymer films sandwiched between the two glass slides were removed to study the morphology, orientation, and crystal structures by optical and electron microscopy. No crystallites were visible when the film was viewed between two crossed polarizers with film plane normal to incident light. However, crystallites appeared in color upon tilting the film.

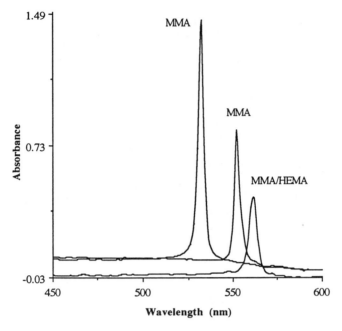

Figure 1. Visible spectra of 152 nm diameter TPM silica dispersed in MMA and MMA/HEMA (65/35 vol %). Data are in Table II. (Adapted from ref 25).

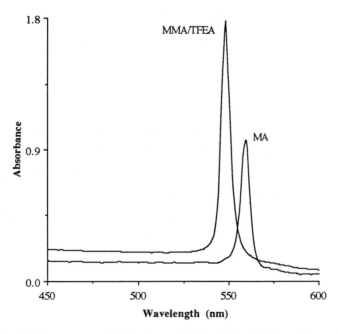

Figure 2. Visible spectra of TPM silica in MMA/TFEA (65/35 vol %) and in MA.

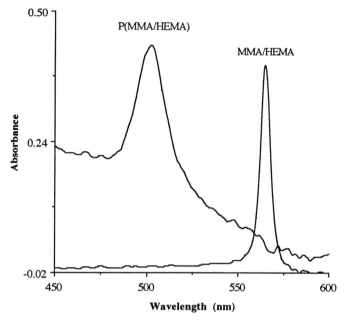

Figure 3. Visible spectra of TPM silica in MMA/HEMA before and after polymerization.

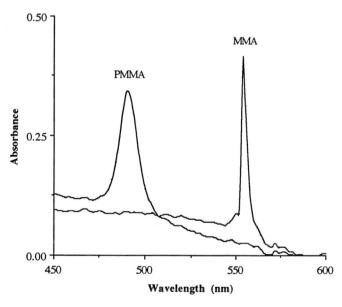

Figure 4. Visible spectra of TPM silica in MMA before and after polymerization. (Adapted from ref 25).

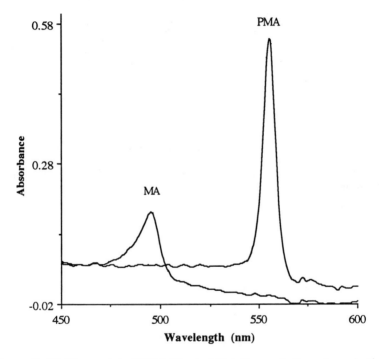

Figure 5. Visible spectra of TPM silica in MA before and after polymerization.

Table III. Comparison of Polymer with Monomer Dispersions

composite	before polymerization		after polymerization		decrease
	λ_{max} (nm)	bandwidth (nm)	λ_{max} (nm)	bandwidth (nm)	in d_{111} (%)
PMMA	554	4.0	490	13.6	14.6
P(MMA-*co*-HEMA)	564	6.0	502	19.0	14.1
P(MMA-*co*-TFEA)	548	5.3	no diffraction		
PMA	556	6.0	496	13.5	14.3

Discussion

Colloidal Crystals in Monomer Films. The 152 nm diameter TPM silica particles in monomers spontaneously formed colloidal crystals. The repulsive forces due to the negative charges on the particle surface, arising from the dissociation of the residual surface hydroxyl groups of the parent silica and of the surface layer from TPM, are responsible for this phase transition. Higher particle number of silica in monomers than of latexes in water is required to form colloidal crystals due to lower charge of the silica and lower dielectric constant of the medium. At high particle numbers the fcc crystal structure is favored.

The silica particles and the monomers do not absorb in the visible range. The sharp peaks in the spectra of the colloidal dispersions (Figures 1 and 2) are due to Bragg diffraction of light by the colloidal crystals. Two different batches of TPM-silica in MMA showed similar crystalline behavior. The d_{111} spacings of fcc crystals calculated from volume fraction, particle diameter, and lattice parameter agree well with those calculated from λ_{max} (= λ_{corr}) by equation 1. The λ_{max} in MMA shifts to longer wavelength as the volume percent of the particles decreases, indicating the wavelength dependence on lattice spacing. The differences of λ_{max} at constant volume fraction in different monomers and monomer mixtures show the dependence of diffracted wavelength on the refractive index of the dispersion. The 3.4-3.6 nm bandwidths of the diffracted peaks from the dispersions of 152 nm diameter silica in MMA agree well with the diffracted bandwidth ($\Delta\lambda_0$) of 3.1 nm calculated for the same volume percent using equation 5. This indicates that the d_{111} planes of all the crystallites in the sample are oriented parallel to the film plane. In general, the diffracted bandwidth depends on the particle diameter and the ratio of the refractive index of the particles to that of the surrounding medium. We observed Bragg diffraction from colloidal dispersions of silica in MMA in which the difference of the refractive indexes (Δn) was only 0.035, whereas Δn for polystyrene latexes in water is 0.26. The diffracted bandwidths for other dispersions are narrow and in the range of 5.0 to 6.0 nm, in close agreement with the calculated bandwidths (Table II).

High degree of ordering of the crystallites, smaller difference between the refractive indexes of the particles and the monomer, and smaller particle diameters narrow the bandwidth of the diffracted light from the dispersions. Others have reported that the degree of ordering of polystyrene latexes in water is influenced by the cell thickness (*17,24*): the thinner the sample, the greater the ordering of crystallites. We observed no significant differences of the bandwidths of apparent absorption peaks between dispersions in 132 and 264 μm cells. The intensity of the diffracted peaks varied with sample composition. The intensity was maximum for MMA/TFEA and minimum for MMA/HEMA (Figure 2), as expected for dependence of diffraction intensity on the difference between the refractive indexes of the particles and the medium. No colloidal crystals could be seen in a dispersion of particles in pure HEMA, either because the particles and the monomer were optically index matched or because the high viscosity of the monomer (5.9 cps at 30 °C) slowed the rate of crystallization. Most of each sample appeared black except at the edges, indicating the crystallites are at extinction. Their (111) planes parallel to the plane of the cell do not change the direction of the polarization vector of transmitted light. The crystallites at the edges were colored indicating that the lattice planes were not normal to the incident plane polarized light.

Colloidal Crystals in Polymer Composite Films. No phase separation of the silica particles from the polymer matrix was observed in 264 µm thick composite films after photopolymerization. The cell containing the colloidal dispersion must be kept horizontal during crystal growth and polymerization to prevent sedimentation of the particles. No diffraction of light was seen from a silica-PMMA film which was polymerized with the cell vertical. This suggests that the orientation and the structure of the crystallites were lost. However, the silica-poly(MMA-co-HEMA) film diffracted visible light (Figure 3), though the cell was vertical during the polymerization. This suggests that the rate of sedimentation of the particles in this mixture was much slower than in MMA alone due to the higher viscosity and density of HEMA.

When the cells containing dispersions in MMA and MA were kept horizontal during polymerization, colloidal particles maintained their order in the polymer matrix, but the diffraction bandwidths increased (Figures 4 and 5 and Table III). No diffraction of light by the crystals in the silica-poly(MMA-co-TFEA) film was seen because of index matching of the particles with the polymer matrix.

No birefringence was observed from most of the film between crossed polarizers when the light beam was normal to the film plane. This suggests a single index of refraction of the polymer composite. When the composite films were rotated at angle 40-55° to the incident light, mostly purple or blue crystals appeared throughout the film. This indicates that sample is not amorphous, but crystalline, and the lattice planes are normal to the incident light.

A scanning electron micrograph of the surface layer of the silica-PMMA composite showed the fcc d_{111} planes (25). However, the surface layer does not prove the presence of the same crystalline structures throughout the sample. The average center-to-center distance between neighboring particles measured from the micrograph was 234 nm, and the average surface-to-surface distance was 82 nm, indicating that the order was due to the long range repulsions rather than hard-sphere packing.

Polymerization caused a blue shift in the rejection wavelength from the composite films, indicating that the colloidal crystals were compressed. The expected 21% decrease in volume of MMA on polymerization accounts for a 6.8% decrease in d-spacing, compared with an observed 14.7% decrease of d, which must be due to an increase of particle number within the crystalline regions. The measured particle distances from the SEM micrograph do not agree with the calculated lattice spacing using equation 1, which suggests that the particle concentration in the bulk may be higher than in the surface layer. The greater bandwidth of the diffraction peaks of the composite films could be due to lower crystalline order of the particles in the polymer films than in the dispersions.

Conclusion

Colloidal crystals of silica in monomer dispersions form solid composite films by polymerizing with UV radiation. The composites selectively filter visible light with a bandwidth less than 20 nm. The silica filled thermoplastic or elastomeric films are more rigid, stable and easy to handle than the previously reported arrays of polystyrene latexes in hydrogels. These composite materials potentially could replace optical rejection filters currently in use.

Acknowledgment. We thank the National Science Foundation for financial support, and Bruce J. Ackerson for helpful discussion.

Literature Cited

1. Pope, E. J. A.; Asami, M.; Mackenzie, J. D. *J. Mater. Res.* **1989**, *4*, 1018.
2. Lin, H.; Day, D. E.; Stoffer, J. O. *Polym. Eng. Sci.* **1992**, *32*, 344.
3. Abramoff, B.; Covino, J.; *J. Appl. Polym. Sci.* **1992**, *46*, 1785.
4. Alfrey, T. Jr.; Bradford, E. B.; Vanderhoff, J. W. *J. Opt. Soc. Am.* **1954**, *44*, 603.
5. Monovoukas, Y.; Gast, A. P. *Phase Transitions* **1990**, 21.
6. Dhont, J. K. G.; Smits, C.; Lekkerkerker, H. N. W. *J. Colloid Interface Sci.* **1992**, *152*, 386.
7. Okubo, T. *Prog. Polym. Sci.* **1993**, *18*, 481.
8. Flaugh, P. L.; O'Donnell, S. E.; Asher, S. A. *Appl. Spectrosc.* **1984**, *38*, 848.
9. Asher, S. A.; Flaugh, P. L.; Washinger, G.; *Spectroscopy* **1986**, *1*, 26.
10. Asher, S. A., U. S. Patents 4 627 689 and 4 632 517 (1986).
11. Clark, N. A.; Hurd, A. J.; Ackerson, B. J. *Nature* **1979**, *281*, 57.
12. Okubo, T. *Colloid Polym. Sci.* **1993**, *271*, 873.
13. Alvarez, J. L., U.S. Patent 5 131 736 (1992).
14. Asher, S. A.; Jagannathan, S., U. S. Patent 5 281 370 (1994).
15. Sunkara, H. B.; Jethmalani, J. M.; Ford, W. T. *ACS Polym. Mat. Sci. Eng. Preprints* **1994**, *74*, 274.
16. Rundquist, P. A.; Photinos, P.; Jagannathan, S.; Asher, S. A. *J. Chem. Phys.* **1989**, *91*, 4932.
17. Monovoukas, Y.; Gast, A. P. *Langmuir* **1991**, *7*, 460.
18. Spry, R. J.; Kosan, D. J. *Appl. Spectrosc.* **1986**, *40*, 782.
19. Stöber, W.; Fink, A.; Bohn, E. *J. Colloid Interface Sci.* **1968**, *26*, 62.
20. Badley, R. D.; Ford, W. T.; McEnroe, F. J.; Assink, R. A. *Langmuir* **1990**, *6*, 792.
21. Philipse, A. P.; Vrij, A. *J. Colloid Interface Sci.* **1988**, *128*, 121.
22. Ford, W. T.; Yu, H.; Lee, J.-J.; El-Hamshary, H. *Langmuir*, **1993**, *9*, 1698.
23. Hiltner, P. A.; Papir, Y. S.; Krieger, I. M. *J. Phys. Chem.* **1971**, *75*, 1881.
24. Van Winkle, D. H.; Murray, C. A. *Phys. Rev.* **1986**, *34*, 562.
25. Sunkara, H. B.; Jethmalani, J. M.; Ford, W. T. *Chem. Mater.* **1994**, *6*, 362.

RECEIVED October 20, 1994

Chapter 15

Thermo-irreversible Gelation and Percolation-Based Mechanical Response via Metal–Olefin Coordination in Diene Polymers

Francis Bossé, Pronab Das, and Laurence A. Belfiore[1]

Department of Chemical Engineering, Polymer Physics and Engineering Laboratory, Colorado State University, Fort Collins, CO 80523

This study focuses on gelation in ternary solutions and solid state properties of films that contain a diene polymer, polybutadiene or polyisoprene, and an inorganic salt, bis(acetonitrile) dichloropalladium (*II*). Upon mixing, effective crosslinks are formed, because the acetonitrile ligands of the palladium salt are displaced by olefinic pendant groups of the polymers. The sol-gel properties of these mixtures are greatly affected by various parameters, i.e., salt content, molecular weight (MW) of the polymer, solvent, and polymer microstructure . In low dielectric media, the gelation process is favored over precipitation. However, gel formation is slower in benzene or toluene relative to gelation times in solvents with higher dielectric constants, i.e., chloroform or tetrahydrofuran. When higher molecular weight fractions of polyisoprene are employed, the ternary phase diagrams of these mixtures clearly show that the gelation phenomenon is favored over precipitation. The mechanical properties of polybutadiene solid films show a dramatic increase of 3 orders of magnitude in Young's modulus of elasticity (0.2 vs. 200 Kpsi) when the palladium salt content is 4 mole %. A percolation description of the concentration dependence of Young's modulus identifies the threshold concentration (p_c) of palladium salt to be in the vicinity of 0.5 mole %. This p_c value is low for a three-dimensional network. The increase in elastic modulus is much smaller (0.2 vs. 2 Kpsi) for polyisoprene. The presence of macrocrystallites in gels and solid films is confirmed by polarized optical microscopy. The equilibrium dimension and distribution of the crystallites in the polymeric matrix depend greatly on the type of solvent, as well as the concentration of the palladium salt.

[1]Corresponding author

The critical phenomenon of gelation in polymeric systems has been the subject of many investigations. It is now well established that solutions containing a crystallizable polymer can form gels via the formation of microcrystallites.[1] On the other hand, gelation of noncrystallizable polymers is not well understood.[2] However, gels possess a common feature that does not depend on their structure or the type of interaction which leads to their formation. This common feature is the presence of crosslinks or severe entanglements.

Over the past few years, studies describing metal ion crosslinked polymer gels formed as a result of chemical bonding between the metal ions and reactive polymer functional groups have been reported in the literature.[3] For the most part, these studies focus on water soluble low MW polymers bearing reactive alcohol or carboxylate functional groups. Another class of polymers which bear olefinic reactive sites in the backbone or as a pendant group may also represent good candidates that can lead to the formation of covalent gels. This was confirmed in a previous publication by Belfiore et al.,[4] in which it was clearly established that gelation could be achieved in solutions of 1,2-polybutadiene and dichloropalladium (*II*). The gelation phenomenon depends upon many factors including the MW and the structure of the polymer, the pH or ionic strength of the solution, the type of solvent, and the concentration of species in solution.[2] In view of these factors, the relations between gel structure and the solid state properties of the residual film after complete solvent evaporation are still unclear. In order to tailor materials with enhanced mechanical properties, structure-property relations need to be established.

This paper is divided into two sections. In the first part, isothermal ternary phase diagrams for systems containing diene polymers crosslinked with a dichloropalladium (*II*) salt will be discussed in terms of sol-gel properties. The investigated parameters are the MW and structure of the polymer, solvent effects, and the concentration of inorganic salt. In the second part, polarized optical microscopy and mechanical testing are employed to analyze the solid state properties of films generated from ternary solutions.

Experimental

Materials. Atactic 1,2-polybutadiene (80% 1,2-PBu, 11% 1,4-trans PBu and 9% 1,4-cis PBu) and 3,4-polyisoprene (43% 1,4-PI, 53% 3,4-PI and 4% 1,2-PI) samples were supplied by Goodyear Tire & Rubber Company (Akron, Ohio), courtesy of Dr. Adel F. Halasa. These samples were used as received without any additional purification. Bis(acetonitrile) dichloropalladium (*II*) is available from several commercial chemical distributors.

Size-Exclusion Chromatography. Molecular weight characterization of the polymer samples was performed by size exclusion chromatography (SEC) on a Varian 5000 LC at 50°C. The solvent used for the standardization curve was tetrahydrofuran (THF, Chemlab spectroscopic grade). The analyses were performed in duplicate and the results averaged. The weight-average MW (M_w), and the polydispersity index for each sample are given in Table I.

Fractionation of the Polyisoprene Sample. The polyisoprene sample was separated by conventional fractional precipitation as follows.[5] The polymer (1.5g) was dissolved in 300 mL of benzene, and 100 mL of methanol was added to achieved a methanol content of 25 % by volume. The resulting cloudy solution was warmed until it cleared and was then allowed to cool slowly to room temperature with stirring. After one day, the insoluble polymer fraction had settled as a gel layer at the bottom of the flask. The clear supernatant liquid was decanted and the gel treated a second time with the same volume and concentration of the benzene/methanol mixture. The gel was then solubilized in THF and precipitated in methanol. The recovered sample was then dried to constant weight in a vacuum oven at 40°C. The molecular weight of various polyisoprene fractions was determined by SEC (see Table I). The average deviation in the MW determination is about 2%.

Table I   MW characterization for polybutadiene and polyisoprene samples

Polymer Sample	Label	$M_w \times 10^{-4}$	Polydispersity Index
Polybutadiene	PBu	50	1.6
Polyisoprene	PI	8	1.6
Polyisoprene*	PI-1	10	1.5
Polyisoprene*	PI-2	30	1.4

* Fractionated sample

Sample Preparation Methods. The polymers (polybutadiene or polyisoprene) and dichloropalladium (*II*) were dissolved separately while stirring in the solvent, e.g., THF, chloroform (CHL), toluene or benzene. The solubility of bis(acetonitrile) dichloropalladium (*II*) in THF (or CHL) is on the order of 0.5 g per 100 mL. However, in less polar solvents like hexane or toluene, the palladium salt is sparingly soluble (< 0.1g per 100 mL). This solubility problem was overcome by adding a small amount of DMF to the palladium salt solutions ($\cong$ 5% wt/wt). The fresh salt and polymer solutions were then mixed in an air tight vial (20 mL). No further stirring was performed after the initial homogenization of the two solutions. Using this procedure, solutions containing 0.3 to 1.75 wt % of solid per mL of solvent were prepared, while the mole fraction of salt relative to the polymer ranged from 0.5 to 8 mole %. Solid state films were prepared from THF and toluene solutions by pouring the highly viscous ternary solutions, before gelation, into a Petri dish to facilitate evaporation of the solvent in a fume hood at ambient temperature.

Measurements.   Polarized optical microscopy was performed using a Nikon Optiphot-Pol microscope.   Temperature variation was accomplished via a convective hot stage (Fluid Inc.), using filtered laboratory air.   Photographs were obtained using an automatic-exposure-based Polaroid camera (model 545). Birefringence measurements were obtained using the following equation:

$$OPD = t \times |n_1 - n_2|$$

where *OPD* is the optical path length difference between the fast wave and the slow wave, *t* is the thickness of the sample, and $|n_1 - n_2|$ is the absolute magnitude of the birefringence.   The optical path difference is obtained using the following procedure.   First, one must find the extinction angle for the crystal which produces darkness in the field of view.   The first-order wave plate (1 λ) is then added.   The color associated with that minimum is noted and translated into a wavelength (nanometers) using a polarization color chart.[20]   The first order plate is then removed, and the crystal is rotated until the maximum intensity in birefringence can be observed.   After adding the first order plate, the primary wavelength of transmitted light can be obtained using the polarization color chart.[20]   The *OPD* value is then easily obtained from a subtraction of these transmitted wavelengths in the presence of the first-order wave plate.

Stress-strain measurements were performed at room temperature using an Instron model 8501 servohydraulic mechanical testing system.   Samples were cut into strips with average dimensions of 45 mm (length), 65 mm (width), and a thickness ranging from 0.25 to 0.65 mm. The strain rate was 50 mm per minute (2 inches/minute).[6]   Using this procedure, at least four solid films were tested at each salt concentration.

Results and Discussion

General Considerations of the Sol-Gel Phase Transition.   Studies of the reactions of low molecular weight model compounds of palladium (*II*) chloride containing acetonitrile ligands showed that acetonitrile is a weak ligand, and that it can be easily displaced by olefinic groups.[7,8]   In an earlier publication,[9] the IR results revealed that acetonitrile ligands were absent in solid films of atactic 1,2-polybutadiene   (1,2-PBu)   with   bis(acetonitrile)   dichloropalladium   (*II*). Coordination of the double bond in the polymeric sidegroup to the palladium metal center to form crosslinks offers an effective mechanism to generate network-like structures.   If this scheme is operative, it should not be surprising that these mixtures bear the potential to form gels under certain conditions.   For ligand substitution reactions, three types of stoichiometric mechanisms have been recognized, i.e., dissociative (SN$_1$), associative (SN$_2$) and interchange.[10] Associative and dissociative reactions are two-step mechanisms in which an intermediate is formed, which is then transformed into the product.   Elucidating which of these ligand substitution reactions is operative is beyond the scope of this paper. However, for the associative or the dissociative mechanism, the dielectric constant of the solvent should have a major effect on gelation phenomena, since it is well known that the acidity or the effective charge of the ions is decreased in low dielectric media.[11] Thus, one expects greater gelation times when a lower dielectric constant solvent is employed. This prediction is observed herein via qualitative gelation kinetics studies for both diene polymers with the bis(acetonitrile) dichloropalladium (*II*) salt in four different solvents.

Part I -- Isothermal Ternary Phase Diagrams for the Diene Polymer, Bis(acetonitrile) Dichloropalladium (*II*) and the Solvent. Ternary phase diagrams can be used to obtain qualitative information about gelation and precipitation. The type of phase diagram used in this study can be represented as a region defined in a two-dimensional plane, since there are two degrees of freedom (concentrations) at constant temperature and pressure. The abscissa of the phase diagram is the mole fraction (%) of the palladium salt with respect to the polymeric repeat unit, while the ordinate is the total mass of solid (in grams) per 100 mL of solvent. Examples of ternary phase diagrams for PBu and the dichloropalladium salt are illustrated in Fig. 1. In Fig. 1 a), THF is used as the solvent, while Fig. 1 b) shows the same system in benzene. From this figure it is clear that gelation occurs throughout a region that is twice as large when benzene is employed as the solvent, relative to the same system in THF. This behavior will be discussed in greater detail in the following section.

Solvent Effects. The results obtained from ternary phase diagrams for polybutadiene and dichloropalladium (*II*) in various solvents are summarized in Table II. The table shows that the solvent must have a dielectric constant greater than 2.2 to observe gelation, and as mentioned above, this can be related to the reduction of ion reactivity in low dielectric media.[11] However, for systems in which gelation can be achieved, it was observed that the composition range over which gelation occurs increases as the dielectric constant of the ternary solution decreases (see Fig. 1). Figure 2 illustrates the effect of the solvent dielectric constant on the equilibrium time required for gelation in two diene polymers that contain olefinic reactive groups, i.e., PBu (filled symbols) and PI (open symbols). Triplicate samples, with a total weight of solid (in grams) per 100 mL of solvent equal to 1.2%, and a palladium mole fraction of 2 %, were used to perform this study. The gels were assumed to have reached steady-state when they were viscous enough to pass the "tilt test" without flow. For both systems, semi-log plots illustrate that the time required for gelation dramatically decreases when the dielectric constant of the solvent increases. From Fig. 2 it is also evident that for the same dichloropalladium (*II*) content, gelation times are much greater for PI relative to PBu. The discrepancy in reactivity between the two polymers is probably related to their structure, since for PI twice as many reactive olefinic groups reside in the backbone (40% vs. 20% for PBu), and may be inaccessible to form coordination crosslinks. In view of the greater time required for gelation, it

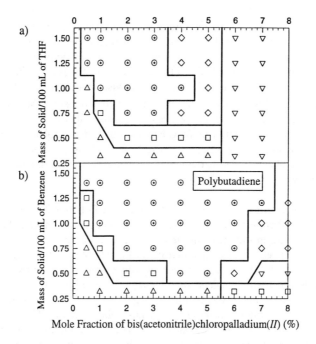

Figure 1   Ternary phase diagrams for polybutadiene and bis(acetonitrile) dichloropalladium (*II*) in two different solvents at room temperature. Fig. 1 a), THF is used as the solvent, while for Fig. 1 b) the solvent is benzene.

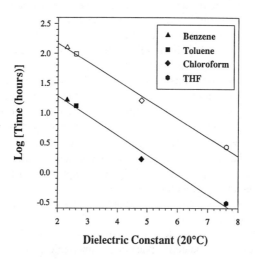

Figure 2   Gelation equilibrium time vs. solvent dielectric constant on semi-log axes for two diene polymers that contain olefinic reactive sidegroups, i.e., PBu (filled symbols) and PI (open symbols). The ternary solutions contained 12 mg of solid per mL of solvent, and the mole fraction of bis(acetonitrile) dichloropalladium (*II*) respective to the polymer repeat unit was constant at 2 %.

is clear that the viscosity of the mixture in low dielectric media increases at a much slower rate relative to the solution viscosity for high dielectric constant solvents. This may offer an explanation for the broader gelation region of the ternary phase diagram (see Fig. 1) with lower dielectric constant solvents. It has been observed in other related systems, that a slow increase in viscosity may favor the formation of interchain crosslinks relative to intrachain interactions.[12] Moreover, since ligand exchange is hampered in low dielectric media, this may lead to a lower crosslink density, which could favor gelation instead of precipitation. If a lower density of crosslinks is present, a "looser" network is generated that is less susceptible to collapse under its own weight. Thus, even if precipitation occurs, it does not imply that crosslinks are not present.

Effects of Palladium (*II*) Salt Concentration and Polymer Molecular Weight on the Sol-Gel Properties. An increase in the concentration of palladium will lead to an increase in the crosslink density which is present in the gel at steady-state. This increase in crosslink density is known to produce segregated gel networks, i.e., isolated regions with a high concentration of polymer chains and other regions which are "pockets" of solvent.[12] When segregation occurs at high salt concentrations, one expects that the gel will precipitate, leading to the formation of a gelatinous material which has "phase-separated" from the pure solvent. On the other hand, if the palladium (*II*) concentration is constant and the molecular weight of the polymer increases, then one expects a lower crosslink density and less segregation. Thus, an increase in the polymer chain length should favor gelation over precipitation. In other words, as the polymer molecular weight increases, precipitation should be observed at higher salt content.

Using the procedure described in the experimental section, ternary phase diagrams for the fractionated and the original polyisoprene samples (see Table I), PI, PI-1 and PI-2, were established. Fig. 3 a) and b) show the ternary phase diagrams for fractionated polyisoprene samples PI-1 ($M_w = 300K$) and PI-2 ($M_w = 100K$), respectively. The MW effect of the polymer on gelation behavior is not as dramatic as the one illustrated in Fig. 1 for solvent effects. However, a consistent trend is observed. As the polyisoprene chain length is increased, gelation preferentially occurs over precipitation at higher salt concentrations. This trend is in good agreement with previous literature observations for other closely related systems,[13] where it was suggested that an increase in the MW of the polymer leads to the formation of less compact structures with less rigid strands between crosslinks. Thus, for the polyisoprene samples, an increase in molecular weight favors gelation over precipitation.

It is worth noting that high palladium (*II*) salt concentrations induce precipitation of the gels. In THF, Fig. 1 a) and Fig. 3, reveal that the gel/precipitate boundary is located between 3 and 4 mole % $PdCl_2$. If (i) the crosslinks are evenly distributed in space on a cubic lattice, (ii) all d^8 metal centers form a bridge between two distinct polymer chains, and (iii) no intramolecular loops are produced, then the mean length of the polymer strands should be on the order of 10 repeat units. Under these conditions, precipitation should not occur,

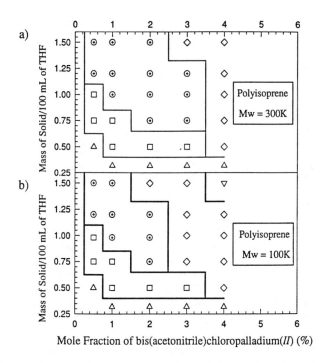

Figure 3 Ternary phase diagrams for polyisoprene and bis(acetonitrile) dichloropalladium (*II*) in THF at room temperature. Fig. 1 a), the MW of the polyisoprene is 300 K, while for Fig. 1 b) the MW is 100 K.

since the network will not be very compact. Moreover, the crosslink density of the network will be low enough to accommodate solvent molecules. Evidently, since precipitation occurs, a large number of defects must be present in the network structure. However, for polybutadiene systems an additional factor enhances segregation, This factor is the crystallization of foreign crystallites in the network, which will be discussed in greater detail in the next section.

Table II   Influence of the solvent on sol-gel behavior for polybutadiene and dichloropalladium (*II*)

Solvent	Dielectric Constants (20°C) [a,b]	Solution Blends Characteristics [c]	Gelation [d]
n-Hexane	1.9	Precipitation	NO
Cyclohexane	2.0	Precipitation	NO
Dioxane	2.2	Precipitation	NO
Benzene	2.3	Clear-Yellow Solutions	YES
Toluene	2.6	Clear-Yellow Solutions	YES
Chloroform	4.8	Clear-Yellow Solutions	YES
Tetrahydrofuran	7.6	Clear-Yellow Solutions	YES

[a]   The solvent dielectric constants were obtained from reference [14].
[b]   A solvent with a dielectric constant greater than THF will not solubilize PBu.
[c]   3 mg of solids per mL of solvent, and bis(acetonitrile) dichloropalladium (*II*) concentration ranging from 1 to 5 mole %.
[d]   Greater than 5 mg of solids per mL of solvent, and bis(acetonitrile) dichloropalladium (*II*) concentration ranging from 1 to 5 mole %.

Part II-- Properties of Solid Films Obtained from Ternary Solutions.
According to the definition given by Ferry more than three decades ago,[15] a dried gel is not considered to be a gel because it is not a dilute system. However, on the basis of more recent percolation[16] and scaling theories[12], dry gels which contain strongly bonded crosslinks (vulcanized rubbers) possess most of the features which are present in dilute systems and should be considered "true" gels. For reversible gels which are formed by physical entanglements or any other mechanism, establishing relations between gels and solid films could be specious. Thus, prior to any discussion of the properties of solid films, one must ascertain that the PBu/palladium salt systems form "strong" chemically bonded gels, i.e., that they are covalent thermo-irreversible gels. There are several experimental observations

which suggest that the systems under investigation can be classified in this manner. First, gelation cannot be induced by cooling viscous ternary mixtures identified by open squares in Fig. 1 and 3. Second, the steady-state gels or solid films cannot be re-dissolved in the pure solvent. Third, using variable-temperature polarized optical microscopy, it was observed that these gels are thermally irreversible. Fourth, the strong dependence of sol-gel properties on the solvent clearly indicates that these are not physical gels.[17] Finally, gels retain their physical properties and appearance when some of the solvent is extracted.

Polarized Optical Microscopy. The presence of crystallites in ionomers[18] and gels[2] has been observed in the past. Crystallinity in gels is usually due to polymer chains that form ordered regions. The crystallites observed in ionomeric systems result from phase separated regions containing a high concentration of ionic material. For both of these systems, typical crystallite sizes are relatively small ($\cong$ 100 nanometers). In this study, the crystallization phenomenon observed for PBu/palladium (*II*) salt systems is of a different type. It corresponds to the growth of crystals that reach impressive dimensions (micron range). These crystals are not observed in the starting solutions (polymer and salt solutions), or in the solid residues of pure polymer or pure salt obtained after solvent evaporation. This suggests that the polymer matrix must be present in order for crystallization to occur. Small crystals are formed immediately after the two solutions are mixed together. The crystals grow rapidly until the equilibrium gel is formed, i.e., when the gel passes the tilt test. Crystal growth is quenched when the equilibrium gel forms in THF, while in toluene a modest crystal growth continues after gelation. For aged gels ($\cong$ 2 months), it was observed that the leached out solvent contains unreacted palladium salt. This indicates that when the gel structure is formed, the microviscosity is high enough to inhibit diffusion of palladium ions[19] and quench the crystal growth process. This observation also implies that if the crosslinking mechanism is diffusion controlled, then it is inhibited when the polymer rich region is well above the critical concentration for entanglement formation (c*).[19] Thus, even though a minimum amount of palladium is required to induce gelation (greater than 0.5 mole % PdCl$_2$ for ternary solutions with a total solids content of 10mg/mL THF), only a few effective crosslinks are required to form gels.

Figure 4 illustrates the type of crystals that were observed in gels and solid films. Fig. 4 a) corresponds to the crystallites that form in THF. These crystallites resemble clear transparent needles whose appearance is independent of salt concentration. The average length of the needles shown in Fig. 4 a) is about 10 microns. Fig 4 b) shows the type of crystallites that form in toluene. These crystallites appear to be red-brown leaflets whose appearance is independent of salt concentration. The appearance and lower decomposition temperature of these structures ($\cong 270°C$ vs. >300°C for bis(acetonitrile) dichloropalladium (*II*)) rule out recrystallization of the original salt. It is also clear that the resulting palladium crystals depend greatly on the solvent. This suggests that the solvent participates in the crystallization process, and that it may also be present in the crystal structure. At this stage, preliminary studies indicate that certain parameters have

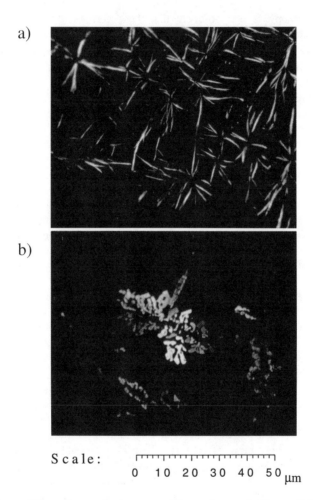

Figure 4   Illustration of the crystallites observed for solid films of polybutadiene with PdCl$_2$. Fig. 4 a) corresponds to the crystallites (transparent needles) formed when THF was employed as a solvent. Fig 4 b) shows the type of crystallites (red-brown leaflets) that were formed when the solvent was toluene. For both films, the palladium salt content relative to the polymer was 2 mole %.

an influence on crystallization behavior. It was observed that a decrease in salt concentration generates smaller crystals which are more uniformly distributed in the gel or the solid media. The rate of salt crystallization is much faster in a high dielectric constant solvent like THF, relative to crystallization rates in a low dielectric constant solvent like toluene. It is worth noting that no crystallization was ever observed for polyisoprene mixtures. Using standard procedures,[20] the intrinsic birefringence was measured for crystals present in the solid films formed using THF as the solvent. Figure 5 summarizes the results obtained for these films (solid circles). From this figure, it is clear that the relation between intrinsic birefringence and salt content is linear. Moreover, the birefringence increases with higher concentrations of the palladium salt. The birefringent pattern for a solid film prepared from toluene is illustrated in Fig. 4 b) (intrinsic birefringence = 0.014 in toluene, i.e., solid square in Fig. 5). At constant thickness, it is well known that an increase in birefringence corresponds to an increase in crystallinity. This suggests that crystals obtained at high salt content are more ordered.[20] The implications of this claim are not fully understood yet. In order to establish more quantitative relations, the current systems are still under investigation. These studies will elucidate the effect of other relevant parameters (polymer microstructure, temperature, and aging) on the crystallization behavior.

Elastic Moduli of Solid Films. Using the procedure described in the experimental section, stress-strain curves were acquired for solid films containing from 0.5 to 4 mole % of the palladium salt. Young's modulus was calculated via Series IX™ software from Instron, using the original stress-strain data without smoothing. Figure 6 summarizes the results obtained for PBu and PI films. Since crystallites are present in the solid films of polybutadiene with $PdCl_2$, two factors can be responsible for this dramatic increase in modulus by approximately three orders of magnitude, i.e., a filler effect and an effect due to palladium chemical crosslinks. For polyisoprene with $PdCl_2$, this increase can only be justified by the presence of chemical crosslinks, based on observations in the optical microscope. If one focuses on the elastic moduli data for polybutadiene films and assumes that (i) all of the starting palladium salt has crystallized, (ii) the density of the resulting salt is 3 g/cm^3, which is a common value for palladium crystals,[21] and (iii) the modulus of the crystal is at least a thousand times greater than that of the polymer matrix (200 vs. 0.2 Kpsi), one can estimate the increase in elastic modulus that the crystallites would produce. For a palladium mole fraction of 4 % of salt with respect to the polymer, the salt occupies 6 % of the total volume. The theoretical elastic modulus for the film can then be obtained using the Kerner equation (depending on the geometry of the dispersion) or the Guth equation (geometry independent).[22] Using the Kerner equation, the maximum increase in modulus should be on the order of 4 times the original PBu modulus if the crystallites are oriented parallel to the stretch direction, but only a 2 fold increase should be observed for random crystallite orientation. Thus, it is clear that a filler-effect is negligible and the 1000-fold increase in modulus is primarily attributed to the presence of chemical crosslinks between palladium (*II*) and the double bonds in the polymer sidegroups. In order to determine the effective contribution of the crystals to the increase in modulus, samples should be analyzed after the crystals are destroyed via heat treatment at 300°C, since it was observed that these crystals do not reform upon cooling. At such high temperatures, thermal degradation of the polymer matrix is highly probable. Therefore, this approach could be unreliable.

The elastic moduli data in Fig. 6 are analyzed in terms of a simple model based on percolation theory in the vicinity of the critical point. The standard percolation equation that relates elastic modulus to the concentration of reactive groups is well known.[23] However, this equation is only valid above the percolation threshold of the network. Macroscopic elastic stiffness of the network can be predicted as follows:

$$\kappa_{exp} = \kappa \left(p - p_c \right)^\tau \text{ or log } \kappa_{exp} = \tau \log \left(p - p_c \right) + \log \kappa, \text{ for } p \geq p_c \quad (1)$$

where κ is the local stiffness constant of a rigid component of the network, κ_{exp} is the resulting macroscopic stiffness for the entire network (Young's modulus of the sample), τ is the critical exponent, while p and p_c represent concentration and critical concentration, respectively. According to reference 23, for a three-dimensional network in which the bonds are randomly distributed throughout the matrix, the value of τ should be equal to 3.6 . If some segregation is present then the value of τ should be lower. In the extreme case in which the crosslinks adopt a random walk geometry, the critical exponent should be equal to 2.9, once again for a three-dimensional network.

Evidently, this relation is not applicable to the whole concentration range illustrated in Fig. 6, because at low salt concentration the modulus is non-zero and slowly increasing below the threshold. This corresponds to a situation which has already been discussed elsewhere,[24] i.e., a material that contains a certain fraction of weak conductor sites to which good ones are slowly added. In this study of mechanical properties, the situation is similar below the critical point, but instead of weak conductors, weak "springs" are present. As the concentration of palladium increases, more and more of these weak springs are replaced by "stronger" ones. An expression that would be valid throughout the entire salt concentration range could be derived using the renormalization group treatment (Chapter 4 in reference 24). However, the derivation of that expression is beyond the scope of the present paper.

Visual inspection of the data in Fig. 6 indicates that the contribution of weak springs to the elastic modulus becomes negligible for a palladium concentration of $\cong 1$ mole % or less. Thus, the equation described above should be valid for salt concentrations greater than 1 mole %. A commercially available non-linear curve fitter (Jandel Scientific) was employed to determine the best values for the three parameters, i.e., τ, p_c and κ. The values determined using this software are $\tau = 3.1$, $p_c = 0.48$, and $\kappa = 0.42$ Kpsi. The percolation threshold, p_c,

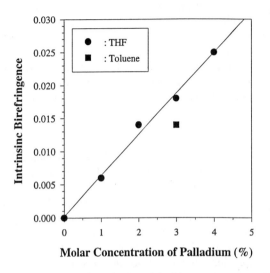

Figure 5 Plot of the intrinsic birefringence vs. bis(acetonitrile) dichloropalladium (*II*) concentration for solid polybutadiene films prepared from ternary solutions prior to gelation.

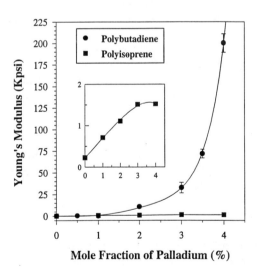

Figure 6 Plots of Young's modulus of elasticity vs. palladium salt concentration for the two diene polymers. The filled circles represent polybutadiene samples, while the filled squares are assigned to polyisoprene samples. The insert is a magnification of the results for polyisoprene films.

is close to the minimum palladium concentration which is required to induce the sol-gel transition. This observation strongly suggests that there is a correlation between the gel and the solid film networks. For PI films, data analysis suggests that these complexes are below the percolation threshold. In comparison with PBu films, this clearly indicates that the crosslink density is much lower for PI. The difference in crosslink density for PBu/PdCl$_2$ vs. PI/PdCl$_2$ films might be explained by the microstructural differences between the two diene polymers.

Conclusion

Atactic 1,2-diene polymers form covalent gels with dichloropalladium (*II*) in a variety of solvents via coordination crosslinks. The gels are thermally irreversible, and the physicochemical interactions are much stronger than those which exist in physical gels. Some of these gels are highly segregated. The sol-gel properties of ternary mixtures are greatly influenced by several factors. The dielectric constant of the solvent has a major effect on sol-gel properties. Greater equilibrium gelation times are required for low dielectric constant solvents, and gelation is favored over precipitation when the dielectric constant of the media is low. The solvent affects gelation thermodynamics displayed in ternary phase diagrams, as well as the kinetics of gelation (Fig. 2). The MW of the polymer thermodynamically influences the gelation phenomenon as evidenced by the fact that gelation is favored when the molecular weight is increased. Palladium salt crystallization was observed for PBu gels and solid films. A linear relationship was found between the intrinsic birefringence of the crystals and the palladium concentration.

The solid films properties depend strongly on salt concentration, as well as the preparation method. For the PBu system, a dramatic increase in Young's modulus of elasticity was observed at relatively low palladium salt concentration (1000 times at 4 mole %), while for the PI films the increase in modulus is moderate (10 times at 4 mole %). For PBu films, the critical concentration, p_c, and the critical percolation exponent, τ, were determined to be 0.48, and 3.1, respectively.

The global mechanism of the gelation process needs to be established. The detailed kinetics of the gelation process will be investigated by dynamic light scattering in a subsequent study.

Acknowledgments

The research described herein is supported by the National Science Foundation's Division of Materials Research (Polymer program) through grant # DMR-9214022, and the donors of the Petroleum Research Fund administered by the American Chemical Society. Professor Don Radford in the Department of Mechanical Engineering at Colorado State University is gratefully acknowledged for obtaining preliminary stress-strain data for polybutadiene and the polymeric palladium complex. The authors benefited from many helpful discussions with

Professor Mark Bradley in the Physics Department at CSU related to applications of percolation theory for the description of elastic moduli in the vicinity of the threshold concentration.

References

1   Domzy; R. C.; Alamo, R.; Edwards, C. O. and Mandelken, L *Macromolecules* 1986, **19**, 310, 1986; Mandelken, L. In *Microdomains in Polymer Solutions*; Dubin, P. L., Ed.; Plenum Press: New York, 1985.

2   Russo, P. S. in *Reversible Polymeric Gels and Related Systems*; Russo, P. S., Ed.; ACS Symposium Series 350; Am. Chem. Soc., Washington, D. C.: Chapter 1 and reference therein, 1987.

3   Menjivar, J. A. In *Water-Soluble Polymers*, Glass, J.E., Ed.; Adv. Chem Ser. **213**, 209, 1986; Moradi-Araghi, A.; Beardmore, D. H. and Stahl, G. A. In *Water-Soluble Polymers for Petroleum Recovery*, Stahl, G. A.; and Schulz, D. N., Eds.; Plenum: New York, 299, 1988;

4   Belfiore, L. A. and McCurdie, M. P; *ACS Proceedings*; Division of Polymeric Material Science and Engineering, 1994, **70** (1), p. 433.

5   Boyer, R. F. *J. Polym. Sci.* 1953, **9**, 197.

6   Peiffer, D. G.; Agarwal, P. K.; Duvdevani, I. and Lundberg ,R. D. *Plastics Engineering* 1987, **43**, 35

7   Maittlis, P. M.; Espinet, P and Russel, M. J. H. in *Comprehensive organometallic chemistry--synthesis, reactions, and structures of organometallic compounds*, Vol. 6, Wilkinson, G; Stone, F. G. A. and Abel, E. W., Eds.; Pergamon Press: Oxford , Ch. 38.6, p. 363-384, 1986.

8   Andrews, M. A.; Chang, T. C. T.; Cheng, C. W. F.; Emge, T. J.; Kelly, K. P. and Koetzle, T. F. *Journal of the American Chemical Society* 1984, **106**, 5913.

9   Belfiore, L. A.; McCurdie, M. P. and Bossé, F. submitted to Polymer International.

10 Shriver, D. F.; Atkins, P. W. and Langford, C. H. in *Inorganic Chemistry*, W. H. Freman: New York, 1990, p.466-468.

11 Jørgensen, C. K. in *Absorption spectra and chemical bonding in complexes*, Pergamon Press: Oxford, 1962.

12 de Gennes, P. G. in *Scaling Concepts in Polymer Physics*, Cornell University Press: London, Chapter 5 and 7, 1979.

13 Illiopoulos, I.; Audebert, R. and Quivoron, C. in *Reversible Polymeric Gels and Related Systems*; Russo, P. S., Ed.; ACS Symposium Series 350; Am. Chem. Soc., Washington, D. C.: Chapter 6, 1987.

14 Bruno, T. J. and Svoronos, P. D .N. in *Handbook of basic tables for chemical analysis*; CRC Press, Inc.: Boca Raton, 1989.

15 Ferry, J. D. in *Viscoelastic Properties of Polymers*; Wiley: New York, 1961, p. 391.

16 Essam , J. W. in *Phase Transitions and Critical Phenomena*, Vol. 2, Domb, C. and Green, M., Eds.; Academic Press: New York, p. 197, 1972; Kirkpatrick, S. *Rev. Mod. Phys.* 1973, **45**, 574

17 Guenet, J. M. in *Thermoreversible Gelation of Polymers and Biopolymers*; Academic Press: London, 1992, p. 47.

18 Duvdevani, I; Lunberg, R. D.; Wood-Cordova, C. and Wilkes, G. L. in *Coulombic Interactions in Macromolecular Systems*; Eisenberg, A. and Bailey F. E., Eds.; ACS Symposium Series 302; Am. Chem. Soc., Washington, D. C.: Chapter 15, 1986.

19 Morawetz, H. in *Macromolecules in Solution*; Wiley, New York, 1975, p 305.

20 Wood A. E. in *Crystals and Light 2nd edition*, Dover Pubication, Inc.: New York, Chapter 10, 1977; Robinson, P. C. and Bradbury S. in *Qualitative Polarized-Light Microscopy -- Microscopy Handbook*, Vol. 9, University Press: Oxford, Chapter 6, 1992;

21 Carmichael, R. S. *Handbook of Physical Properties of Rocks*, CRC Press: Boca Raton, Florida, 1984.

22 Nielsen, L. E. in *Mechanical Properties of Polymers and Composites,* Vol. 2, Chapter 7, Marcel Dekker, Inc.: New York, 1974.

23 Kantor Y. and Webman I *Phys. Rev. Lett.* 1984, **52**, 1891.

24 Stauffer D. in *Introduction to Percolation Theory*, Taylor & Francis: London, Chapter 5, 1985

RECEIVED October 26, 1994

Bonding and Other Interactions

Chapter 16

Hybrid Organic–Inorganic Silica Materials
Chemical Evidence for Organization in the Solid

G. Cerveau, C. Chorro, R. Corriu[1], C. Lepeytre, J. P. Lère-Porte, J. Moreau, P. Thepot, and M. Wong Chi Man

Précurseurs Organométalliques de Matériaux, Université Montpellier II, Case 007, Place E. Bataillon, 34095 Montpellier Cedex 5, France

The synthesis and chemical reactivity of hybrid silica gels containing reactive organic fragments have been explored. The reactions which occured in the solid state provide evidence for a partial organization of the organic moities within the silica network.

Interactions between the organic molecules incorporated in the silica network allow facile polyaddition or polycondensation of the organic units.

As shown upon metal complexation the location of the organic fragments at the surface or in the core of the amorphous solid is highly dependent on the organic structures.

Organic-inorganic hybrid materials are a field of growing interest, expanding very quickly *(1) (2)* . Hybrid solids consisting of polymers or organic molecules included into inorganic oxide networks are receiving much attention.

Three possibilities can be used for the introduction of an organic molecule in an inorganic network :

(i) Organic molecules can be incorporated by mixing them with an inorganic component (without covalent bond between the organic and inorganic fragments). The formation of an alloy is possible when some molecular interactions occur.

(ii) Surface interactions, either hydrogen bonding or covalent bond, allow to attach organic molecules at the surface of solids.

(iii) The incorporation in the core of an amorphous solid requires at least two covalent bonds between the organic molecule and the solid.

Our interest lies in the preparation of hybrids in which organic molecules are covalently bound to a silica *(3) (4) (5)* .

In this research, silica-based hybrid materials seem the most favorable case for the following reasons :

(i)   The covalent SiC bond is very chemically stable towards air, bases, acids, etc...

(ii) There are many known routes for the attachement of silyl groups to organic molecules.

(iii) The sol gel processes are well adapted to the preparation of oxide networks under reaction conditions which do not cleave Si-C bonds.

(iv) Finally the industry of silicones provides many organosilicon molecules which permit to work at a reasonnable scale.

[1]Corresponding author

Besides the intrinsic physical properties (electronic, optical, magnetic, etc.) which are possibly afforded by the organic molecules embedded in the solid, there are two interesting questions to consider :
1) Is the texture of the solid (specific surface area, porosity, etc.) related to the molecular structure of the organic moiety ?
2) Does the molecular structure give rise to some organization in the amorphous solid?

Concerning the first question, there is some evidence that the texture of the solid is not controlled by the molecular structure *(4)* . The molecular structure appears as a parameter, but the specific surface area is highly function of the experimental conditions and particularly of the solvent. The first table illustrates very well this dependence.

Table I

Exp.	Précursors	Conc. Ml^{-1}	Solvent	Catalyst (conc.)	Gelation time (mn)	Specific surface area (m^2/g)
1	Si(OMe)$_3$	3	MeOH	NH$_4$F(10^{-3})	10^{-1}	1030
2		1	MeOH	NH$_4$F(10^{-3})	10	516
3		1	THF	NH$_4$F(10^{-3})	5	548
4		0,4	THF	NH$_4$OH(2,6)	45	1262
5	Si(OMe)$_3$	0,4	THF	HCl(5,10^{-3})	600	129
6		1	MeOH	NH$_4$F(10^{-3})	15	685
7	(MeO)$_3$Si〜S〜Si(OMe)$_3$	1	THF	NH$_4$F(10^{-3})	15	19
8		3	MeOH	NH$_4$F(10^{-3})	60	325
9	(MeO)$_3$Si〜〜Si(OMe)$_3$	1	MeOH	NH$_4$F(10^{-3})	24.10^3	0
10		0,1	MeOH	NH$_4$OH(2,6)	30	977

This paper is mainly devoted to the second question. We have developped hybrid solids in which the reactivity between the organic moieties could provide some chemical evidence for the presence of a partial organization in the amorphous hybrid silica network.

In all cases, discussed below, the organic fragments included are covalently bound to silica and the organic molecules choosen present an unambiguous reactivity toward each other under well define conditions.

Scheme 1 presents the synthesis of the organic precursor *(6)* and the preparation of the hybrid gel by hydrolysis. The conservation of the molecular structure of the organic fragment in the gel was controlled by spectroscopy.

The diyne organic moiety was choosen since an appropriate geometry is required in the polymerisation of the bis acetylenic units into enynes *(7) (8)* . As illustrated in schema 2 , this reaction takes place with organisation requirements. The 1-4 addition implies a parallelism between the bis-acetylenic units *(9)* .

Bis Acetylenic moieties

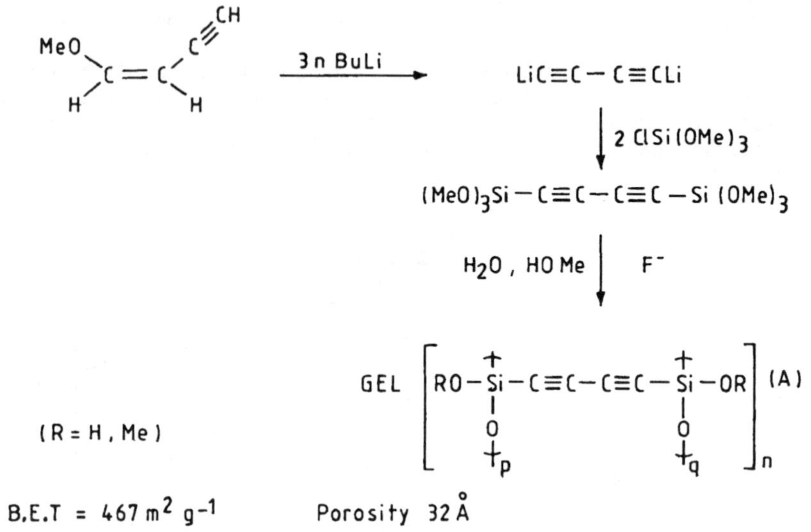

$$MeO{\diagdown}{\underset{H}{C}}={\underset{H}{C}}{\diagup}C{\equiv}CH \xrightarrow{\text{3 n BuLi}} LiC{\equiv}C-C{\equiv}CLi$$

$$\downarrow\ 2\ ClSi(OMe)_3$$

$$(MeO)_3Si-C{\equiv}C-C{\equiv}C-Si(OMe)_3$$

$$H_2O,\ HO\ Me\ \Big\downarrow\quad F^-$$

(R = H, Me)

$$GEL\ \left[\overset{+}{RO-Si}-C{\equiv}C-C{\equiv}C-\overset{+}{Si}-OR\right]_n\ (A)$$

with $\underset{O}{|}$, $\overset{+}{\underset{p}{}}$ and $\overset{+}{\underset{q}{}}$ groups

B.E.T = 467 m^2 g^{-1}　　　　Porosity 32 Å

Scheme 1

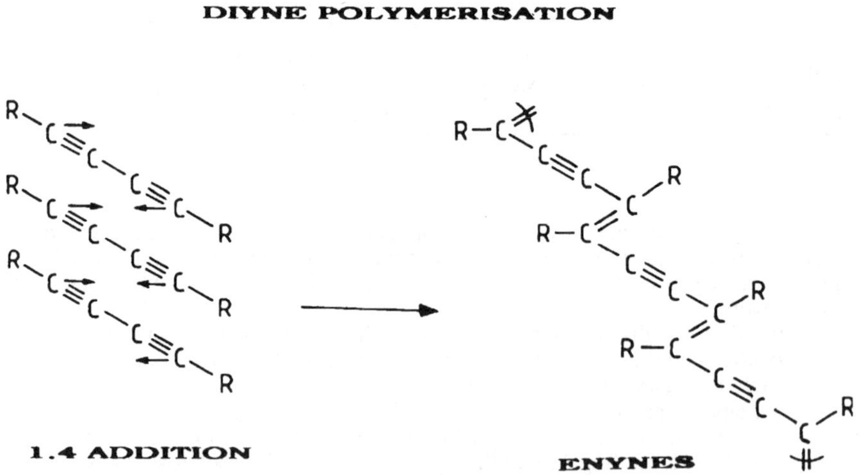

DIYNE POLYMERISATION

1.4 ADDITION　　　　　　　　　**ENYNES**

Scheme 2

Upon heating the gel at 200°C under Argon , it is possible to detect the formation of enynes by spectroscopy *(10)*.

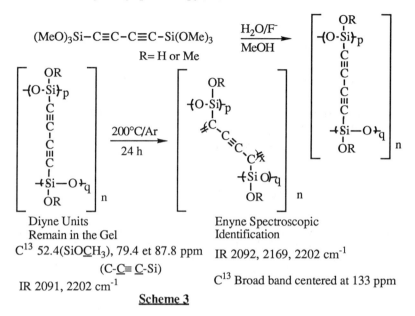

Diyne Units
Remain in the Gel

C^{13} 52.4(SiO$\underline{C}$H$_3$), 79.4 et 87.8 ppm

(C-$\underline{C}$≡ $\underline{C}$-Si)

IR 2091, 2202 cm^{-1}

Enyne Spectroscopic
Identification

IR 2092, 2169, 2202 cm^{-1}

C^{13} Broad band centered at 133 ppm

Scheme 3

The TDA curve exhibits an exothermic transformation between 120 and 350°C with a maximum at 250°C which can be assigned to the diyne polymerisation in the gel.

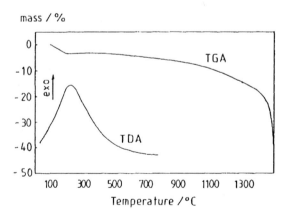

Scheme 4

There is a comparable situation in the case of polymers in which the bis-acetylenic units are bound to SiR$_2$ units *(8)*. The cross-linkage due to the bis-acetylenic units reaction appears even at lower temperature in the case of the gel.

$+(SiR2-C\equiv C-C\equiv C)_n$

Degree of   (R = Me) > (R = Et) > (R=Ph) > (R = SiMe3)
cristallinity: ($\searrow$)

Texo°C: ($\nearrow$)   200 < 225   < 255, 315 < 350(broad)

Exothermic Activity: 181   -   125   70
(ΔH cal. g-1)($\rightsquigarrow$)

<u>Scheme 5</u>

Finally, the same spectroscopic features have been found when the gelation is performed before the cross- linkage (route A) or when the cross-linkage is induced by irradiation before the gelation (route B).

The easy polymerisation at a relatively low temperature of the diyne fragment within the amorphous gel may be indicative of a favorable arrangement of the diyne moities. These results suggest the formation, in the silica network, of substructures which permit the enyne formation : this observation involves parallelism between several bis-acetylenic units.

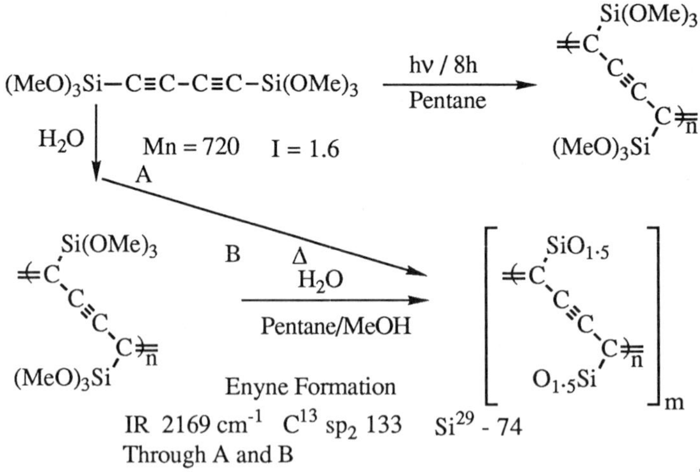

$(MeO)_3Si-C\equiv C-C\equiv C-Si(OMe)_3$ $\xrightarrow[\text{Pentane}]{hv / 8h}$

H_2O | Mn = 720   I = 1.6

A

B $\xrightarrow[\text{Pentane/MeOH}]{\Delta \quad H_2O}$

Enyne Formation

IR 2169 cm-1 C^{13} sp2 133 Si^{29} - 74

Through A and B

<u>Scheme 6</u>

Complexation of aromatic groups bound to silica

Aromatic organosilanes easily form Cr(CO)3 complexes upon reaction with Cr(CO)6. This reaction was used to test the accessibility of phenyl groups attached to a silica network (11) (12). The scheme 7 shows the case of monosilylated phenyl systems for which the introduction of Cr(CO)3 is possible after gelation. The accessibility is a good evidence that most of the phenyl groups are located at the surface of silica.

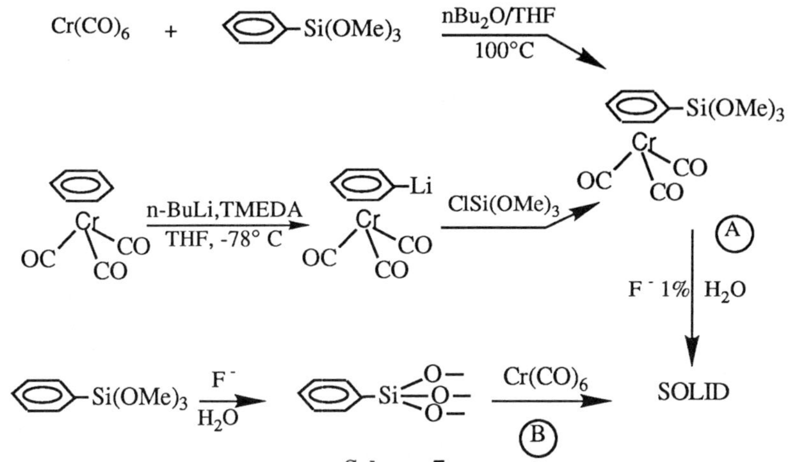

Scheme 7

At the opposite in the case of the bis-silylated phenylene groups *(3) (4)* , it is possible to prepare gels containing the Ar-Cr(CO)3 units first by complexation followed by gelation. However the operation is not commutative : when the gelation is performed fisrt, it is impossible to introduce the Cr(CO)3 moieties on the phenylene ring . We can conclude that most of the phenyl groups are located in the core of the solid and are not accessible for reaction. This is in agreement with the high hydrophilicity observed for the gels containing p-phenylene units.

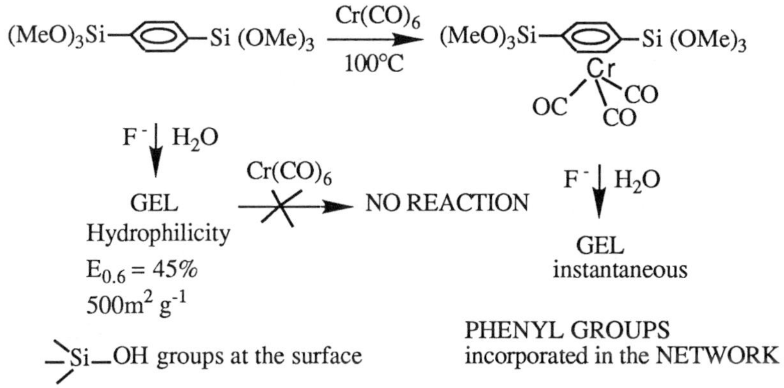

Scheme 8

However as shown on scheme 9 this conclusion cannot be generalized, since very similar compounds do not behave similarly.

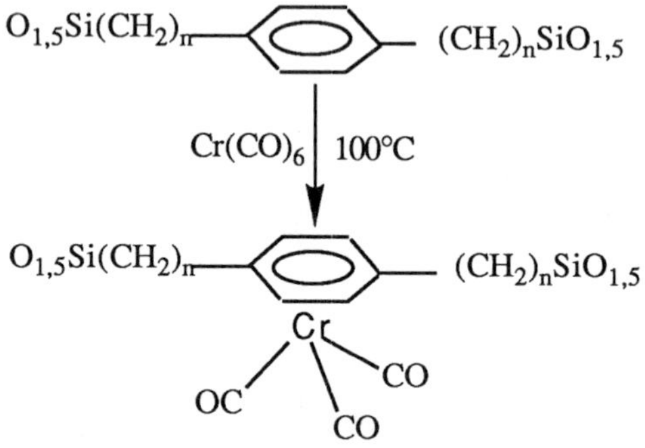

$O_{1,5}Si(CH_2)_n$ —⬡— $(CH_2)_nSiO_{1,5}$

$Cr(CO)_6$ | 100°C

$O_{1,5}Si(CH_2)_n$ —⬡— $(CH_2)_nSiO_{1,5}$

Cr

OC / | \ CO

CO

Complexation characterized by spectroscopy and % of complexation determined by T.G. analysis

n=0	0%	$E_{0,6} = 45\%$
n=1	40%	$E_{0,6} = 7\%$
n=2	52%	$E_{0,6} = 5\%$

Scheme 9

This difference in reactivity illustrates very well that interactions between aromatic groups during the gelation process are completely different in the case of the similar molecules reported in scheme 8 and scheme 9.

Polymerisation of thiophenes units inside the matrix

Polythiophenes are obtained by electropolymerisation of thiophene *(13)* . Silylated thiophenes have been shown to be very efficient precursors for this polymerisation *(14) (15)* leading to polymers with a good regularity *(16)* . We have prepared gels containing bis- silylated mono, bis and ter-thiophene. (Scheme 10).

The spectroscopic features show that molecular structure is maintained in the solid. The electrochemical oxydation of the gels deposited as films at the electrode leads to a fast polymerisation reaction *(17)* . (Scheme 11).

29SiCPMAS NMR Spectra of xerogels

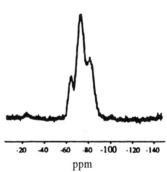

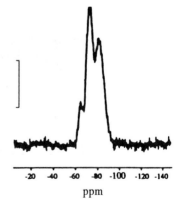

Scheme 10

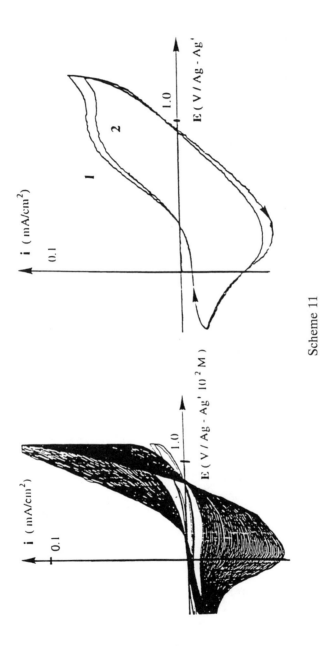

Scheme 11

The polymer is identified in the film by spectroscopy *(17)* .

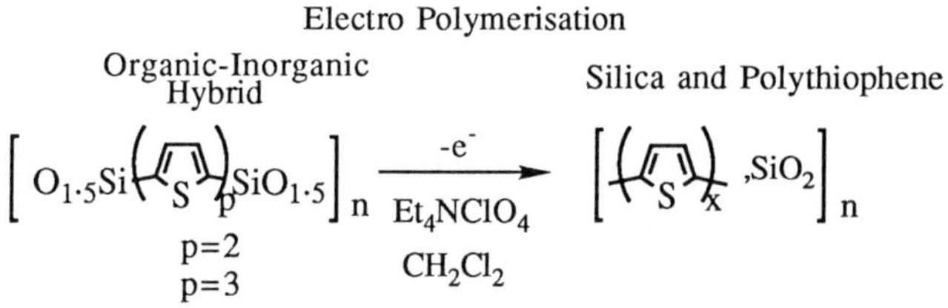

Electro Polymerisation

Organic-Inorganic Hybrid   Silica and Polythiophene

$$\left[O_{1.5}Si\left(\underset{S}{\overbrace{}} \right)_p SiO_{1.5} \right]_n \xrightarrow[\substack{Et_4NClO_4 \\ CH_2Cl_2}]{-e^-} \left[\left(\underset{S}{\overbrace{}} \right)_x , SiO_2 \right]_n$$

p=2
p=3

Spectroscopic Identification of Polythiophene formation

Raman Resonance Spectra

652, <u>682</u>, <u>700</u>, 1047, 1171, 1220 and 1458 cm-1

Absorption spectrum 490 nm (420 nm gel)

Scheme 12

The same reaction is possible by chemical oxydation performed on the gel.

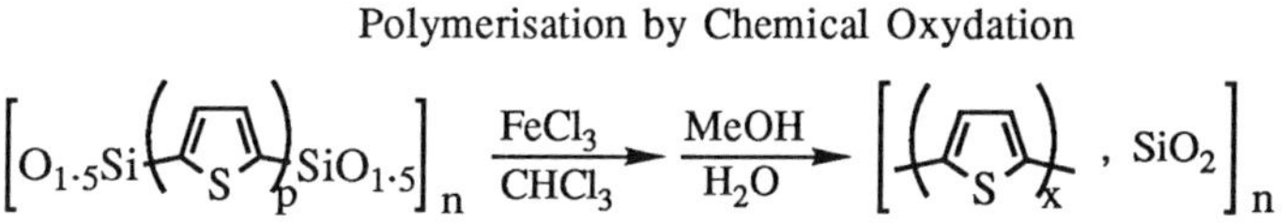

Polymerisation by Chemical Oxydation

$$\left[O_{1.5}Si\left(\underset{S}{\overbrace{}} \right)_p SiO_{1.5} \right]_n \xrightarrow[CHCl_3]{FeCl_3} \xrightarrow[H_2O]{MeOH} \left[\left(\underset{S}{\overbrace{}} \right)_x , SiO_2 \right]_n$$

p=1 (685 m^2.g^{-1})
p=3 (188 m^2.g^{-1})

Scheme 13

The polymer can be idenfied by the same way. However it is impossible to separate the polymer obtained from the silica which formed during the polymerisation, although the ^{29}Si solid state NMR shows that most of the Si-C bonds are cleaved into SiO bonds, as observed by Si^{29} CP MAS NMR spectroscopy. (Scheme 14).

The mechanism *(16)* of oxydative polymerisation of silylated thiophene implies the thiophene units are close one to each other. Thus, the observation of fast polymerisation inside the gels suggests a partial organization in the amorphous solid. The thiophene units are more or less aggregated and oxydation permits the formation of polymeric segments by coupling of proximal thiophene units.

Since the mobility is restricted in the solid, and since the polymerisation takes place by coupling of two radicals, we can conclude that the polymer formation occurs inside thiophene aggregates.

The interaction between units is also illustrated by the possible inclusion of charge transfer complexes, formed between terthiophene and TCNQ, in the gels *(17)* .

These complexes have been included in the gel upon hydrolysis of a charge-transfer complex precursor. They remain unchanged after gelation as shown by light

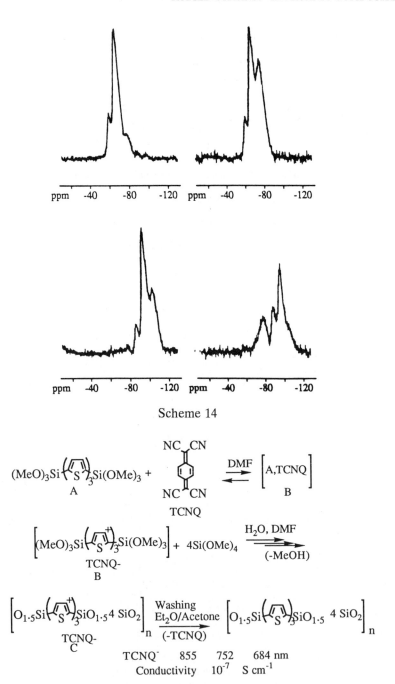

Scheme 14

Scheme 15

absorption and by the conductivity in the solid. Even if these interactions are very weak they are strong enough for controlling a structure in the amorphous solid.

Control of solid formation by intermolecular interactions due to a template effect

We examined the sol-gel polycondensation of an olefinic trialkoxysilane $(MeO)_3Si$-CH_2-CH= $CHCH_2Si(OMe)_3$ *(4)* in the presence of Pd (II) complexes.

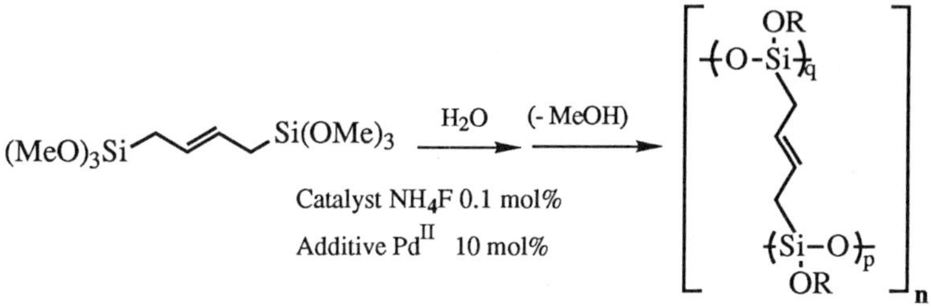

$$(MeO)_3Si \diagdown \diagup Si(OMe)_3 \xrightarrow[\text{Catalyst } NH_4F \ 0.1 \text{ mol\%}]{H_2O \quad (-MeOH)} \left[\begin{array}{c} OR \\ +O\text{-}\overset{.}{S}i\overset{\downarrow}{]_q} \\ \diagup \diagdown \\ +\overset{.}{S}i\text{-}O\overset{\downarrow}{]_p} \\ OR \end{array} \right]_n$$

Additive Pd^{II} 10 mol%

R = H or Me   Scheme 16

This system was choosen because of the high ability of -CH=CH- bonds for coordination around Pd (template effect) *(18)* . The sol-gel polymerisation was performed in THF using various reaction conditions. The results are summarized in Table II.

Table II

Precursor	Additive	Additive conc mol (eq.)	Gel time (min)
 $(MeO)_3Si \diagdown\diagup Si(OMe)_3$	-	-	29×10^3
	Li_2PdCl_4	10^{-1}	30
	$LiCl$	10^{-1}	50×10^3
	$(CH_3CN)_2PdCl_2$	5×10^{-2}	30

Catalyst : NH_4F 10^{-3} mole/l

Transparent gel formed in all cases. ^{13}C and ^{29}Si NMR analysis of xerogels clearly established that the Si-C bond is retained in the solid.

Whereas differences in the experimental conditions did not produce significant changes in the hybrid network at the molecular level, drastic differences appeared in the kinetic of gelation : in the presence of Pd (II) the gel time was reduced by almost three order of magnitude.

In order to establish the nature of the Pd activation in the sol-gel process, the polycondensation of two other related monomers was studied. The monomer containing a double bond with non-allylic silicon atoms and the non-coordinating hydrocarbon monomer were hydrolysed under similar reaction conditions.

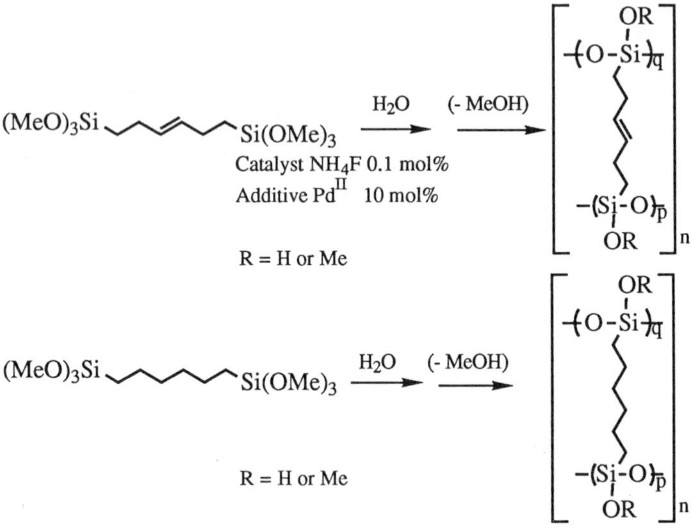

Scheme 17

The hybrid gel which formed exhibited spectroscopic characteristics consistent with the mean structures presented in scheme 17. Again here, important differences appeared in the kinetic of gelation according to the reaction conditions. As shown in the Table III, Pd (II) acts as a very efficient condensation catalyst for monomers containing unsaturated units. The C_6 unsaturated monomer underwent very rapid condensation in the presence of Pd, whereas no change was observed for the saturated C_6.

Table III

Precursor	Additive	Additive conc mol (eq.)	Gel time (min)
 (MeO)$_3$Si~~~~Si(OMe)$_3$	-	-	10^4
	Li$_2$PdCl$_4$	5×10^{-2}	120
	Li$_2$PdCl$_4$	2×10^{-1}	120
(MeO)$_3$Si~~~~Si(OMe)$_3$	-	-	2.9×10^3
	Li$_2$PdCl$_4$	5×10^{-2}	2.9×10^3

Catalyst : NH$_4$F 10^{-3} mole/l

Since similar observations were made for the two olefinic monomers, most probably palladium activation proceeds by a template effect of the metal. As shown in scheme 18, the formation of η^2 olefin complex with unsaturated units allows an intramolecular condensation within the coordination sphere of the metal. While held in the complex, the monomer are ideally positioned for condensation and the siloxane bridge formation. The acceleration of the polycondensation can thus be attributed to the template effect of Pd.

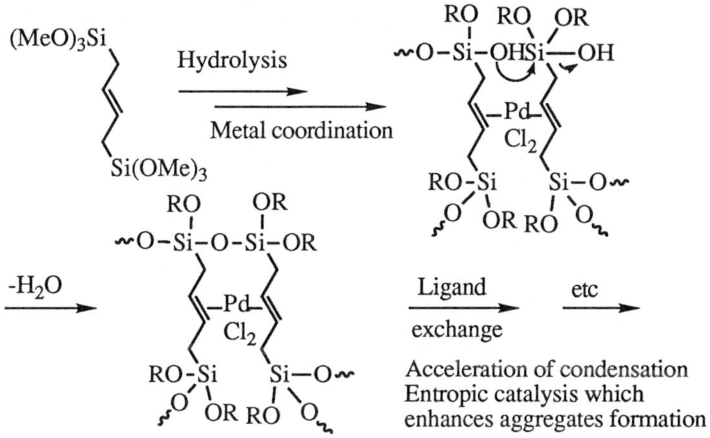

Scheme 18

We also examined the influence of Pd on the composition and properties of the resulting gel.

As summarized in scheme 19 the main features are as follows :

1) For the gel prepared from unsaturated monomer, 80-100% of Pd was trapped in the hybrid network and remained in the xerogel. Conversely only a few Pd was retained in the gel prepared from the saturated C_6 monomers.

2) The gels containing Pd revealed no catalytic activity in hydrogenation reactions.

This contrasts with the quite active related hydrogenation catalyst consisting of Pd supported on a siloxane framework *(19)* . All these results are consistent with a dispersion of Pd within the bulk of the material and no Pd at the surface.

This is confirmed by the experiment shown in scheme 19. The condensation performed in MeOH instead of THF resulted in a quite different material.

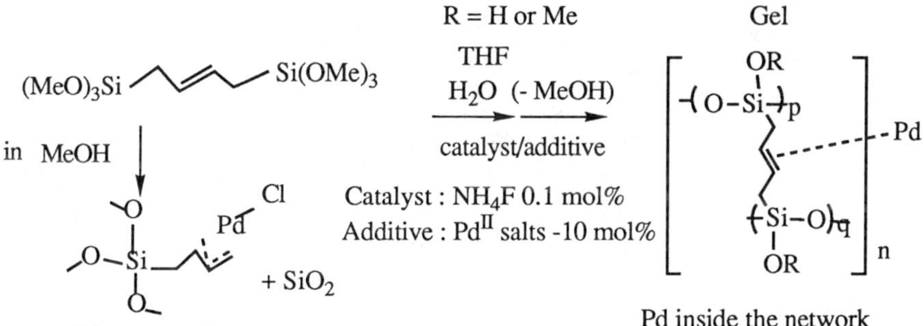

Pd at the surface
Catalytic Effect

Scheme 19

Pd inside the network
No Catalytic Effect

In this case, formation of η^3-allyl Pd species and cleavage of the Si-C bond was established *(20)* . The resulting solid with η^3 allyl-Pd complexes at the surface behave as a very efficient catalyst.

Conclusion

The results presented here illustrate the possibility of interactions between the organic molecules incorporated in a silica network leading to the formation of a substructure in the solid. The location of organic units at the surface or inside of the solid is also highly depending on their molecular structure.

Literature Cited

1- Mac Kenzie J.D., *J.Sol-Gel Sci.Tech.* in press and references therein.
2- (a) Schmidt H.K., *Mater.Res.Soc.Symp.Proc.* **1984**, 32, 327 ; (b) Schmidt H.K., *Inorganic and Organometallic Polymers, ACS Symp.*, ser.n°360, American Chemical Society : Washington, DC **1988**, p 333 ; (c) Schmidt H.K., *Mater.Res.Soc.Symp.Proc.* **1990**, 180, 961 and references therein.
3- (a) Shea K.J., Loy D.A.,Webster O.W., *Chem.Mater* **1989**, 1, 574 ; (b) Shea K.J., Loy D.A.,Webster O.W., *J.Am.Chem.Soc.* **1992**, 114, 6700, and references therein.
4- (a) Corriu R.J.P., Moreau J.J.E., Thepot P., Wong Chi Man M., *Chem.Mater* **1992**, 4, 1217-1224 ; (b) ib. authors, *J.Sol-Gel Sci.Tech.,* in press.
5- Cerveau G., Corriu R.J.P., Costa N., *J.of Non Cryst.Solids* **1993**, 163, 226.
6- Brefort J.L., Corriu R.J.P., Gerbier Ph., Guerin C., Henner B.J.L., Jean A., Kulhman T., Garnier F., Yassar A., *Organometallics* **1992**, 11, 2500.
7- (a) Bässler H., Enkelman V., Sixl H., *Adv.Polym.Sci.* **1984**, 63 ; (b) Chance R.R., Pattel G.N., *J.Polym.Sci.Polym.Phys.Ed.* **1978**, 16, 859 ; (c) Chance R.R., Pattel G.N., Turi E.A., Khanna Y.P., *J.Am.Chem.Soc.* **1978**, 100, 1307; (d) Patil A.O., Desmpande D.D., Talwar S.S., Biswas A.B., *J.Polym.Sci.Polym.Chem..Ed.* **1981**, 19. 1155 ; (e) Kato J., Nakamura K., Yamazakis S., Amano T., *J.Polym.Sci.Part A Polym.Chem..* **1989**, 27, 1853.
8- (a) Corriu R.J.P., Guerin C., Henner B.J.L., Jean A., Mutin H., *J.Organomet. Chem.* **1990**, 396, C3 ; (b) Corriu R.J.P., Gerbier Ph., Guerin C., Henner B.J.L., Jean A., Mutin H., *Organometallics* **1992**, 11, 2507.
9- Corriu R.J.P., Gerbier Ph., Guerin C., Henner B.J.L., Fourcade R.J., *Organometal.Chem.* **1993**, 449, 111.
10- Melreyer A.J., Banghmann R.H., *J.Polymer Sci.Polymer Phys.Ed.* **1989**, 27 1853, and references therein.
11- Moran M., Cuadrado I., Pascual M.C., Casado C.M., Losada J., *Organometallics* **1992**, 11, 1210.
12- Kyung Moon Choi, Shea K.J., *Chem.Mater* **1993**, 5, 1067.
13- (a) Garnier F., *Angew.Chem.Int.Ed.Engl.* **1989**, 28, 513 ; (b) Roncali F., *Chem.Rev.* **1992**, 92,711 and references therein.
14- (a) Lemaire M., Büchner W., Garreau R., Hoa H., Guy A., Roncali J., *J.Electroanal.Chem.* **1991**, 312, 547 ; (b) Lemaire M., Büchner W., Garreau R., Hoa H., Guy A., Roncali J., *J.Electroanal.Chem.* **1990**, 281, 293 ; (c) Roncali J., Guy A., Lemaire M., Garreau R., Hoa H., *J.Electroanal.Chem.* **1991**, 312, 277.
15- (a) Ritter S.K., Noftle R.E., *Chem.Mater 4* **1992**, 872 ; (b) Matsuda H., Taniki Y., Kaeriyama K., *J.Polym.Sci.*, A, 30, **1992**, 1667 ; (c) Guay J., Diaz A., Wu R., Tour J.M., Dao L.H., *Chem.Mater 4* **1992**, 254.
16- Sauvajol J.L., Chorro C., Lère-Porte J.P., Corriu R.J.P., Moreau J.J.E., Thépot P., Wong Chi Man M., *Synth.Metals* **1994**, 62, 233.

17- Corriu R.J.P., Moreau J.J.E., Thépot P., Wong Chi Man M., Chorro C., Lère-Porte J.P., Sauvajol J.L., *Chem.Mater* **1994**, 6, 15.
18- (a) Schrauzer G.N., Eichler S., *Chem.Ber.* **1962**, 95, 550 ; (b) Schrauzer G.N.,Glockner P., Eichler S., *Angew.Chem.Int.Ed.* **1964**, 3, 185 ; (c) Ward J.S., Pettit R., *J.Am.Chem.Soc.* **1971**, 93, 262 ; (d) Humphries A.P., Knox S.A.R., J.Chem.Soc.Dalton **1978**, 1514 ; (e) Kochi J.K., *Organometallics Mechanism and catalysis*, Academec Press, London **1978**, p 429.
19- (a) Schubert U., Ambert-Schwab S., Breitscheidel B., *Chem.Mater* **1989**, 1, 516, and references therein ; (b) Tour J.M., Cooper J.P., Pendalwar S.L., *J.Org.Chem.* **1990**, 55, 3452 ; (c) Ferrari C., Predieri G., Tiripicchio A., Costa M., *Chem.Mater* **1992**, 4, 243 and references therein.
20- (a) Kliegman J.M., *J.Organometal.Chem.* **1971**, 29, 73 ; (b) Hayashi T., Ito H., Kumada M., *Tetrahedron lett.* **1982**, 23, 4605 ; (c) Hayashi T., Konishi M., Kumada M., *J.Chem.Soc.Chem.Commun.* **1983**, 736 ; (d) Corriu R.J.P., Escudie N., Guérin C., *J.Organometal.Chem.* **1984**, 271, C7 ; (e) Hayashi T., Yamamoto A., Iwata T., Ito Y., *J.Chem.Soc.Chem.Commun.* **1987**, 398 ; (f) Fugani K., Ohsima K., Utimoto K., Nozaki H., *Bull.Chem.Soc.Jpn* **1987**, 60, 2509 ; (g) Ogoshi S., Yoshida W., Ohe K., Murai S., *Organometallics* **1992**, 12, 578.

RECEIVED November 4, 1994

Chapter 17

Hybrid Organic–Inorganic Materials
The Sol–Gel Approach

J. D. Mackenzie

Department of Materials Science and Engineering,
University of California, Los Angeles, CA 90024–1595

Many crystalline and non-crystalline ceramic oxides have now been prepared via the sol-gel method. The most recent exploitation of the Sol-Gel method is the preparation of hybrid organic inorganic materials which can be broadly divided into three types. In Type 1, the porous oxide gel is impregnated with organics to form a nanocomposite. In Type 2, an organic is added to the liquid sol-gel solution as a mixture. After gelation, the organic is trapped in the porous oxide. In Type 3, the organic reacts with the precursors of the sol-gel liquid solution such that after gelation, it is chemically bonded to the inorganic oxide. The properties and structures of these new materials are reviewed in this Chapter.

Crystalline ceramic oxides and oxide glasses are commonly made by reacting raw materials at temperatures above 1000°C. High temperature reactions have many disadvantages including losses from volatilization, undesirable reactions with containers and stresses associated with shrinkage during the cooling of the product. In the past decade, the sol-gel process has generated a great deal of interest as a potentially viable low temperature technique for the preparation of crystalline and glassy oxides (1,3). This process involves room temperature reactions of metal-organic precursors (commonly, metal alkoxides) in liquid solutions to form amorphous porous gels. The gels are then dried and heat-treated to give crystalline or glassy oxide. A most common example is the use of tetraethoxysilane (TEOS) to give silica glass. Fully dense silica glass of excellent optical properties can be made at temperatures around 1000°C whereas the melting approach would necessitate temperatures in excess of 2000°C. The sol-gel method further permits the formation of very homogeneous ceramics because of the use of liquid solutions. It is particularly suited for the fabrication of thin films. One serious disadvantage of the sol-gel method is that the porous gel is mechanically

0097–6156/95/0585–0226$12.00/0

very weak and shrinkages during drying and high temperature heat-treatment often lead to brittle fracture. The presence of non-bridging organic groups in the metal-organic precursors, for instance C_2H_5-$Si(OEt)_3$ to replace TEOS would suggest that brittle fracture may be minimized since the Si has only three interconnecting Si-O bonds rather than four. The dried gel would now contain one C_2H_5-group and thus an "organically modified silicate" (Ormosil) is formed.

The most recent exploitation of the sol-gel method is the preparation of hybrid organic inorganic materials which can be conveniently divided into three types. The Ormosil mentioned above contains a chemical bond between the organic (C_2H_5) and the inorganic (SiO_2) components A chemical bond can also be formed between an organic such as a long chain polymer and oxide groups in the starting liquid solutions. These form one family of organic inorganic hybrids. Another type of hybrids can be made by dissolving or dispersing organics in a sol-gel liquid solution. On gelation, the organic is trapped in the porous inorganic amorphous oxide. A third type involves the impregnation of an organic into the continuous ultrafine pores of a solid oxide gel to form a nanocomposite. The processing of these three types of organic-inorganic hybrid materials is represented in Figure 1. The main features of the preparation of these materials, their structures and microstructures and their properties are summarized in this review.

Type I - Impregnated Hybrids. Oxide gels are usually highly porous prior to drying and the pores are interconnecting with diameters ranging from a few angstroms to few thousand angstroms depending on the chemistry and the processing conditions. After drying and heat-treatment to temperatures above 500°C, the porous gel is now mechanically strong enough for further treatment. SiO_2 gel has been the most widely studied material of this class. Organic monomers of various types have been impregnated into SiO_2 gels and polymerized in situ to give transparent nanocomposites (4,5). Polymethylmethacrylate (PMMA) - SiO_2 nanocomposites containing a wide range of PMMA were found to have properties obeying ideal mixture rules. Thus, transparent solids which are lighter than soda-lime silicate glass but much harder and stronger than PMMA can be fabricated. Table I gives a comparison between the mechanical properties of common inorganic glasses, PMMA and a PMMA-SiO_2 nanocomposite. Organic dyes, for instance, can also be dissolved in the monomer prior to polymerization. In the case of PMMA, its refractive index is almost identical to that of SiO_2 and in addition, because impregnation can be almost 100% complete, a highly transparent nanocomposite can be formed. More recently, L.L. Hench and coworkers were able to prepare SiO_2 gel with very narrow pore size distributions in the diameter range of 14 to 90Å. Various organic dyes such as Rhodamine 6G and 4PyPO-MePTS were impregnated into the pores to give optically active materials for laser applications (6,7). Such

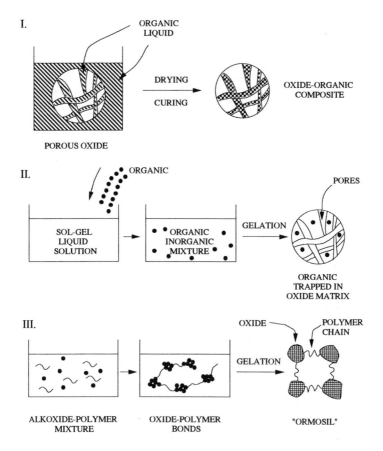

Figure 1.   Three general types of Organic-Inorganic Hybrid Materials.

dye-SiO_2 composites have been shown to perform better than the conventional liquid dye systems in dye lasers (*8,9*). SiO_2 gels have been fabricated into microspheres with diameters ranging from 20 to 70 µm and impregnated with dyes (*10*). The use of such microspheres as components for optical display panels were demonstrated and their applications as sensors, slow release media for medicines, fragrances and other chemicals suggested (*10*). It is entirely feasible that polycrystalline porous oxides such as TiO_2 and ZrO_2 can be used as the host for organic impregnates and thus widening the opportunities for further exploitation of these organic-inorganic hybrids.

It is obvious that opportunities abound for the use of these Type I nanocomposites for many future applications. For example, the pores need not be completely filled with the organics with the first impregnation but instead, can be narrowed by a coating of the organic. Thereafter, a second impregnation with another organic can be made. Another possibility is the incorporation of optically active transition metal ions into the oxide phase to create the possibility of interactions between these ions and an impregnated organic dye.

Type II - Entrapped Organics Hybrids. In 1984, Avnir, Levy and Reisfeld first stirred organic dyes into a SiO_2 sol-gel solution and prepared a gel containing Rhodamine 6G and suggested the use of such materials as dye lasers (*11*). Since that time, many organic materials have been impregnated into primarily SiO_2 gel matrices. A partial list of such impregnates is shown in Table II. There is a great deal of developmental activity at present to exploit the applications of these nanocomposites. The hard colored coatings using organic pigments instead of dye molecules dispersed in a SiO_2 gel is apparently a successful commercial development in high definition television. A highly promising version of a chemical sensor is shown in Figure 2. The tip of a common optical fiber is stripped of the oxide cladding and replaced by a thin coating of a dye-oxide sensor fabricated via sol-gel liquid solution. The evanescent wave is modified when the sensor is exposed to gases to which the dye is sensitive. Small sensors only three inches in length and reasonable in cost have apparently been developed in Ireland (B.D. MacCraith, et. al., J. Sol-Gel; Sci and Tech., in press) for the detection of a variety of gases. The liquid crystal device developed by Levy et. al. (*15*) also holds promise. In these devices, the liquid crystal droplets are dispersed in a suitable oxide gel matrix sandwiched between two glass slides with transparent indium-tin oxide electrodes. In the absence of an electric field, the liquid crystal droplets would scatter light. However, a field can align the liquid crystals in the droplet and gives transparency. In many publications on these Type II hybrid materials, some critical questions can be raised and currently answers are absent. The scientific knowledge of entrapped organics in an inorganic gel matrix is still in its infancy. For instance, the SiO_2 gel formed

Table I. Comparison of Mechanical Properties of Various Transparent Solids (5)

Materials	Density g/cc	Relative Abrasion (10^{-3} mm^3/cycle)	Vickers Hardness (kg/mm^2)	Modulus of Rupture (kpsi)
SiO$_2$ glass	2.20	12	700	15,500
Soda-lime glass	2.50	23	450	12,000
SiO$_2$-33% PMMA	1.85	35	220	12,000
PMMA	1.20	350	30	8,000

Table II. Examples of Entrapped Organics in Oxides

Entrapped Organics	Potential Applications	Reference
Rhodamine 6G	Dye laser	Avnir, Levy and Reisfeld (11)
Quinizarin	Optical data storage	Tani, Namikawa, Arai and Makishima (12)
Poly-pophenylene vinylene	Third Order NLO	Prasad (13)
o-phenanthrolin	Chemical sensors	Zusman, Rottman, Ottolenghi and Avnir (14)
Liquid crystals	Display systems	Levy, Pena, Serma and Oton (15)
Spiropyrane	Photochromics	Levy, Einhorn and Avnir (16)
Phenoxazinium	Hard colored coatings	Nakazumi and Amano (17)

even after heat-treatment to 200°C, still contain OH groups and probably unreacted alkoxides. The gel is very porous and mechanically very weak. The porosity and its related low mechanical strength can be partially mitigated by filling the pores with polymers or gel solutions as described above. However, there is still little knowledge at present regarding the locations of the entrapped organics and their interactions with the oxide matrix. For instance, the organics can be entirely surrounded by the oxide, partially surrounded or exposed in the pore. Figure 3 shows where a relatively large organic molecule such as EDTA can be trapped within a porous SiO_2 gel. EDTA is known to form chelates with heavy metal ions and hence the composite can be potentially useful as an ultrafilter and/or sensor. However, its usefulness must be dependent on where and how the EDTA is trapped and at present, such information is lacking. Another uncertainly is whether a dispersed organic will have any effects on the gelation mechanism and hence on the microstructure of the oxide gel itself. Thus, a great deal more scientific studies must still be made before one can have confidence of the proposed applications of these Type II hybrid materials.

Type III - Chemically Bonded Organic - Inorganic Hybrids. In the Introduction section, it has been mentioned that organic groups such as C_2H_5 can be incorporated in a SiO_2 gel via the starting precursor material. The porous gel so formed cannot be heat-treated as for a pure oxide gel because the organic group would decompose. However, if a high temperature stable gel is not required or if the organic group can be subsequently removed at high temperatures without causing brittle fracture, then these gels containing organics can be highly desirable during drying. Figure 4(a) depicts a purely inorganic oxide gel and Figure 4(b) that of a Type III hybrid. Since network connectivity terminates at every R group, the ability of this hybrid to relieve stresses during shrinkage in drying is improved and fracture tendencies are minimized.

It is also possible to induce the formation of chemical bonds between the inorganic components in a gel and organic components during the gelation process. The most common example is the reaction between TEOS and polydimethylsiloxane (PDMS):

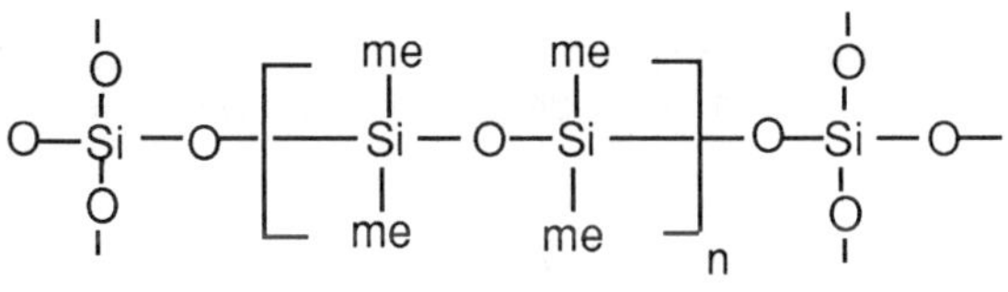

Depending on the ratio of TEOS to PDMS, very hard "Ormosils" (organically modified silicates) or rubbery solids can be prepared (18,19).

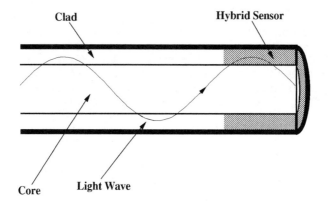

Figure 2.   Fiber optic sensor developed by B.D. MacCraith,
             et al (J. Sol-Gel Science and Technology, in press).

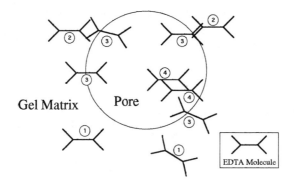

Figure 3.   Various positions where an EDTA molecule can be
             trapped in an oxide gel. The pore size is 30Å.

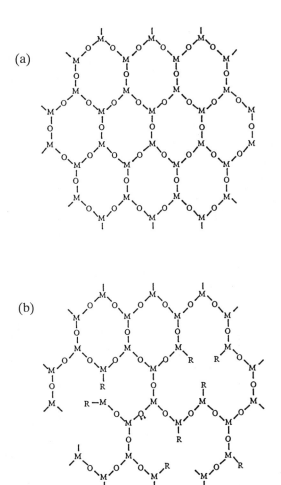

Figure 4a. and b.   Presence of organic (R) group can minimize shrinkage stresses during the drying of gels.

Table III.   A Comparison of Some Rubbery ORMOSILs and Commercial Rubbers in Mechanical Properties

Materials	Density (g/cm³)	Tensile Strength (psi/MPa)	Ultimate Elongation (%)	Young's Modulus (psi/MPa)	Resilience (%)	Relative Damping (%)
Ormosil I	0.50	310/2.2	25	2800/19.2	50	28.9
Ormosil II	0.83	230/1.6	40	1000/6.9	72	15.2
Ormosil III	0.80	265/1.8	50	150010.0	70	16.3
Silicone rubber (red)	1.24	700/4.8	200	60 0/4.1	49	29.6
Neoprene rubber (black)	1.50	1000/6.9	300	900/6.2	26	47.3

Note:
Ormosil I: TEOS/PDMS$_{1700}$.
Ormosil II: TEOS/AEROSIL 90/PDMS$_{1700}$.
Ormosil III: TEOS/AEROSIL 90/PDMS$_{4200}$.

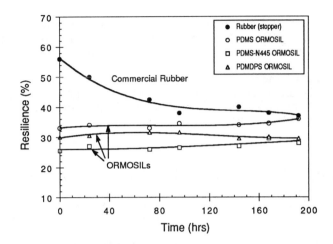

Figure 5.   Variation of resilience of various rubbers as a function of time at 150°C in air.

The chemical bonding between the organic and the inorganic was shown by NMR studies. Such reactions were first carried out by the pioneering research of Schmidt (20) and Wilkes (21) in 1985. The rubbery behavior of these "Ormosils" is caused by the coiling and uncoiling of the PDMS chain when the solid is under externally applied stress since the Si-O-Si angle along the chain is about 150°. When the PDMS concentration is less than about 10% by volume, very hard "Ormosils" can be prepared with Vickers Hardness Numbers approaching 200 Kg/mm^2 versus the values of around 25 Kg/mm^2 exhibited by the common hard organic plastics.

The hard coatings may be useful for organic plastics. Samples containing as much as 80 weight percent of SiO_2 can still exhibit rubbery behavior which is somewhat surprising. The proposed microstructure of a rubbery Ormosil is shown in Figure 1. Table III gives a comparison between the mechanical properties of Ormosils and various types of solids. Since the rubbery Ormosils can contain as much as 80 weight percent of SiO_2 and their moduli and resilience are comparable to common organic rubbers, they may truly form a new class of rubbery solids. The variation of resilience of the Ormosils after heating in air at 150°C for various times is shown in Figure 5 in comparison with a sample of commercially available rubber. In view of the fact that at this early stage of research when attempts to further improving their properties such as the use of fillers have not yet been made, the rubbery Ormosils are certainly promising candidates for future development.

Future Projections. At present, there is a great deal of research performed on Ormosils in general. The most promising developments are in the areas of sensors, especially biosensors, hard coatings and new rubbery solids. Another interesting future activity is the preparation of organic-inorganic hybrids in which both phases are optically active, for example, dyes and rare earth ions. The possibility of energy transfer between the organic and inorganic species can create yet another new dimension for the exploitation of these new materials.

Acknowledgment. The author is grateful to the Air Force Office of Scientific Research, Directorate of Chemistry and Materials Science and to SDIO/IST for the support of his research under Grants No. AFOSR-91-0096 and AFOSR-91-0317.

Literature Cited

1. Brinker, C.J.; Scherer, G.W., *Sol-Gel Science*, Academic Press: New York, New York, 1990.
2. *Sol-Gel Technology for Thin films, Fibers, Preforms, Electronics and Specialty Shapes*; Klein, L.C., Ed.; Noyes Publishing, Park Ridge, New Jersey, 1988.
3. *Ultrastructure Processing of Advanced Ceramics*; Mackenzie, J.D.; Ulrich, D.R. Eds.; John Wiley and Sons, New York, New York, 1988.
4. Pope, E.J.A.; Mackenzie, J.D.; MRS Bull., Vol. 12, p. 29, 1988.
5. Pope, E.J.A.; Asami, A.; Mackenzie, J.D., J. Mater. Res., Vol. 4, p. 1018, 1989.
6. Hench, L.L.; West, J.K.; Zhu, B.F.; Ochos, R.; SPIE Proc., Vol. 1328, p. 230, 1990.
7. Hench, L.L.; LaTorre, G.P.; Donovan, S.; Marotta, J; Valliere, E.; SPIE Proc., Vol. 1758, p. 94, 1992.
8. Dunn, B.; Mackenzie, J.D.; Zink, J.I.; Stafsudd, O.M.; SPIE Proc., Vol. 1328, p. 174, 1990
9. Lin, H.T.; Bescher, E; Mackenzie, J.D.; Dai, H; Stafsudd, O.M., J. Matl. Sci. Vol. 27, p. 264, 1992.
10. Pope, E.J.A., SPIE Proc. Proc., Vol. 1758, p. 860, 1992.
11. Avnir, D; Levy, D; Reisfeld, R, J. Phys. Chem, Vol. 88, p. 5956, 1984.
12. Tanik, T.; Namikawa, H.; Arai, K.; Makishima, A., J. Appl. Phys, Vol. 58, p. 3569, 1985.
13. Prasad, P.N., SPIE Proc., Vol. 1328, p. 168, 1990.
14. Zusman, R.; Rottman, C.; Ottolenghi, M.; Avnir, D., J. Non-Cryst. Solids, Vol. 122, p. 107, pp. 1990.
15. Levy, D.; Pena, J.M.S.; Serna, C.J.; Oton, J.M.; J. Non-Cryst. Solids, Vol. 147, p. 646, 1992.
16. Levy, D.; Einhorn, S.; Avnir, D.; J. Non-Cryst. Solids, Vol. 113, p. 137, 1989.
17. Nakazumi, H.; Amano, S., J. Chem. Soc. Chem. Comm., p. 1079, 1992.
18. Mackenzie, J.D.; Chung, Y.J.; Hu, Y., J. Non-Cryst. Solids, Vol. 147, p. 271, 1992.
19. Iwamoto, T.; Morita, K.; Mackenzie, J.D., J. Non-Cryst. Solids, Vol. 159, p. 65, 1993.
20. Schmidt, H., J. Non-Cryst. Solids, 73, p. 681, 1985.
21. Wilkes, G.L.; Orter, B.; Huang, H, Polymer Prep., Vol. 26, p. 300, 1985.

RECEIVED December 6, 1994

Chapter 18

Sol–Gel-Derived Silica–Siloxane Composite Materials

Effect of Reaction Conditions in Polymer-Rich Systems

E. P. Black[1], T. A. Ulibarri[1,3], G. Beaucage[1], D. W. Schaefer[1], Roger A. Assink[1], D. F. Bergstrom[2], P. A. Giwa-Agbomeirele[2], and G. T. Burns[2]

[1]Sandia National Laboratories, Albuquerque, NM 87185
[2]Dow Corning Corporation, Midland, MI 48686

In situ sol-gel techniques were used to prepare silica/siloxane composite materials. Tetraethylorthosilicate (TEOS) was reacted to simultaneously produce the silica-filler phase and crosslink the silanol end-capped polydimethylsiloxane (PDMS). By systematically changing the reaction conditions, the phase separation within the materials was varied. The resulting materials were characterized by scanning electron microscopy (SEM), Small-Angle Neutron Scattering (SANS), solid state ^{29}Si NMR and mechanical testing. A correlation between the molecular weight of the matrix, M, and the size of the resulting silica domains (varies with $M^{0.4}$) was found. It was also shown that the catalyst amount and activity affects the phase separation characteristics, with evidence for chain extension at low catalyst amounts. Increasing the catalyst activity increased the number of micron-size particles within the system. ^{29}Si NMR indicates that the observed changes in phase separation were not due to changes in the molecular environments present.

Although silicones are inherently weak at room temperature due to their low glass transition temperature and lack of stress crystallization, methods have been developed to reinforce these materials and they now represent a multibillion dollar industry. Current reinforcement techniques center around blending pyrogenic silicas with polydimethylsiloxane (PDMS) polymers *(1-4)*. While these conventional filling/blending techniques have proven successful, further advances in silicone-based materials production require the development of new methods of reinforcement.

It has been determined that silica particles with high surface areas and ramified structures are the most effective at providing reinforcement. While silica particles of this type are not necessarily the only means of reinforcement, a reasonable approach to obtaining new high performance silicone-based materials lies in developing new methods for producing inorganic particles of this nature. One possible method of producing highly dispersed inorganic fillers which interact strongly with the PDMS matrix is the use of *in situ* sol-gel techniques.

[3]Corresponding author

Currently, the production of *in situ* reinforcement in polymeric systems by sol-gel methods is undergoing rapid development *(5-20)*. However, despite this activity, a detailed understanding of synthesis/structure/property relationships is still lacking. In order to produce sol-gel derived composite materials with sufficient mechanical properties for commercial applications, this deficit of information must be addressed. Utilizing solid state ^{29}Si NMR, scanning electron microscopy (SEM), mechanical testing and Small-Angle Neutron Scattering (SANS), we have completed a detailed investigation on the effect of a number of reaction conditions on *in situ* silica growth in polydimethylsiloxane (PDMS)/tetraethylorthosilicate (TEOS) systems. In this chapter, we summarize our findings on some of the factors which affect silica domain formation, such as polymer molecular weight, catalyst activity and silica loading.

Experimental Details.

Materials and Instrumentation. Hydroxyl terminated polydimethylsiloxane was obtained from Dow Corning Corporation and Petrarch/Huls America. The number average molecular weight of the PDMS for the molecular weight study were 2,000; 4,200; 18,000; 36,000; 77,000; 150,000; 310,000 g mol^{-1}. TEOS (Reagent Grade) was obtained from Fischer Scientific and Aldrich Chemical Company and was used without further purification. Tin octoate (TO), [CH$_3$(CH$_2$)$_3$CH(C$_2$H$_5$)CO$_2$]$_2$Sn, was obtained from Huls as a 50% solution in PDMS or in neat form from Pfaltz & Bauer. Dibutyltin dilaurate (DBTDL), [CH$_3$(CH$_2$)$_{10}$CO$_2$]$_2$Sn[(CH$_2$)$_3$CH$_3$]$_2$, was obtained from Hüls as a 25% solution in PDMS.

The ^{29}Si NMR spectra were recorded at 39.6 MHz on a Chemagnetics console interfaced to a General Electric 1280 data station and pulse programmer. The samples were spun about the magic angle at 2.5 kHz. A direct polarization sequence was used with pulse delay times of 120s (a factor of 4 times the longest observed T$_1$) so the resonance areas are expected to be quantitative. Neutron scattering was performed at the High Flux Isotope Reactor at Oak Ridge National Laboratory, the Manuel Lujan Neutron Scattering Center at Los Alamos National Laboratory and at the Cold Neutron Research Facility at the National Institute of Standards and Technology. Data from different q ranges were matched by an arbitrary vertical shift factor. Scanning Electron Micrographs (SEM) were obtained on Au/Pd coated samples using a JEOL 6400 scanning microscope.

Ultimate tensile strength, 50% and 100% stress values, and elongations were obtained using an Instron Model 1122 Tester. Dogbone samples of dimensions 1 3/4 in (4.5 cm) long by 3/8 in (1 cm) wide were pulled to rupture at a speed of 500 mm/min using the procedure defined by ASTM Standard D412-87. Values reported are the average of five tensile samples. Hardness values were obtained on a Type A Durometer according to ASTM Standard D2290. Sample pieces were stacked to a thickness of about 1 cm and hardness measurements were done at three different positions on each side of the sample. The reported values are the average of the six numbers.

General Procedure. TEOS and a tin catalyst (TO or DBTDL) were added to the hydroxyl terminated polydimethylsiloxane in amounts sufficient to both cross-link and fill the matrix with silica to the mole percent desired. The solution was stirred until it became homogeneous. The mixture was then poured into a glass petri dish which was previously treated with a mold release agent. The sample was degassed, placed into a humidity cabinet and cured at 60% relative humidity (R$_H$) for 1 week.

Results and Discussion

Molecular Weight Dependence. In order to investigate the relationship between the molecular weight of the PDMS matrix and the resulting size of the silica domains generated, we prepared two series of silica/siloxane composite materials *(19)*. Within each series, the molecular weight of the PDMS matrix was varied from 2,000 to 310,000 average molecular weight. The two series of materials differed only in the amount of catalyst used with one series containing TO in a catalyst/TEOS ratio of 1/300 and the other a 1/70 ratio.

Neutron scattering data (Figure 1) shows a distinct shift to smaller q with increasing polymer molecular weight. This shift is consistent with the idea that the elasticity of the network controls the domain size since the molecular weight between crosslinks determines the elastic modulus. For both of the catalyst amounts, the domain size, R, was calculated from least-square fits to a five parameter model for interacting particles (e.g., Figure 2) and plotted versus the elastomer molecular weight (Figure 3). The observed slopes of 0.43 and 0.45 ± 0.10 are consistent with de Gennes' prediction of 0.5 for domain growth limited by network elasticity *(21, 22)*.

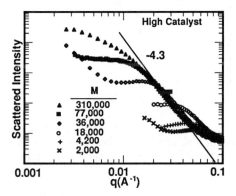

Figure 1. SANS data for high catalyst series (reproduced with permission from ref. 19. Copyright 1992 Materials Research Society.).

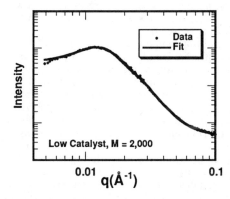

Figure 2. Fit to the SANS data (reproduced from ref. 19. Copyright 1992 Materials Research Society.).

By SANS, the high and low catalyst ratios show identical behavior except for a prefactor shift to larger length scales for the low-catalyst system (Figure 3). If network elasticity limits the domain growth as proposed by de Gennes, this shift indicates that the lower catalyst amount results in a material with a lower effective modulus, consistent with the occurrence of chain extension of the PDMS oligomers during the crosslinking reaction. Lower catalyst ratios will lead to slower hydrolysis of the TEOS; therefore, in the initial stages of the reaction, there will be more silanol groups due to polymer end groups than from TEOS hydrolysis products and chain extension reactions will be statistically more prevalent. Evidence for enhanced chain extension at low catalyst amounts has been previously noted *(23)*.

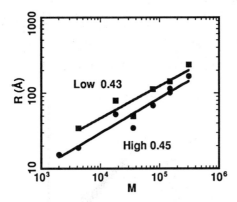

Figure 3. Dependence of the domain size (R) on molecular weight (M) of the elastomer precursor for high and low catalyst systems (slope given) (reproduced from ref. 19. Copyright 1992 Materials Research Society.).

Loading and Catalyst Activity. In order to investigate the effect of loading and catalyst activity, two additional series of silica/siloxane composite materials were synthesized *(20)*. The first series of samples were made using tin octoate (TO), which is the highly reactive tin(II) catalyst used in the molecular weight study. The second series used dibutyltin dilaurate (DBTDL), which is a tin(IV) catalyst of moderate activity. With each catalyst, a series of samples were made at different nominal silica loading levels (5, 10, 20 and 40 mol% in the final product with respect to mols of siloxane Me_2SiO).

Examining the resulting samples by SANS and SEM revealed the growth of silica domains on two different length scales. SANS was used to study the 70-1000 Å, small-scale silica domain growth within these materials, while the micron size range, large-scale structure was investigated using SEM. All of the DBTDL-catalyzed samples showed no obvious silica-containing domains on the micron size scale by SEM (Figure 4), while the highly loaded TO-catalyzed samples (>20 mol%) contain domains a few microns in size (Figure 5). Not only are there micron size particles apparent in the TO-catalyzed samples, but there is also a definite increase in the number of particles per unit area observed when going from 20% to 40% loading. The high q tail associated with scattering from these large-scale domains accounts for the large intensity at small q for the TO system.

In general, SANS indicates that in both the DBTDL and TO-catalyzed samples small-scale domains are produced at higher loading levels (Figures 6 and 7). For the TO system, increased loading leads to greater signal intensity, indicating greater

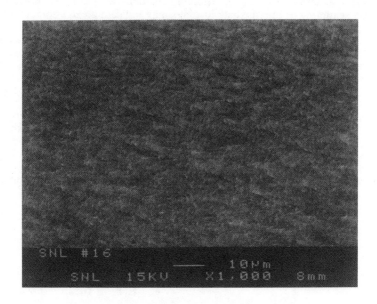

Figure 4. SEM of 20% DBTDL catalyzed sample.

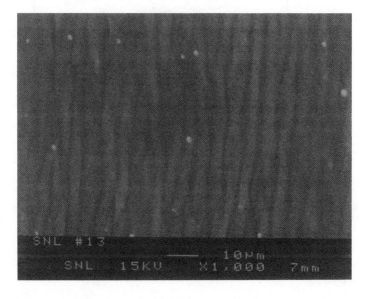

Figure 5. SEM of 20% TO catalyzed sample.

amplitude associated with the composition fluctuations. For DBTDL, however, the enhanced scattering at intermediate loadings may indicate greater compatibility as expected from typical polymer phase diagrams.

Although interpretation of the data are compromised by residual scattering by the micron-scale domains, the DBTDL system does not demonstrate a change in the correlation length with loading, while the TO system does. For TO, the data follows a monotonic trend with the correlation length moving to larger length scales as the loading increases. In addition, analyzing the SANS data indicates that DBTDL catalysis leads to smaller ångstrom range domains than TO catalysis.

Examining the low q SANS data in greater detail indicates the presence of micron-size domains, which also differ for the two catalyst systems. As shown in Figure 8, there are two possible explanations for the difference in the low q region of the SANS data between the two catalytic systems. For DBTDL (Figure 6), the scattering due to the large-scale domains is pushed to very small q, indicating either very few large-scale domains exist or they are large compared to the TO system.

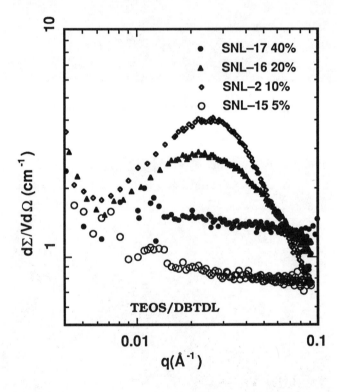

Figure 6. SANS data for 5-40% loaded TEOS/DBTDL samples (Reproduced from ref. 20.).

The strength of the tail from the large-scale domains in the TO system (Figure 7) suggests that either a large number of these domains are present or they are of smaller size than those produced with DBTDL. However, while two explanations exist for the SANS large-scale structure data, the presence of SEM-observable domains only in the TO system indicates that the best explanation for the SANS data is that there are few large-scale domains with DBTDL and many large-scale domains with TO (e.g., the average size of the large scale domains is the same, they only differ in number). Since the development of ceramics by sol-gel chemistry depends on the relative rates of hydrolysis and condensation within the system *(24)*, the catalyst activity should affect the silica domain growth. Hence, our observation that the differences in the catalyst activity lead to changes in the silica growth patterns is reasonable.

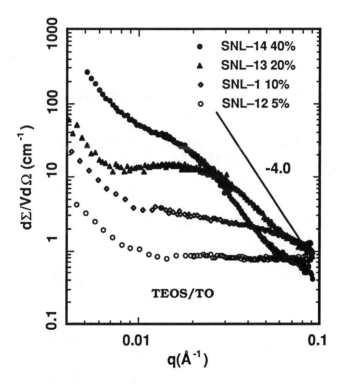

Figure 7. SANS data for 5-40% loaded TEOS/TO samples (Reproduced from ref. 20.).

Examination of the mechanical properties shows that, as expected, the tensile strength (50% and 100% stress) and durometer measurements generally increase with higher loading, while the elongation at break decreases (Table I). These trends are attributed to an averaging of the properties of the two phases although enhanced crosslinking of the base polymer may also contribute. An increase in the crosslink density leads to greater incompatiblity of the two phases and, therefore, to more distinct domains. This enhanced phase separation with loading is supported by the

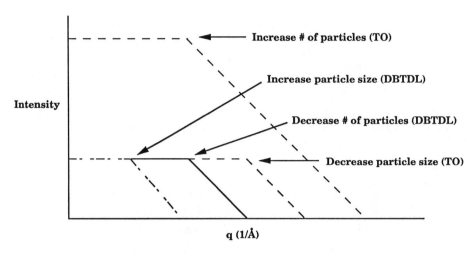

Figure 8. Possible explanations for changes in SANS low q data.

Table I.   Mechanical measurements for TEOS/DBTDL and TEOS/TO samples as a function of loading [a,b]

Sample/Cat.	Loading	Duro.	Elong	Tensile[c]	Stress 50% (100%)
1/TO	5%	34	204	648	53 (165)
2/TO	10%	45	106	529	255 (527)
3/TO	20%	51	----[d]	----	--------
4/TO	40%	43	----	----	--------
5/DBTDL	5%	34	194	647	51 (133)
6/DBTDL	10%	42	174	925	74 (306)
7/DBTDL	20%	47	92	677	259 (574)
8/DBTDL	40%	55	----	----	--------

[a]   Reprinted from ref. 20.
[b]   Durometer (Type A), Elongation (%), Tensile and Stress (psi).
[c]   Poor correlation between loading and tensile strength exists with these samples because of the flaw-dependent nature of this measurement.
[d]   Due to poor mechanical properties of the higher loaded samples, mechanical measurements were not obtained.

Table II. ^{29}Si NMR of Q distribution and extent of reaction for 20% loaded TEOS/DBTDL and TEOS/TO [a]

Sample/Cat.	Q^2	Q^3	Q^4	%Q	Extent of Reaction
3/TO	12.1	44.1	43.8	22.0	0.83
7/DBTDL	11.3	45.3	43.1	20.8	0.83

[a]   Reprinted from ref. 20.

aforementioned increase in intensity of the SANS data and the increased growth of large domains, as indicated by SEM.

In order to determine whether the differences in the phase separation characteristics of the two systems are beacuse of changes in the molecular environments present due to differences in the actual state of hydrolysis and condensation or are caused by changes in the connectivity of the system, the two catalyst systems were examined by ^{29}Si NMR. Using the 20% loaded materials as representative samples, ^{29}Si NMR indicates that while the phase separation is markedly different, the extent of the reaction and the Q distribution are the same (Table II). This similarity indicates that the differences are not due to changes in the molecular environments, but are due to changes in the overall growth or connectivity of the system.

Conclusion

Systematically changing the reaction conditions can control the phase separation in *in situ* silica-filled siloxane materials. For example, there is a direct correlation between the molecular weight of the matrix, M, and the size of the resulting silica domains (varies with $M^{0.4}$). It has also been shown that the catalyst amount and activity affects the phase separation characteristics. Lower amounts of catalyst lead to chain extension reactions, increasing the effective molecular weight of the matrix polymer and concomitantly the silica particle size. Increasing the catalyst activity increases the number of micron-size particles within the system. Finally, all of the observed changes in phase separation appear to be the result of changes in connectivity and not due to changes in the molecular environments present.

Acknowledgments

We thank G. L. Zender for technical assistance in obtaining the SEM data. We thank Ann Norris and Randy Schmidt for assistance in mechanical data interpretation. SANS experiments were conducted with the supervision and assistance of George Wignall, Phil Seeger and Charlie Glinka. This work is sponsored by the U. S. Department of Energy under contract number DE-AC04-94AL8500 and was performed under CRADA CR91/1028.

References

1.   Polmanteer, K. E.; Lentz, C. W. *Rubber Chem. Technol.* **1975**, *48*, 795.
2.   Warrick, E. L.; Pierce, O. R.; Polmanteer, K. E.; Saam, J. C. *Rubber Chem. Technol.* **1979**, *52*, 437.
3.   Boonstra, B. B. *Polymer*, **1979**, *20*, 691.
4.   Iler, R. K., *The Chemistry of Silica*; Wiley, New York 1979.
5.   Wilkes, G. L.; Huang, H.; Glaser, R.H. in Silicon-Based Polymer Science (Advances in Chemistry Series 224); Zeigler, J. M.; Feardon, F. W. Eds; American Chemical Society: Washington, DC 1990; pp 207-226.
6.   Wang, B.; Wilkes, G. L.; Hedrick, J. C.; Liptak, S. C.; McGrath, J. E. *Macromolecules*, **1991**, *24*, 3449.
7.   Schmidt, H. *Mat. Res. Soc. Symp. Proc.*, **1990**, *171*, 3.
8.   Schmidt, H.; Wolter, H. *J. Non-Cryst. Solids*, **1990**, *121*, 428.
9.   Novak, B. M. *Adv. Mater.*, **1993**, *5(6)*, 422.
10. Novak, B. M.; Ellsworth, M. W. *Polym. Prepr. (Am. Chem. Soc., Div. Polym. Chem.)* **1992**, *33(1)*, 1088.

11. Wei, Y.; Bakthavatchalam, R.; Yang, D.; Whitecar, C. K. *Polym. Prepr. (Am. Chem. Soc., Div. Polym. Chem.)* **1991**, *32(3)*, 503.
12. Pope, E. J. A.; Asami, M.; MacKenie, J. D. *J. Mater. Res.*, **1989**, *4(4)*, 1018.
13. Chujo, Y.; Ihara, E.; Kure, S.; Saegusa, T. *Macromolecules*, **1993**, *26*, 5681.
14. Landry, C. J. T.; Coltrain, B. K.; Landry, M. R.; Fitzgerald, J. J.; Long, V. K. *Macromolecules*, **1993**, *26*, 3702.
15. Landry, C. J. T.; Coltrain, B. K.; Wesson, J. A.; Zumbulyadis, N.; Lippert, J. L. *Polymer*, **1992**, *33(7)*, 1496.
16. Mark, J. *Chemtech*, **1989**, *19(4)*, 230.
17. Sun, C. -C.; Mark, J. E. *Polymer*, **1989**, *30*, 104.
18. Schaefer, D. W.; Mark, J. E.; McCarthy, D. W.; Jian, L.; Ning, C. -Y.; Spooner, S. in Ultrastructure Processing of Ceramics, Glasses and Composites; Ulrich, D. R.; Uhlman, D. R. Eds., John Wiley & Sons, New York, 1992.
19. Ulibarri, T. A.; Beaucage, G.; Schaefer, D. W.; Olivier, B. J.; Assink, R. A. *Mat. Res. Soc. Symp. Proc.*, **1992**, *274*, 85.
20. Black, E. P.; Ulibarri, T. A.; Beaucage, G.; Schaefer, D. W.; Assink, R. A.; Bergstrom, D. F.; Giwa-Agbomeirele, P. A.; Burns, G. T., *Proceedings of the American Chemical Society, Division of Polymeric Materials: Science and Eng.*, **1994**, *70*, 382.
21. deGennes, P. G. *J. de Phys.*, **1979**, *40*, 69.
22. Bettachy, A.; Derouiche, A.; Benhamou, M.; Daoud, M. *J. Phys.*, **1991**, *1*, 153.
23. Huang, H.; Glaser, R. H.; Wilkes, G. L. *Polym. Prepr. (Am. Chem. Soc., Div. Polym. Chem.)* **1987**, *28(1)*, 434.
24. Keefer, K. D. *Mat. Res. Soc. Symp. Proc.*, **1984**, *32*, 15.

RECEIVED October 25, 1994

LADDER AND BRIDGED STRUCTURES

Chapter 19

Hypervalent Spiro Polysiliconate and Polygermylate Ionomers

Novel Ladder and Network Materials

James H. Small[1], Kenneth J. Shea[1,3], Douglas A. Loy[2],
and Gregory M. Jamison[2]

[1]Department of Chemistry, University of California, Irvine, CA 92717
[2]Properties of Organic Materials Department, Sandia National
Laboratories, Albuquerque, NM 87185

Novel penta- and hexacoordinate spiro- polysilconate and polygermylate ionomeric materials have been synthesized. These materials are
representatives of a new family of hybrid organic-inorganic ionomeric
materials that contain penta- and hexacovalent silicon and germanium
atoms in the main chain of the polymer. The materials are air stable,
non-hygroscopic, amorphous powders. The incorporation of the
hypervalent silicon and germanium functionality has been verified by
^{29}Si and ^{13}C solid state and solution NMR, ^{1}H solution NMR, IR, and
elemental analysis. The materials are thermally stable to ~325°C
(TGA and DSC) and are of a non-porous nature (0.3-8.5 m^2g^{-1}) (BET,
N_2).

Silica and silicates comprise the bulk of the earth's crust. The natural abundance of
silicon and oxygen, 26% and 50% respectively, represent approximately three-fourths
of the earth's surface by weight (1). Silica and silicates have been utilized as raw
materials by mankind as sources of primitive weapons, building materials, and
ceramics for hundreds of thousands of years. Recent developments in the electronics
industry over the last quarter century have revealed the importance of silicon metal as
a semiconductor. Silicon metal, the primary intermediate from which most modern
applications of silicon are derived, is prepared by the reduction of silica with carbon
in an electric arc furnace at 3000°C. Due to the extensive resources of inexpensive
silica available for the production of silicon metal, a tremendous amount of research
has been conducted on the preparation of both crystalline and amorphous silicate
materials, silicone rubbers, specialty silicon composites, and various other
applications (Figure 1) (2).

Naturally occuring siliconate materials are limited to minerals created under
extreme pressures. One such example is the mineral stishovite, a polymorph of quartz
that contains the hexacoordinate dianionic siliconate functionality (3-4). While the
vast majority of silicon-containing materials are tetracoordinate (i.e. silica, silicate
glasses, and silicone polymers) stable penta- and hexacoordinate siliconate materials
have recently been prepared (5-7). The synthetic materials recently prepared are
members of a new class of novel hybrid organic-inorganic materials and are unique
examples of backbone-containing ionomers. Although a wide variety of ion-

[3]Corresponding author

0097–6156/95/0585–0248$12.00/0
© 1995 American Chemical Society

containing polymers have been prepared to date, relatively few examples of backbone-containing ionomers exist in the literature (*8*). These materials may find applications as solid and solution polymer electrolytes (*9*), ion-exchange resins (*10*), preceramic materials (*11*), or as non-linear optical (NLO) materials (*12*).

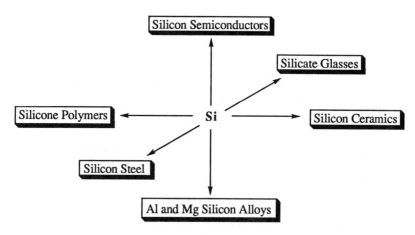

Figure 1. Some common applications of silicon derived materials.

We wish to report further research on the synthesis and characterization of penta- and hexacoordinate anionic polysiliconate ladder and network materials. Specifically, we describe the preparation of unusual spiro- hypervalent polysiliconates. In addition, we also report the synthesis and characterization of spiro- anionic polygermylate network materials. These materials represent some of the first examples of a hypervalent polygermylate material.

Background

Hypervalent siliconates have been known for some time (*13-15*). Penta- and hexacoordinate siliconate compounds exhibit interesting structural characteristics, unique reactivity, a variety of potential synthetic applications, and have been proposed as intermediates and transition states in nucleophilic substitution reactions at silicon (*16*).

Stable hypervalent siliconate compounds contain electron-withdrawing substituents (F^-) and/or multidentate oxo or aza ligands (i.e. aryl 1,2-diols) which aid in the stabilization of the hypervalent silicon by inductive delocalization of the electron density at silicon. Rosenheim first discovered that finely divided silica gel could be depolymerized in anhydrous basic solution to afford the symmetric hexacoordinate tris(catechol) siliconate **1** (Scheme 1) (*17*). Interestingly, recent developments by Corriu and co-workers have shown that the reverse reaction, the polymerization of silica from **1**, done under mild aqueous acidic conditions, yields silica materials with unique morphological characteristics (Scheme 1) (*18*).

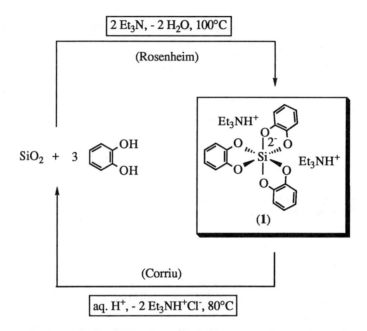

Scheme 1. Depolymerization-polymerization of silica with catechol to form the dianionic hexacoordinate siliconate **1**.

A similar strategy was employed by Frye for the preparation of the phenyl substituted pentacoordinate siliconate **2** by the condensation of phenyltriethoxysilane with catechol in the presence of triethylamine (Scheme 2) (*19*). Additional alkyl- and aryl-substituted pentacoordinate siliconate compounds have been prepared in an analogous fashion by the condensation of the appropriate trialkoxysilane with catechol utilizing a variety of bases (*20-22*).

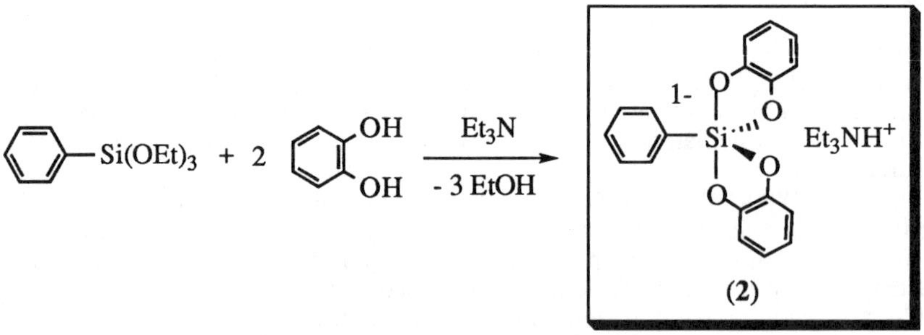

Scheme 2. Condensation of phenyltriethoxysilane with catechol to form an anionic pentacoordinate siliconate **2**.

We have previously reported the synthesis and characterization of bis(triethoxysilyl) arylene, ethynylene, and alkylene derivatives. These compounds were converted by sol-gel methodology to polysilsesquioxanes, a novel family of hybrid organic-inorganic materials (*23-26*). An example is shown below for the synthesis of a 1,4-phenylene-bridged polysilsesquioxane material (Scheme 3). These materials have unique morphological properties with surface areas varying from ~ 1-2 m^2g^{-1} (non-porous) to ~1800 m^2g^{-1} (highly porous) (BET, N_2). In addition, the average pore diameter is primarily confined in the micropore (< 20 Å) to low mesopore region (20-60 Å). Most importantly, we have established that the type of bridging organic component, as well as the reaction conditions, will influence the resulting surface area and porosity of the materials.

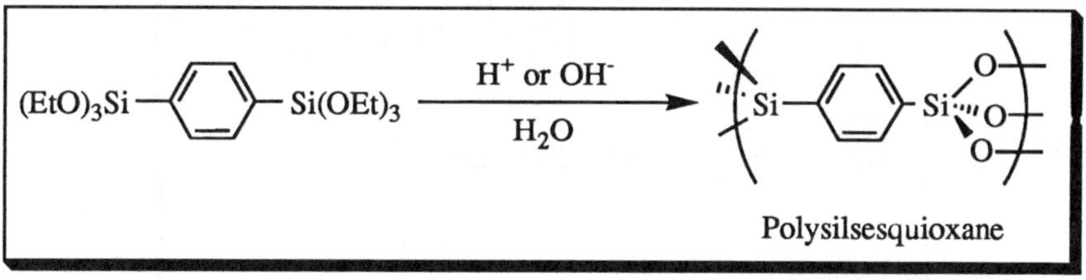

Scheme 3. Synthesis of a 1,4-phenylene-bridged polysilsesquioxane.

Due to the facile synthesis of stable hypervalent siliconates, and the availability of bis-trialkoxysilanes, we extended the methodology of Frye to prepare bridged bis-siliconates (Scheme 4) (*27*). Bis-siliconates, while found in the literature, are relatively few in number (*28-31*). These arylene- and alkylene-bridged bis-siliconates increase the number of bis-siliconates known, but more importantly, serve as model "building blocks" for the synthesis of anionic polysiliconate ladder and network ionomeric materials.

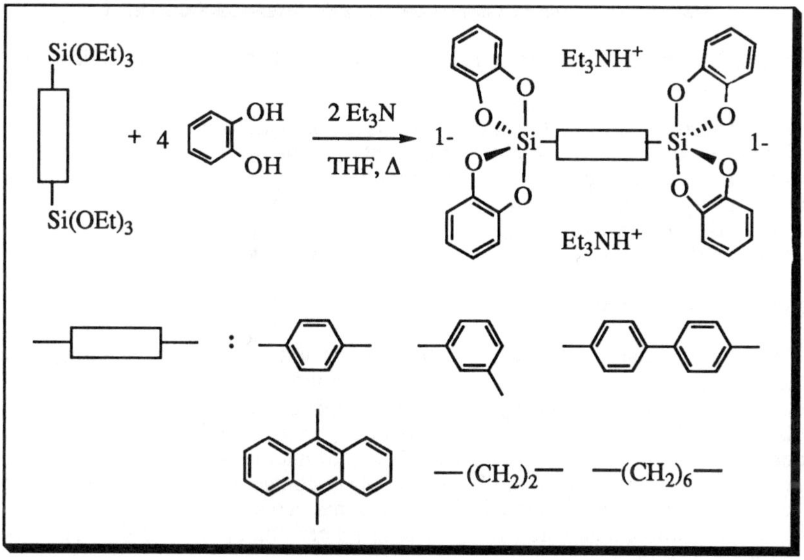

Scheme 4. Synthesis of arylene- and alkylene-bridged bis-siliconates.

The general strategy for the synthesis of these materials is to utilize a tetrafunctional condensing reagent to couple the hypervalent silicon or germanium functionalities. This is accomplished by linking the penta- or hexacoordinate functionalities by a bis(aryl 1,2-diol) ligand as shown for a pentacoordinate ladder anionic polysiliconate (Figure 2).

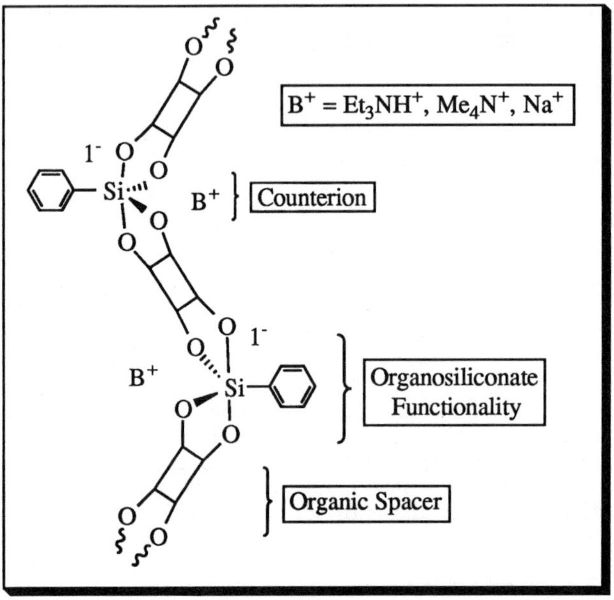

Figure 2. Strategy for the preparation of an anionic hypervalent polysiliconate.

Our approach allows for the synthesis of a wide variety of materials by manipulation of the three main components of the ionomeric framework: the organosiliconate functionality, the tetrafunctional organic spacer, or the cationic counterion. Systematic variation of these components will generate a diverse family of materials. The alteration of these components and their effect on the bulk properties of the hypervalent materials will be discussed in this report.

Results and Discussion

We have previously reported the first synthesis and characterization of penta- and hexacoordinate anionic polysiliconate ionomeric materials (5-7). The synthesis of these materials was achieved by the condensation of 1,2,4,5-tetrahydroxybenzene (THB) (3) with the corresponding alkoxysilane to yield either a pentacoordinate ladder, a pentacoordinate network, or a hexacoordinate network ionomeric material (Scheme 5). The first generation of anionic polysiliconate materials, which contained the tetrahydroxybenzene functionality, were low molecular weight ionomers, prone to end-group oxidation, and were slowly hydrolyzed resulting in loss of the hypervalent silicon functionality.

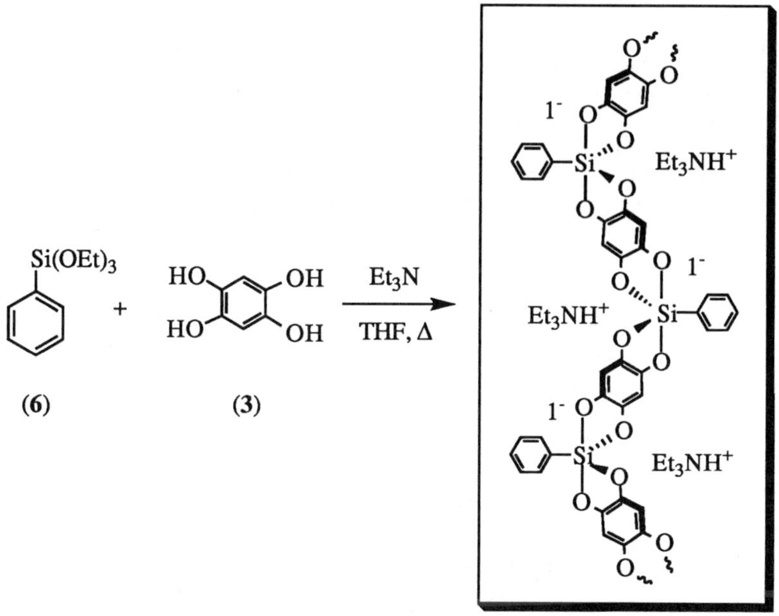

Scheme 5. Synthesis of a pentacoordinate ladder polysiliconate.

A more robust condensing reagent, (±)-5,5',6,6'-tetrahydroxy-1,1'-spiro-bis(indane) (spirocatechol=SCAT) (**4**), was chosen for the second generation polysiliconate materials (Scheme 6) (*32-34*). This compound was expected to enhance the formation of higher molecular weight materials due to an increase in solubility in organic solvents. Furthermore, use of SCAT should alleviate the problem of end-group oxidation due to the separation of the catechol moieties with saturated carbon centers.

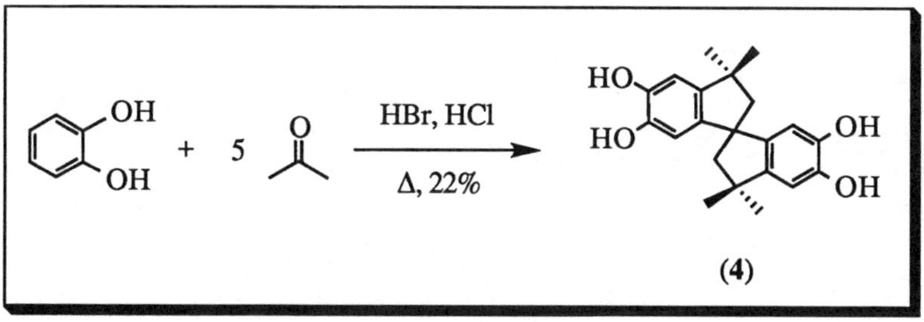

Scheme 6. Synthesis of (±)-5,5',6,6'-tetrahydroxy-1,1'-spiro-bis(indane) (SCAT) (**4**).

Monomer Synthesis: To broaden the scope of this chemistry we have examined a variety of alkoxysilane monomers for the preparation of penta- and hexacoordinate anionic polysiliconate materials (Scheme 7). Tetraethoxysilane (TEOS) (**5**), phenyltriethoxysilane (PTES) (**6**), 1,6-bis(trimethoxysilyl) hexane (BMSH) (**7**), and

1,2-bis(trimethoxysilyl) ethane (BMSE) (**8**) were purchased from commercial suppliers. 1,4-Bis(triethoxysilyl) phenylene (BESP) (**9**) and 4,4'-bis(triethoxysilyl) biphenylene (BESB) (**10**) were prepared by Barbier-Grignard methodology from the corresponding arylene-dibromides (*23-25*). 1,2-Bis(triethoxygermyl) ethane (BEGE) (**11**), 1,6-bis(triethoxygermyl) hexane (BEGH) (**12**), and 1,10-bis(triethoxygermyl) decane (BEGD) (**13**) were prepared by the hydrogermylation of the appropriate unsaturated organic compounds (*35*). The monomers were purified by distillation and verified by ^1H, ^{13}C, and ^{29}Si nuclear magnetic resonance, infra-red spectroscopy, mass spectrometry, and gas chromatography.

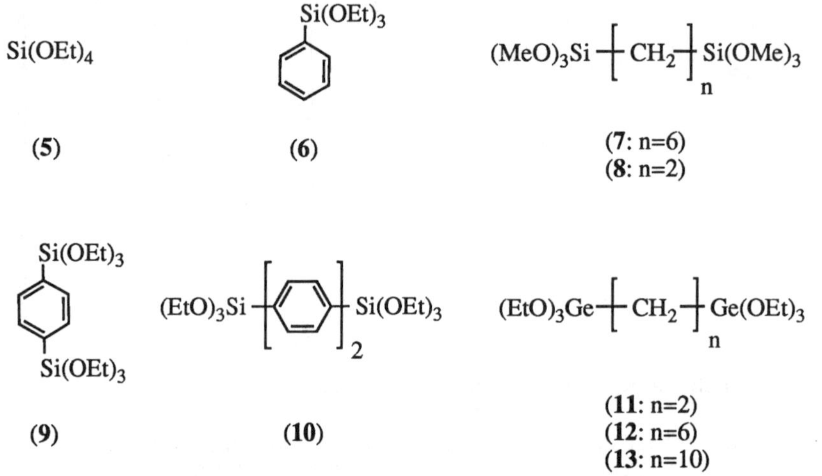

$Si(OEt)_4$

(**5**)

$Si(OEt)_3$ (phenyl ring)

(**6**)

$(MeO)_3Si\!-\!\!\left[CH_2\right]_n\!\!-\!Si(OMe)_3$

(**7**: n=6)
(**8**: n=2)

$Si(OEt)_3$ (phenyl ring) $Si(OEt)_3$

(**9**)

$(EtO)_3Si\!-\!\!\left[\!\!\left(\text{phenyl}\right)\!\!\right]_2\!\!-\!Si(OEt)_3$

(**10**)

$(EtO)_3Ge\!-\!\!\left[CH_2\right]_n\!\!-\!Ge(OEt)_3$

(**11**: n=2)
(**12**: n=6)
(**13**: n=10)

Scheme 7. Alkoxysilyl (**5-10**) and germyl (**11-13**) monomers.

Polysiliconate Synthesis: A typical synthesis and characterization of an anionic polysiliconate material is given for the synthesis of the 1,4-phenylene-bridged pentacoordinate anionic polysiliconate with triethylamine as base (BESP-SCAT-Et_3NH^+, **X-9**) (Figure 3) To a refluxing solution of BESP (0.90 g, 2.22 mmol) and SCAT (1.50 g, 4.41 mmol) in THF (9.0 mL) was added Et_3N (2.00 mL, 14.4 mmol). After 16 h a white solid had formed. The solution was refluxed for an additional 4 h and then cooled slowly to room temperature. The solids were filtered, washed with THF (3 X 25 mL), and dried under dynamic vacuum (< 1mTorr, 6 h) at room temperature to afford a tan powder (1.84 g, 83%). ^1H NMR (300 MHz, DMSO-d_6) δ 8.7-8.5 (br d, $(CH_3CH_2)_3NH^+$), 7.7-7.1 (br m, Ph-*H*), 6.6-6.2 (br s, SCAT), 6.0-5.7 (br s, SCAT), 3.8-3.7 (br s, OC*H*$_2$CH$_3$), 3.7-3.5 (br s, THF), 3.0-2.8 (br s, $^+$HN(C*H*$_2$CH$_3$)$_3$), 2.3-1.8 (br m, CH$_2$), 1.8-1.65 (br s, THF), 1.5-1.1 (br m, C*H*$_3$), 1.1-0.7 (br m, $^+$HN(CH$_2$C*H*$_3$); ^{13}C Solid State NMR (50.29 MHz, CP MAS) δ 149.0 (SCAT), 141.9 (SCAT), 133.5 (Ph), 104.9 (SCAT), 68.0 (THF), 60.4 (*C*H$_2$), 57.8 (O*C*H$_2$CH$_3$), 47.2 (N*C*H$_2$CH$_3$), 43.6 (Q), 31.6 (*C*H$_3$), 25.8 (THF), 18.5 (OCH$_2$*C*H$_3$), 8.8 (NCH$_2$*C*H$_3$); ^{29}Si Solid State NMR (39.73 MHz, CP MAS) δ -57.6, -87.1; ^{29}Si Solid State NMR (39.73 MHz, Bloch Decay) δ -87.0; FTIR (KBr) υ 3446, 3037, 2951, 2860, 1637, 1485, 1360, 1279, 1220, 1141, 878, 695, 600 cm^{-1}; Elemental analysis calcd. for $C_{60}H_{76}N_2O_8Si_2$: C 71.39, H 7.59, Si 5.56, found: C 69.77, H

7.49, Si 5.62. This material, which is typical of the spirocatechol polysiliconates, is an air stable powder and only slightly soluble in polar solvents (DMSO, DMF, MeOH). The alkylene-bridged polysiliconates, **X-7** and **X-8**, and the ladder material, **X-6**, are soluble in polar organic solvents.

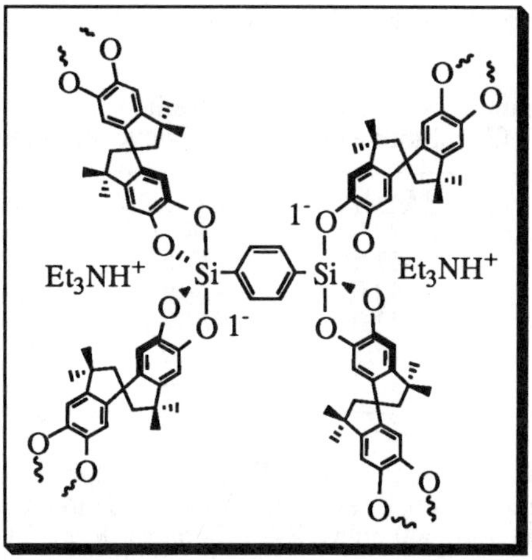

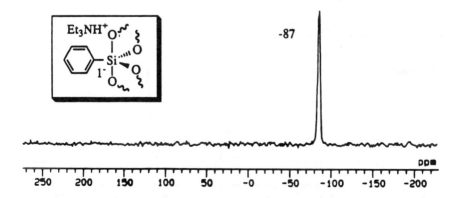

Figure 3. 1,4-Phenylene-bridged polysiliconate BESP-SCAT-Et₃NH⁺ (**X-9**).

Polysiliconate Characterization: ^{29}Si NMR has been utilized as the primary diagnostic in the evaluation of the valence at silicon (*14*). The presence of the hypervalent siliconate functionality within the materials was verified by ^{29}Si solid state cross-polarization magic angle spinning (CP MAS) NMR. Figure 4 shows the ^{29}Si solid state CP MAS spectra of the phenyl-substituted ladder polysiliconate (PTES-SCAT-Et₃NH⁺) (**X-6**).

Figure 4. ^{29}Si Solid stateCP MAS NMR spectra of PTES-SCAT-Et₃NH⁺ (**X-6**).

The resonance centered at -87 ppm is characteristic of a mono aryl-substituted pentacoordinate siliconate and corresponds well with the known "model" catecholate value (16). The lack of an absorption at -57 ppm, due to residual tetracoordinate silicon, verifies the presence of only the pentacoordinate siliconate functionality. The ^{29}Si NMR chemical shifts of the monomers 5-10, their respective catecholates (27), and their respective polysiliconates (X-5 to X-10) are given in Table 1.

Table 1. ^{29}Si NMR data of catecholates and polysiliconates 5-10. Values are in ppm relative to TMS (solution) and hexamethylcyclotrisiloxane (solid).

Compound	Monomer	Catecholate	Polysiliconate
5	-99.5	-140.0	-99, -140
6	-57.5	-87.5	-86, -87
7	-42.0	-71.9	-49, -59, -74
8	-42.0	-73.7	-54, -72, -98
9	-57.7	-87.0	-58, -87
10	-58.0	-87.4	-58, -87

The most intense resonance in all cases is the hypervalent siliconate functionality. An extremely small resonance, attributable to tetracoordinate silicon, is seen for X-5 (-99 ppm), X-7 (-49 ppm), X-8 (-54 ppm), X-9 (-58 ppm), and X-10 (-58 ppm). Interestingly, the 1,2-ethylene-bridged polysiliconate (X-8) shows a slight fragmentation of the organosilicon bond in the hydrocarbon bridging component of the material (-98 ppm, (RO)$_4$Si). ^{29}Si solution NMR spectra, also acquired for X-6, X-7, and X-8, were identical to spectra obtained from solid state NMR experiments showing their stability in polar organic solvents (DMSO).

The materials were analyzed by ^{13}C solid state CP MAS NMR to verify the incorporation of SCAT , the stability of the organic bridging spacer, and the degree of condensation. An example is shown in Figure 5 for the 1,4-phenylene-bridged polysiliconate BESP-SCAT-Et$_3$NH$^+$ (X-9).

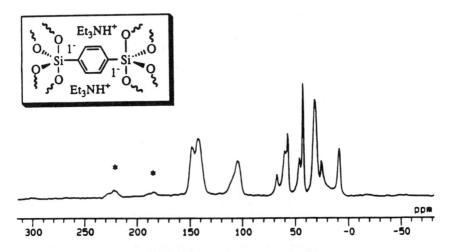

Figure 5. ^{13}C solid state CP MAS NMR of BESP-SCAT-Et$_3$NH$^+$ (X-9). Asterisks (*) denote spinning side bands.

Analysis of the ^{13}C solid state NMR data indicates that both the bridging hydrocarbon moiety and the SCAT component have been successfully introduced into the polysiliconate material (Table 2). It was not possible to resolve the alkylene-bridged carbon resonances in **X-7** and **X-8** due to coincidental overlap.

Table 2. ^{13}C Solid state CP MAS NMR data of polysiliconates **X-5** to **X-10**. Values are in ppm relative to ^{13}C solid state NMR standard hexamethylbenzene

Material	Aromatic	Quarternary	CH$_2$	CH$_3$	Bridge	$^+$HNCH$_2$CH$_3$
SCAT	144.6, 142.1, 140.6, 110.0 108.3	55.3, 42.4	59.6	31.6, 30.5	-------	-------------
X-5	150.6, 144.4, 104.7	58.0, 43.5	60.7	32.2	-------	46.9, 9.1
X-6	151.2, 142.4, 105.0	43.6	59.2	34.4	126.9	47.8, 8.3
X-7	148.6, 142.7, 105.1	57.8, 43.3	60.4	31.9	****	46.6, 9.0
X-8	149.0, 143.9,	58.2, 43.8	60.9	32.2	****	47.5, 9.4
X-9	149.0, 141.9, 104.7	43.6	60.4	31.6	133.5	47.2, 8.8
X-10	149.0, 142.4,	57.6, 43.0	60.1	32.2	125.0	46.6, 7.9

**** Unable to accurately resolve alkylene-bridging carbons.

Analysis of the ^1H solution NMR spectra, FT-IR, and elemental analysis data support the proposed structures for polysiliconates **X-5** to **X-10**. Examination of the ^1H NMR data verify the incorporation of both the SCAT component and the bridging organic moiety in the material (Table 3). The characteristic resonances attributable to the triethylammonium counterion are clearly observed at 8.75, 2.95, and 1.04 ppm and correlate with known catecholate values (27). The presence of residual hydroxy resonances at ~8.5 ppm as end-groups in **X-5**, **X-7**, **X-8**, and **X-10** imply incomplete condensation of each SCAT molecule.

Table 3. ^1H NMR data of polysiliconates X-5 to X-10

Material	Aromatic	CH$_2$	CH$_3$	Bridge	$^+$HNCH$_2$CH$_3$
SCAT	8.52, 6.49, 6.06	2.16, 1.99	1.25, 1.19	-------	---------------
X-5	8.4-8.1, 6.5-6.9	2.2-1.9	1.3-1.1	-------	8.7-8.4, 3.0-2.7, 1.05
X-6	6.36, 6.34, 5.84, 5.82	2.12, 2.03, 1.95, 1.90	1.24, 2.03,	7.47, 7.09	8.75, 2.95, 1.04
X-7	8.4-8.3, 6.5-6.4, 6.4-6.2, 6.2-6.0, 5.9-5.7	2.2-1.9	1.3-1.1	0.85, 0.40	8.7-8.4, 3.0-2.8, 1.07
X-8	8.5-8.4, 6.5-6.4, 6.2-6.1, 6.0, 5.7	2.1-1.9	1.3-1.0	0.3-0.5	8.7-8.4, 2.95, 1.10
X-9	6.6-6.2, 6.0-5.7,	2.3-1.8	1.5-1.1	7.7-7.1	8.7-8.5, 3.0-2.8, 1.1-0.7
X-10	8.4-8.3, 6.5-6.4, 6.4-6.2, 6.0-5.8	2.2-1.9	1.4-1.1	7.7-7.2	8.7-8.4, 3.0-2.7, 1.1-0.7

Due to the network formation of the bridging pentacoordinate polysiliconates (**X-7** to **X-10**) and the hexacoordinate polysiliconate (**X-5**), it was thought that these ionomeric materials may possess significant surface areas. The materials were analyzed by N_2 adsorption porosimetry at 77K and the surface areas calculated by the BET method (*36*). The observed surface areas were between 0.3-8.5 m^2g^{-1} indicative of non-porous materials.

Thermal analysis, differential scanning calorimetry (DSC) and thermogravimetric analysis (TGA), indicates an onset of decomposition ranging from 265-338°C (DSC) corresponding to a 62-76% weight loss by 900°C (TGA) (Figure 6). The TGA of SCAT reveals an onset of decomposition at 325°C and complete weight loss by 600°C. A correlation between the onset of decomposition for SCAT and the polysiliconates seems likely. This infers that the hypervalent siliconate functionality is thermally stable to temperatures greater than ~350°C.

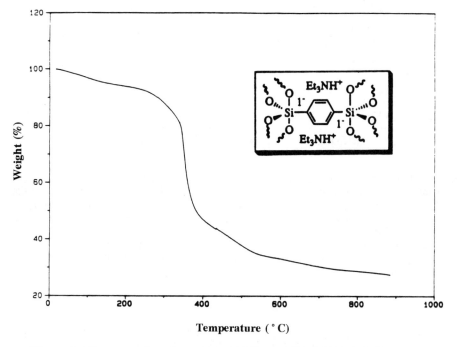

Figure 6. Thermogravimetric analysis (TGA) of BESP-SCAT-Et$_3$NH$^+$ (**X-9**).

Alternative Counterion Polysiliconates

To investigate the effect of alteration of the counterion on the bulk properties of the materials, we have synthesized the tetramethylammonium (Me$_4$N$^+$) and sodium (Na$^+$) counterion derivatives. Neither counterion contains an acidic proton, in contrast to the triethylammonium counterion with a pK_a=~10-11, and should further aid in the stability of the hypervalent siliconate functionality.

The materials were prepared by the condensation of the appropriate alkoxysilane **5-10** with either (25 wt% Me$_4$NOH in MeOH) or (NaOMe in MeOH) to afford the corresponding sodium or tetramethylammonium polysiliconate. The tetramethylammonium materials [**X-5 to X-10(Me$_4$N$^+$)**] were completely insoluble [except **X-6(Me$_4$N$^+$)**], air stable, and non-hygroscopic powders. In contrast, the sodium materials prepared thus far [**X-5, X-6, and X-9(Na$^+$)**] were slightly soluble and somewhat moisture sensitive powders. The sodium materials slowly hydrolyzed upon exposure to moisture with subsequent loss of the hypervalent siliconate functionality.

The ^{29}Si solid state NMR spectra of these materials verifies the presence of the hypervalent siliconate functionality. An example of the ^{29}Si solid state NMR spectra for PTES-SCAT-Na$^+$ [**X-6(Na$^+$)**] and PTES-SCAT-Me$_4$N$^+$ [**X-6(Me$_4$N$^+$)**] is given in Figure 7. Both materials show a single resonance at ~ -87 ppm corresponding to the pentacoordinate siliconate functionality. As with the triethylammonium material [**X-6(Et$_3$NH$^+$)**], the lack of an absorbance at -57 ppm indicates complete formation of the pentacoordinate siliconate functionality. In addition, ^{13}C solid state NMR spectroscopy, IR spectroscopy, and elemental analysis experiments support the structures proposed. A more detailed evaluation of these alternative counterion materials will be published elsewhere.

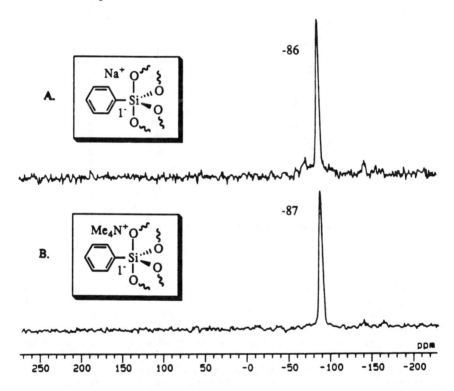

Figure 7. ^{29}Si NMR spectra of A). PTES-SCAT-Na$^+$, and B). PTES-SCAT-Me$_4$N$^+$.

Polygermylate Synthesis: Hypervalent germanium materials could also be prepared using the preceeding methods. For example, a typical synthesis and characterization of an anionic polygermylate material is given for the 1,6-hexylene-bridged polygermylate BEGH-SCAT-Et$_3$NH$^+$ (**X-12**) (Figure 8). The polygermylate materials **X-11** to **X-13** were formed immediately upon addition of base. This is believed to occur due to the enhanced reactivity of the germanium monomers to nucleophilic substitution (*35*). ^{1}H NMR (500 MHz, DMSO-d$_6$) δ 8.85-8.75 (br s, (CH$_3$CH$_2$)$_3$N*H*+), 6.3-6.1 (br s, SCAT), 5.9-5.7 (br s, SCAT), 2.95 (q, (CH$_3$C*H*$_2$)$_3$NH+), 2.2-1.9 (br s, C*H*$_2$), 1.35-1.15 (br m, C*H*$_3$ and GeCH$_2$C*H*$_2$CH$_2$-), 1.10-0.95 (br t, (C*H*$_3$CH$_2$)$_3$NH+ and GeC*H*$_2$CH$_2$CH$_2$-); ^{13}C Solid State NMR (50.29 MHz, CP MAS) δ 149.5 (SCAT), 140.9 (SCAT), 105.9 (SCAT), 68.5 (THF), 60.8 (*C*H$_2$), 57.9 (O*C*H$_2$CH$_3$), 47.4 (N*C*H$_2$CH$_3$), 43.5 (Q), 32.1 (*C*H$_3$ and GeCH$_2$CH$_2$CH$_2$-), 28-20 (THF, OCH$_2$CH$_3$, and Ge*C*H$_2$*C*H$_2$*C*H$_2$-), 9-6 (NCH$_2$*C*H$_3$); FTIR (KBr) υ 3413, 2951, 2858, 1481, 1346, 1277, 1219, 860, 822, 694, 613, 513 cm^{-1}; Elemental analysis calcd. for C$_{60}$H$_{84}$N$_2$O$_8$Ge$_2$: C 66.59, H 5.59, N 2.59, Ge 13.41, found: C 62.51, H 6.40, N 2.23, Ge 12.59.

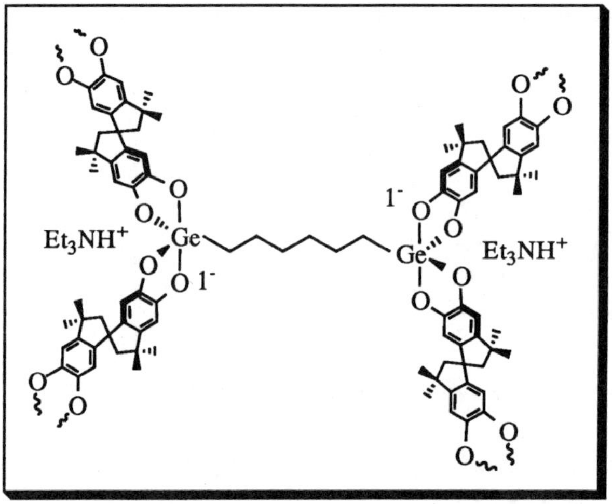

Figure 8. 1,6-Hexylene-bridged polygermylate BEGH-SCAT-Et$_3$NH+ (**X-12**).

Polygermylate Characterization: ^{73}Ge solution NMR has been shown to be extremely difficult to obtain due to low sensitivity, low natural abundance, and the inherent quadrupolar nature of the ^{73}Ge nucleus (*37-38*). ^{73}Ge solution NMR was attempted on monomers **11-13** but proved unsuccessful. Therefore, verification of the hypervalent germanium functionality was established using ^{1}H solution NMR chemical shift values. We have previously observed an upfied shift of ~0.3 ppm of the aromatic catechol protons in the "model" catecholate siliconates (*27*). ^{29}Si NMR spectroscopy has verified the presence of the hypervalent siliconate functionality in the model catecholates as well as in the polysiliconates **X-5** to **X-10**. A similar trend was seen for the germanium catecholates. This permitted indirect verification of the presence of the hypervalent germanium functionality. The polygermylate materials **X-11** to **X-13** also exhibit the characteristic upfield shift in the aromatic spirocatcatecholate protons supporting the presence of the pentacoordinate germylate functionality.

^{13}C Solid state CP MAS NMR was conducted on **X-11, X-12**, and **X-13** and verified that the SCAT and hydrocarbon-bridging components have been incorporated into the polygermylate materials. An example of a ^{13}C solid state CP MAS NMR spectra of a polygermylate material is given for the 1,6-hexylene-bridged polygermylate material BEGH-SCAT-Et$_3$NH$^+$ (**X-12**) (Figure 9). Resolution of the aliphatic-bridging carbon resonances was extremely difficult due to coincidental overlap with SCAT aliphatic resonances. The values reported are best estimates based upon correlation with model catechol germylate values. In addition, the materials were also analyzed by infra-red spectroscopy and elemental analysis. The resulting data correspond well with expected values. A more detailed analysis of the germylate model compounds and polygermylate materials will be published elsewhere.

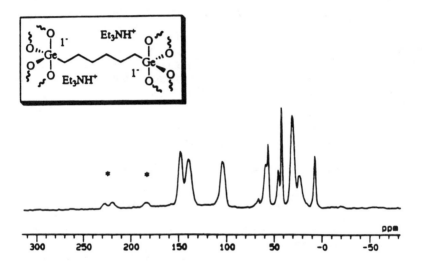

Figure 9. ^{13}C Solid state CP MAS NMR spectra of BEGH-SCAT-Et$_3$NH$^+$ (**X-12**).

Conclusion

The preparation of novel hybrid organic-inorganic materials with the ionic functionality contained within the backbone of the polymer has been achieved. We have synthesized and characterized a new family of spiro- anionic penta- and hexacoordinate polysiliconate materials incorporating both organic and alkali metal counterions. We have also prepared the *first* members of a new family of unique spiro- anionic pentacoordinate polygermylate materials. These materials were thermally stable to moderate temperatures and are predominantly air and moisture stable. Analysis of the materials by solution and solid state NMR, IR, and elemental analysis verified the presence of the hypervalent silicon and germanium functionalities within the material. Further examination of the effects of alteration of the three components of the hypervalent materials is presently underway in our lab.

Acknowledgments

We would like to thank Dr. Jiejun Wu for his helpful discussions on the analysis of the solid state and solution NMR data. We would also like to recognize the Division of Materials Research of the National Science Foundation and the Air Force Office of Scientific Research for financial support. The research conducted at Sandia National Laboratories was supported by the U.S. Dept. of Energy, Contract Number DE-AC04-94AL85000.

Literature Cited

1. Iler, R.K. *The Chemistry of Silica*; Wiley: New York, NY, 1979.
2. Rochow, E.G. *Silicon and Silicones*; Springer-Verlag: Berlin, Germany; 1987; pp 11.
3. Thomas, J. M.; Gonzales, J.M.; Fyfe, C. A.; Gobbi, G.C.; Nicol, M. *Geophys. Lett.* **1983**, *10*, 91.
4. Edge, R. A.; Taylor, H.F.W. *Nature* 1969, *224*, 363.
5. Shea, K.J.; Loy, D.A.; Small, J.H. *Chem. Mater.* **1992**, *4*, 255.
6. Shea, K.J.; Loy, D.A.; Small, J.H. In *Better Ceramics Through Chemistry V*; Klemperer, W.G.; Brinker, C.J.; Hampden-Smith, M., Eds.; Mater. Res. Soc. Proc. 271; Materials Research Society: Pittsburgh, PA, 1992, pp 711-718.
7. Small, J.H.; Shea, K.J.; Loy, D.A.; Jamison, G.M. *Polym. Prepr., Am. Chem. Soc. Div. Polym. Mater: Sci. Eng.* **1994**, *70*, 378.
8. Holliday, L. *Ionic Polymers;* Applied Science Publishers, LTD: London, 1975; Chap. 1-2.
9. Gray, F.M. *Solid Polymer Electrolytes;* VCH Publishers, Inc: New York, NY, 1991.
10. Small, H. *Ion Chromatography;* Plenum Press: New York, NY, 1989, pp 41.
11. Schwartz, M.M. *Handbook of Structural Ceramics;* McGraw Hill, Inc: New York, NY, 1992.
12. Prasad, P.N. In *Contemporary Nonlinear Optics*; Agrawal, G.P.; Boyd, R.W., Eds.; Academic Press: San Diego, CA, 1992, pp 265-294.
13. Tandura, S.N.; Voronkov, M.G.; Alekseev, N.V. *Top. Stereochem.* **1986**, *131*, 99.
14. Holmes, R.R. *Chem. Rev.* **1990**, *90*, 17.
15. Corriu, R.J.P.; Young, J.C. In *The Chemistry of Organic Silicon Compounds*; Patai, S., Rappoport, Z., Eds.; Wiley: Chichester, U.K.; 1989, Part 2, Chap. 20.
16. Chuit, C.; Corriu, R.J.P.; Reye, C.; Young, J.C. *Chem Rev.* **1993**, *93*, 1371-1448, and citations therein.
17. Rosenheim, A.; Baibmann, B.; Schendel, G.Z. *Anorg. All. Chem.* **1931**, *196*, 160.
18. Cerveau, G.; Corriu, R.J.P. *C. R. Acad. Sci., Ser. II*, **1993**, *316*, 35.
19. Frye, C.L. *J. Am. Chem. Soc.* **1964**, *86*, 3170.
20. Corriu, R.J.P.; Kpoton, A.; Poirer, M.; Royo, G. *J. Organomet. Chem.* 1984, 277.
21. Stevenson, W.H.; Wilson, S.; Martin, J.C.; Farnham, W.B. *J. Am. Chem. Soc.* **1985**, *107*, 6340.
22. Corriu, R.J.P. *Phosphorous and Sulfur* **1986**, 27, 1.
23. Shea, K.J.; Loy, D.A.; Webster, O.W. *Chem. Mater.* **1989**, *1*, 572.
24. Shea, K.J.; Loy, D.A.; Webster, O.W. *J. Am. Chem. Soc.* **1992**, *114*, 6700.
25. Small, J.H.; Shea, K.J.; Loy, D.A. *J. Non-Cryst. Sol.* **1993**, *160*, 234.

26. Oviatt, H.W.; Shea, K.J.; Small, J.H. *Chem. Mater.* **1993**, *5*, 943.
27. Loy, D.A.; Small, J.H.; Shea, K.J. *Organometallics* **1993**, *12*, 1484.
28. Johnson, S.E.; Day, R.O.; Holmes, R.R. *Inorg. Chem.* **1989**, *28*, 3182.
29. Cerveau, G.; Chuit, C.; Colomer, E.; Corriu, R.J.P.; Reye, C. *Organometallics* **1990**, *9*, 2415.
30. Breliere, C.; Corriu, R.J.P.; Royo, G.; Chi Man, M.W.; Zwecker, J. *C. R. Acad. Sci., Paris Ser. II*, **1991**, 1527.
31. Tamao, K.; Hayashi, T.; Ito, Y. *Organometallics* **1992**, *11*, 2099.
32. Goto, H. *Chem. Abstr.* **1992**, *114*, 163609s.
33. Munoz, A.; Wolf, P.; Despax, B.; Brossas, J.; Clouet, G. *Eur. Poly. Jour.* **1979**, *15*, 631.
34. Backer, W. *J. Chem. Soc.* **1934**, 1678.
35. Jamison, G.M.; Loy, D.A.; Shea, K.J. *Chem. Mater.* **1993**, *5*, 1193.
36. Gregg, S.J.; Sing, K.S.W. *Adsorption, Surface Area, and Porosity*, 2nd Ed.; Academic Press: London, U.K.; 1982; pp 25.
37. Liepins, E.; Zicmane, I.; Lukevics, E. *J. Organomet. Chem.* **1988**, *341*, 315.
38. Wilkins, A.L.; Watkinson, P.J.; MacKay, K.M.; *J. Chem. Soc., Dalt. Trans.* **1987**, 2365.

RECEIVED September 15, 1994

Chapter 20

Hexylene- and Phenylene-Bridged Polysiloxane Network Materials

Douglas A. Loy[1], Gregory M. Jamison[1], Roger A. Assink[1],
Sharon Myers[1], and Kenneth J. Shea[2]

[1]Properties of Organic Materials Department, Sandia National
Laboratories, Albuquerque, NM 87185–0367
[2]Department of Chemistry, University of California, Irvine, CA 92717

Hexylene- and phenylene-bridged polymethylsiloxane xerogels and
aerogels were prepared by sol-gel polymerizations. The materials
were prepared by the hydrolysis and condensation of 1, 6-
bis(diethoxymethylsilyl)hexane **1** and 1, 4-bis(diethoxymethylsilyl)benzene **3** under acidic and basic conditions. The
polymerizations afforded network polymers in the form of gels within
several hours. The gels were dried to afford both xerogels and
aerogels. The materials were characterized by solid state ^{29}Si and ^{13}C
CP MAS NMR spectroscopy, nitrogen sorption porosimetry, thermal
gravimetric analysis, and scanning electron microscopy.

Polymers built by making siloxane (Si-O-Si) bonds are an important and diverse
body of materials ranging from inorganic silicates and silica gels on one extreme to
the linear polysiloxanes on the other (1). In between lies a spectrum of siloxane-
based hybrid organic-inorganic materials with an incredible variety of organic
substituents attached to the silicon atoms. Depending on the potential number of
siloxane linkages to each silicon atom, the resulting material can be a dimer (I), a
linear siloxane (II), or crosslinked siloxanes (III and IV) when fully condensed.
Silica (IV) is a siloxane polymer with a maximum of four siloxane bonds to other
monomers in the network that is formed by the polymerization of tetraethoxysilane.

A consequence of this highly crosslinked structure is that during the sol-gel
process (2), the silica polymer molecules become large enough for gelation to occur.
Reaction conditions (type and concentration of monomer and catalyst, amount of
water, temperature, etc.) affect the structure of the resulting material which in turn
greatly impacts the properties of the final material.

$$[RSiO_{1.5}]_n$$

I   II   III   IV

0097–6156/95/0585–0264$12.00/0

alkoxide substituents in the monomer will alter the kinetics of polymerization and the final structure in the resulting polymer. An alternative way is to organically modify the monomer with a substituent directly attached to the silicon atom through a silicon-carbon bond. For example, if one of the ethoxy groups is replaced with a single hydrocarbon substituent (R = aryl, alkyl, alkenyl, or alkynyl group), a monomer capable of forming a silsesquioxane polymer (III) is obtained. Silsesquioxanes are a class of organic-inorganic hybrids (3, 4) with three siloxane linkages interconnecting the silicon atoms in the network. This is theoretically sufficient for constructing a crosslinked polymer. However, the R group also restricts the siloxane polymer's growth in one of the four possible directions; when R is sterically bulky, cyclization to cubic oligosilsesquioxanes is preferred (Scheme 1). Sol-gel polymerization often leads to soluble oils thought to be composed of lightly interconnected oligosilsesquioxanes cubes. Relatively harsh conditions are required to generate gels.

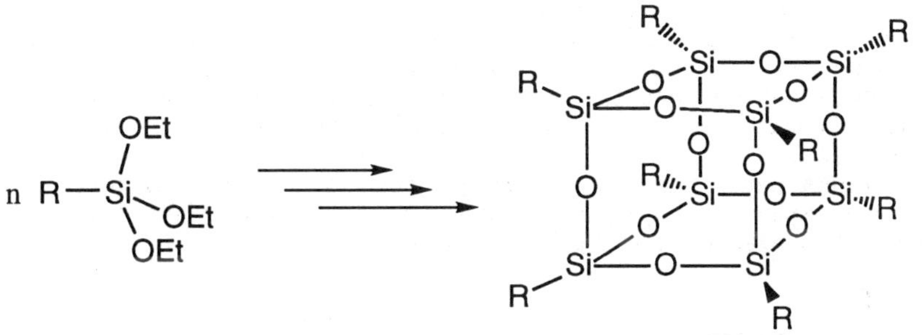

Scheme 1. Formation of Cubic Oligosilsesquioxanes.

Bridged Polysilsesquioxanes. The impediment to forming network polymers based on silsesquioxanes was removed when we placed trichlorosilyl or triethoxysilyl groups about a bridging hydrocarbon group (Scheme 2). With this bridged silsesquioxane building block (5-10) there is a maximum of six siloxane bonds possible; positioning silsesquioxane groups on either side of the bridging group facilitates the formation of *network* structure rather than simple oligomeric cubes. With this strategy, bridged siloxane network polymers can be easily prepared as *gels* by hydrolyzing and condensing the monomers at concentrations as low as 0.01 M. The bridged silsesquioxane motif allows a large variety of bridging groups to be placed into the network polymers. In fact, we have been able to demostrate the versatility of this method for engineering the physical and chemical properties of materials at a molecular level.

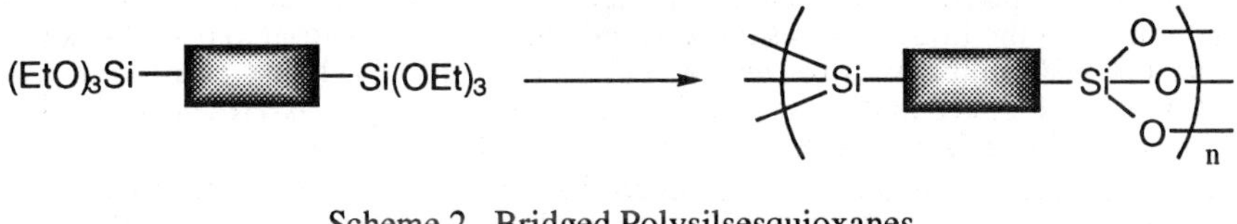

Scheme 2. Bridged Polysilsesquioxanes.

Bridged Polysiloxanes. In our effort to understand the origin of porosity (or its absence) in bridged polysilsesquioxanes, we became interested in the effect of the number of siloxane bonds, or crosslink density, in the bridged polymers on the formation of networks capable of gelation. As the number of siloxane groups per monomer repeat unit is reduced, a point will be reached where a gel will not form.

Traditionally, diethoxysilanes polymerize to afford linear or cyclic polysiloxanes (Scheme 3). Crosslinking can be introduced by placing a polymerizable group (such as a vinyl) on one of the R substituents; this method has been employed to make siloxane based interpenetrating networks or other hybrid organic-inorganic materials.

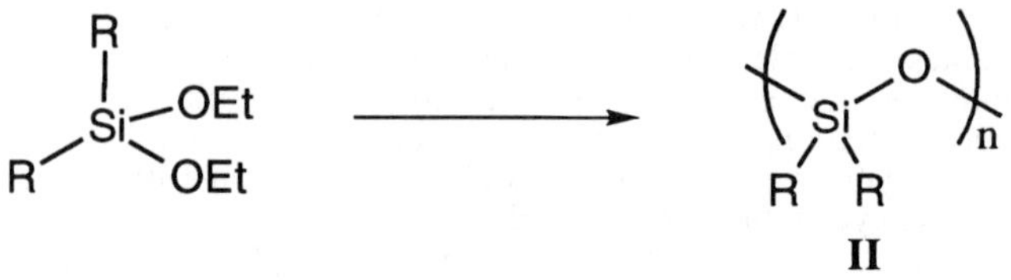

II

Scheme 3. Polysiloxane Polymers.

Bridged polysiloxanes are made from monomers with diethoxysilyl groups attached to either side of a hydrocarbon group, providing a maximum of four potential siloxane bonds. These bridged monomers have the same potential connectivity (number of siloxane linkages) as found in silica, but with less steric crowding at a given silicon atom and an added dimension of variability in the choice of the organic substituents.

In this paper we describe the synthesis and characterization of 1, 6-bis(diethoxymethylsilyl)hexane **1** and 1, 4-bis(diethoxymethylsilyl)benzene **3** and their polymerization to afford gels (Scheme 4) (11).

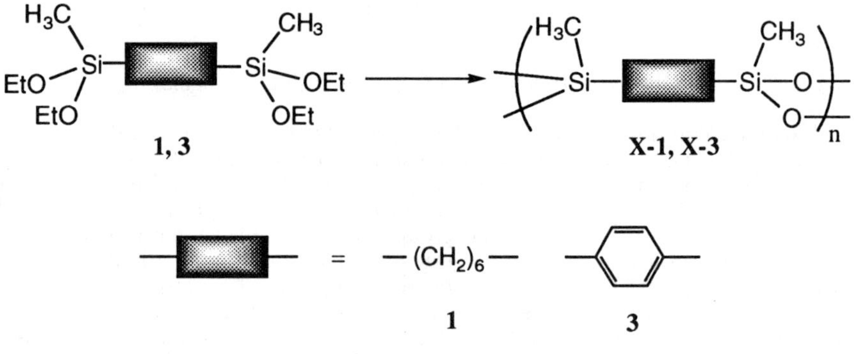

Scheme 4. Bridged Polysiloxanes.

In these monomers, an ethoxy group on each silicon of polysilsesquioxane monomers **2** and **4** is replaced with a methyl group. For purposes of comparison, we also polymerized the bridged silsesquioxane monomers, 1, 6-bis(triethoxysilyl)hexane **2** and 1, 4-bis(triethoxysilyl)benzene **4**. Polymerizations of **1-4** were carried out at 0.4 M monomer concentration in absolute ethanol with aqueous HCl and NaOH as catalysts. The gels were generally processed to afford xerogels by an aqueous work-up. The dried xerogels were characterized by ^{13}C and ^{29}Si CP MAS NMR spectroscopy to provide details of the molecular structures, scanning electron microscopy to observe larger scale structure, nitrogen sorption porosimetry to determine surface areas and pore sizes, and thermal analysis to determine the thermal stability of the materials.

Experimental.

All reactions were run under dry argon using standard Schlenk techniques unless otherwise noted. Solvents were dried and distilled under argon by standard methods; alternatively, anhydrous solvents were used as obtained from Aldrich Chemicals, Inc., as were 1, 5-hexadiene and 1, 4-dibromobenzene. Diethoxymethylsilane was used as obtained from Hüls, Inc. Methyltriethoxysilane was distilled from CaH_2.

Infrared spectra were obtained from a Perkin-Elmer 1750 Fourier Transform Infrared spectrometer. ^1H, ^{13}C and ^{29}Si solution NMR spectra were recorded on Bruker 300 and 400 MHz spectrometers; ^{13}C CP MAS NMR spectra were obtained with a Bruker 400 MHz spectrometer at 100.63 MHz. ^{29}Si solid state CP MAS NMR spectra were obtained on a Chemagnetics console interfaced to a Nicolet 1280 data station at 50.17 MHz.

Thermal analyses of the network materials were performed on a Perkin Elmer Differential Scanning Calorimeter and a Thermal Gravimetric Analyzer. Scanning electron micrographs were obtained with a JEOL Scanning Electron Microscope. Porosimetry measurements were determined using a Quantachrome Autosorb6 multiport nitrogen porosimeter.

Synthesis of $(EtO)_2MeSi-(CH_2)_6-SiMe(OEt)_2$ [1] An oven-dried 250-mL three necked round bottom flask was fitted with a 60 mL addition funnel and an efficient reflux condenser. The flask was charged with 7.82 g (95.2 mmol) of 1, 5-hexadiene, 100 mL of freshly distilled benzene and 0.05 g (0.1% based on hexadiene) of $H_2PtCl_6 \cdot 6H_2O$. Upon addition of the catalyst, a small amount of gas evolution was observed. The addition funnel was charged with 33.16 g (247 mmol) of diethoxymethylsilane. Dropwise addition of the silane with stirring under argon results in the formation, with warming, of a yellow solution. Following complete addition, the mixture was stirred overnight at 40 °C. When the progress of the reaction had stopped (as determined by GC analysis), the volatiles were removed by rotary evaporation under vacuum, and the residue twice distilled to yield 16.31 g (_% yield) of 1, 6-bis-(diethoxymethylsilyl)hexane as a clear, viscous fluid. bp.: 113-116 °C @ 200 mtorr. IR(neat): 2973, 2925, 2878, 2734, 1391, 1257, 1166, 1109, 1083, 952, 800 cm^{-1}. ^1H NMR(CDCl$_3$): δ = 3.74 (q, 8H, (CH$_3$C$\underline{H}_2$O)$_2$MeSi-(CH$_2$)$_6$-SiMe(OC$\underline{H}_2$CH$_3$)$_2$, $^3J_{H-H}$ = 7.0 Hz), 1.30, 0.59 (m, 8H; m, 4H, (CH$_3$CH$_2$O)$_2$MeSi-(C$\underline{H}_2$)$_6$-SiMe(OCH$_2$CH$_3$)$_2$), 1.19 (t, 12H, (C$\underline{H}_3$CH$_2$O)$_2$MeSi-(CH$_2$)$_6$-SiMe(OCH$_2$C$\underline{H}_3$)$_2$, ^3JH-H = 7.0 Hz), 0.07 (s, 6H, (CH$_3$C H$_2$O)$_2$M$\underline{e}$Si-(CH$_2$)$_6$-Si$\underline{Me}$(OCH$_2$CH$_3$)$_2$). $^{13}C\{^1H\}$ NMR (CDCl$_3$): δ = 58.0 ((CH$_3$$\underline{C}H_2$O)$_2$MeSi-(CH$_2$)$_6$-SiMe(O$\underline{C}H_2CH_3$)$_2$), 32.9, 22.7, 13.8 ((C$\underline{H}$$_3CH_2$O)$_2$MeSi-(C$\underline{H}_2$)$_6$-SiMe(OCH$_2CH_3$)$_2$), 18.34 (($\underline{C}H_3CH_2$O)$_2$MeSi-(CH$_2$)$_6$-SiMe(OCH$_2$$\underline{C}H_3$)), -4.9 ((CH$_3CH_2$O)$_2M\underline{e}$Si-(CH$_2$)$_6$-Si$\underline{Me}$(OCH$_2CH_3$)$_2$). ^{29}Si NMR (CDCl$_3$): δ = -4.25. LRMS: 343 amu. HRMS: calcd: 342.5816, found: 343.1768.

1,4-$(EtO)_2MeSi-C_6H_4-SiMe(OEt)_2$ [3] (12). An oven-dried 2000 mL three necked flask with magnetic stir bar was fitted with a 250 mL addition funnel and an efficient reflux condenser, cooled under vacuum and backfilled with dry argon. The flask was charged with 15.4 g (634 mmol) of oven-dried magnesium turnings, 400 mL of anhydrous THF and 189 g (1060 mmol) of methyltriethoxysilane. After charging the addition funnel with a 125 mL THF solution of 1, 4-dibromobenzene, approximately 5 mL of the dibromobenzene solution was added, and the flask mixture stirred with warming until reflux was achieved. The heat was removed and the remainder of the dibromobenzene solution was added dropwise over a 60 minute period (at a rate sufficient to maintain reflux). When the addition was complete, the flask contents was refluxed with stirring for 16 h. The resulting salt suspension was cooled, and volatiles removed under vacuum until the flask contents was reduced to approximately 40% of its original volume. Anhydrous hexanes were added and the suspension stirred for 2h at room temperature to precipitate additional magnesium salts. The suspension was filtered through a medium fritted funnel, and the filtrate volume reduced by rotary evaporation. Distillation of the residue under vacuum yielded 28.8 g (40% yield) of 98% pure (by GC analysis) 1,4-bis-(diethoxymethylsilyl)benzene; an additional 18.77 g (26% product) of 96% pure material can be obtained by redistillation of the impure fractions. bp.: 102-104 °C @ 17 mtorr. IR(neat): 3054, 2974, 2926, 2880, 1391, 1259, 1166, 1142, 1105, 1078, 956, 826, 788, 765 cm^{-1}. ^{1}H NMR(CDCl$_3$): δ = 7.65 (s, 4H, 1,4-(CH$_3$CH$_2$O)$_2$MeSi-C$_6$H$_4$-SiMe(OCH$_2$CH$_3$)$_2$), 3.82 (q, 8H, 1,4-(CH$_3$CH$_2$O)$_2$MeSi-C$_6$H$_4$-SiMe(OCH$_2$CH$_3$)$_2$), 1.24 (t, 12H, 1,4-(CH$_3$CH$_2$O)$_2$MeSi-C$_6$H$_4$-SiMe(OCH$_2$CH$_3$)$_2$), 0.36 (s, 6H, 1,4-(CH$_3$CH$_2$O)$_2$MeSi-C$_6$H$_4$-SiMe(OCH$_2$CH$_3$)$_2$). ^{13}C {^{1}H} NMR (CDCl$_3$): δ = 136.6, 133.2 (1,4-(CH$_3$CH$_2$O)$_2$MeSi-C$_6$H$_4$-SiMe(OCH$_2$CH$_3$)$_2$), 58.5 (1,4-(CH$_3$CH$_2$O)$_2$MeSi-C$_6$H$_4$-SiMe(OCH$_2$CH$_3$)$_2$), 18.2 (1,4-(CH$_3$CH$_2$O)$_2$MeSi-C$_6$H$_4$-SiMe(OCH$_2$CH$_3$)$_2$), -4.3 (1,4-(CH$_3$CH$_2$O)$_2$MeSi-C$_6$H$_4$-SiMe(OCH$_2$CH$_3$)$_2$). ^{29}Si NMR (CDCl$_3$): δ = -18.0. LRMS: 343 amu. HRMS: calcd: 342.5816, found: 343.1768.

Polymerizations. Sol-gel polymerizations of **1** and **3** were carried out at 0.4 M concentrations in 10 mL ethanol. The monomers were dissolved in anhydrous ethanol (5 mL) in a volumetric flask. Then aqueous catalyst (10.8 mol% catalyst, 4 H$_2$O) in ethanol (4 mL) was added with stirring and the volume of the solution adjusted to 10 mL. The resulting solutions were clear and colorless. The solutions were immediately transferred to scintillation vials that were then tightly sealed. Once the gels formed they were allowed to age for two weeks before further processing. Final work-up of the gels consisted of crushing the gels in water (100 mL), washing with water (100 mL), then diethyl ether (100 mL). After air drying, the xerogels were placed under vacuum at 100 °C for 18 hours.

1, 6-Hexylene-Bridged Polymethysiloxane (X-1-NaOH). An oven-dried 25 mL volumertric flask was charged with 3.52 g (10 mmol, 0.4 M) **1** and diluted with approximately 10 ml dry absolute ethanol. In a separate vessel was mixed 1.08 mL of a 1N aqueous NaOH solution and approximately 10 mL absolute ethanol, which was then added to the ethanolic solution of **1** with thorough mixing. The total volume was adjusted to 25 mL with absolute ethanol and thoroughly mixed before transferring to a dry scintillation vial. Gelation occurred within 4 hr. Upon standing for 14 days, the wet gel was crushed in 100 mL distilled water and filtered. The solids were washed with a second 100 mL volume of water, filtered and washed with 100 mL of dry diethyl ether. After standing overnight, the white powder was heated under vacuum for 18 hr. at 100 °C. 2.05 g (101% based on complete condensation of **1**) of **X-1** was recovered. IR(KBr): 3691, 2926, 1260, 1088, 796 cm^{-1}. ^{13}C CP MAS NMR: δ = 58.0 , 33.0, 23.5, 18.7, 0.5. ^{29}Si CP MAS NMR: δ = -12.6 (D^1), -22.0 (D^2). Nitrogen sorption surface area (BET): 297 m^2/g.

1, 6-Hexylene-Bridged Polymethysiloxane (X-1-HCl). An oven-dried 25 mL volumertric flask was charged with 3.51 g (10 mmol, 0.4 M) **1** and diluted with approximately 10 ml dry absolute ethanol. In a separate vessel was mixed 1.08 mL of a 1N aqueous HCl solution and approximately 10 mL absolute ethanol, which was then added to the ethanolic solution of **1** with thorough mixing. The total volume was adjusted to 25 mL with absolute ethanol and thoroughly mixed before transferring to a dry scintillation vial. Gelation occurred within 4 hr. Workup as described above yielded 1.94 g (103% based on complete condensation) of **X-1** was recovered.

IR(KBr): cm^{-1}. ^{13}C CP MAS NMR: δ = 33.7, 23.8, 18.1, 0.3. ^{29}Si CP MAS NMR: δ = -12.6 (D^1), -22.0 (D^2). Nitrogen sorption surface area (BET): 3 m^2/g.

1, 4-Phenylene-Bridged Polymethylsiloxane (X-3-NaOH). An oven-dried 25 mL volumertric flask was charged with 3.43 g (10 mmol, 0.4 M) **3** and diluted with approximately 10 ml dry absolute ethanol. In a separate vessel was mixed 1.08 mL of a 1N aqueous NaOH solution and approximately 10 mL absolute ethanol, which was then added to the ethanolic solution of **3** with thorough mixing. The total volume was adjusted to 25 mL with absolute ethanol and thoroughly mixed before transferring to a dry scintillation vial. Gelation occurred within 30 min.; the wet gel scatters light, as a pink tint was observed upon aging for 14 days. Workup as described above yielded 1.24 g of **X-3** as a white powder. IR(KBr): 3436, 3057, 2970, 1386, 1265, 1146, 1088, 777.0, 516.0, 490.0 cm^{-1}. ^{13}C CP MAS NMR: δ = 186.8 (spinning side band), 180.4 (spinning side band), 139.1, 132.7, 85.0 (spinning side band), -1.6. ^{29}Si CP MAS NMR: δ = -23.5 (D^1), -32.7 (D^2). Nitrogen sorption surface area (BET): 1024 m^2/g.

1, 4-Phenylene-Bridged Polymethylsiloxane (X-3-HCl). An oven-dried 25 mL volumertric flask was charged with 3.43 g (10 mmol, 0.4 M) **3** and diluted with approximately 10 ml dry absolute ethanol. In a separate vessel was mixed 1.08 mL of a 1N aqueous HCl solution and approximately 10 mL absolute ethanol, which was then added to the ethanolic solution of **3** with thorough mixing. The total volume was adjusted to 25 mL with absolute ethanol and thoroughly mixed before transferring to a dry scintillation vial. Gelation occurred within 24 hr.; the wet gel scatters light, as a blue caste was observed upon aging for 14 days. Workup as described above yielded 1.33 g of white xerogel. IR(KBr): 3649, 3056, 2967, 1637, 1386, 1266, 1146, 1083, 1019, 775.0, 601.0, 515.0, 488.0 cm^{-1}. ^{13}C CP MAS NMR: δ = 187.0 (spinning side band), 180.6 (spinning side band), 139.2, 132.8, 85.0 (spinning side band), -1.8. ^{29}Si CP MAS NMR: δ = -24.0 (D^1), -30.9 (D^2). Nitrogen sorption surface area (BET): 676 m^2/g.

Results and Discussion.

The bridged polysiloxane and polysilsesquioxane polymers were prepared by sol-gel processing (Scheme 7). First the monomers had to be synthesized, purified and characterized; none of the monomers **1-4** are commercially available. Then the monomers were hydrolyzed and condensed to afford gels. After aging the gels for two weeks they were typically processed to afford aerogels. However, monolithic aerogels were prepared from hexylene-bridged polysilsesquioxanes and polysiloxanes. Each of these steps will be described in more detail below along with characterization data.

Monomer Synthesis. Hexylene-bridged monomers were prepared in a mild one step transformation from 1, 5-hexadiene and diethoxymethylsilane to give 1, 6-bis(diethoxymethylsilyl)hexane **1** or triethoxysilane to give 1, 6-bis(triethoxysilyl)hexane **2** in good yields (Scheme 5). The hydrosilation reactions were carried out in benzene using chloroplatinic acid as catalyst at room temperature or with gentle heating (40 °C). Some isomerization of terminal olefinic bonds was observed, leading to mixtures of intermediate 6-silylhex-1-ene's and 6-silylhex-2-ene's. However, monomers **1** and **2** could be purified by simple distillations to afford single regio-isomers. Hydrosilation has proven to be a versitile method for synthesizing numerous alkane bridged ethoxysilane monomers in our group.

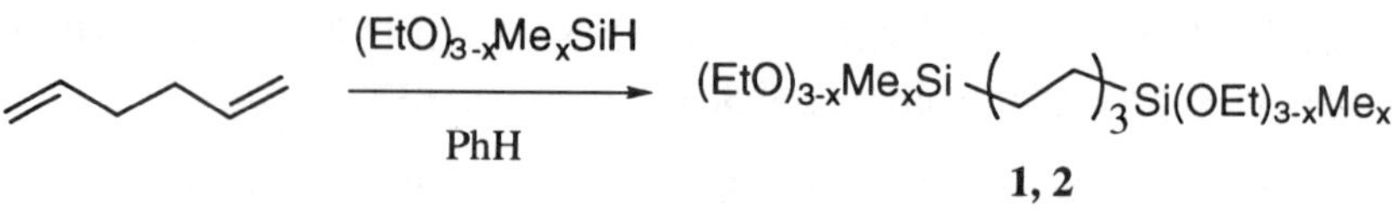

$$(EtO)_{3-x}Me_xSi\left(\begin{array}{c}\\\end{array}\right)_3 Si(OEt)_{3-x}Me_x$$

1, 2

Scheme 5. Hydrosilation of 1, 5-Hexadiene.

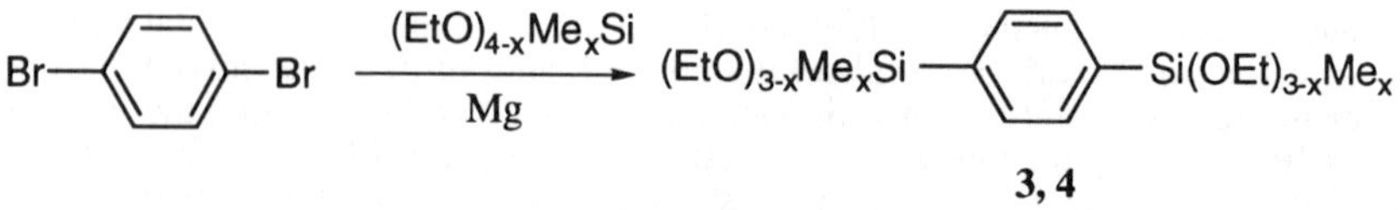

3, 4

Scheme 6. Barbier-Grignard Synthesis of Bridged Silane Monomers.

1, 4-Phenylene-bridged monomers **3** and **4** were prepared from 1, 4-dibromobenzene by a Barbier-Grignard reaction with methyltriethoxysilane and tetraethoxysilane, respectively (Scheme 6). As always with Grignard reactions, care had to be exercised with the addition of the dibromobenzene to the reaction due to the strongly exothermic nature of the Barbier-Grignard chemistry. Fair yields of 1, 4-bis(diethoxymethylsilyl)benzene **3** and 1, 4-bis(triethoxysilyl)benzene **4** were obtained by distillation. Losses were primarily due to multiple substitutions on the ethoxysilane precursors leading to "dimer" and polymer formation.

Sol-Gel Polymerizations. Polymerizations of the ethoxysilane monomers were carried out at room temperature in ethanol. Six equivalents of water were used with the triethoxysilyl monomers **2** and **4**; four equivalents were used with the diethoxymethylsilyl monomers **1** and **3**. Both acidic (HCl) and basic (NaOH) catalysts were used to facilitate hydrolysis and condensation of the monomers. While we knew that the triethoxysilyl monomers would form gels, it was uncertain that monomers **1** and **3**, with only four hydrolyzable ethoxy groups, would undergo sufficient condensation to allow gelation to occur. We were surprised when **1** and **3** formed gels within a few hours (at approximately the same rate as the triethoxysilyl monomers). Apparently, the increased reactivity of the diethoxymethylsilanes over triethoxysilanes compensated for the reduced number of ethoxy substituents available for hydrolysis and condensation. The gels quickly became rigid, acquired a faint blue tint, and rang (or vibrated) when lightly tapped. In all cases the ringing subsided within a day or two after gelation. During this time the gels shrank slightly from the walls of their containers. The gels derived from **1-4** were aged for two weeks before being crushed and rinsed with water and then with diethyl ether. The resulting white polymers were dried at 100 °C under vacuum. They were insoluble in organic solvents, brittle and easily ground with mortar and pestle into fine powders for analysis.

Characterization . Due to the intractable nature of these materials it was necessary to use solid state techniques for their characterization. Since sol-gel processed ethoxysilanes are amorphous, x-ray powder diffraction studies yield little information. Small angle scattering has proven to be a useful method for obtaining structural information from these materials in the size regime between micron and tens of nanometers, but will not be discussed here. Without solid state nuclear magnetic resonance techniques and convenient gas sorption porosimetry analysis, characterization of these materials would be difficult. This may in fact be one of the

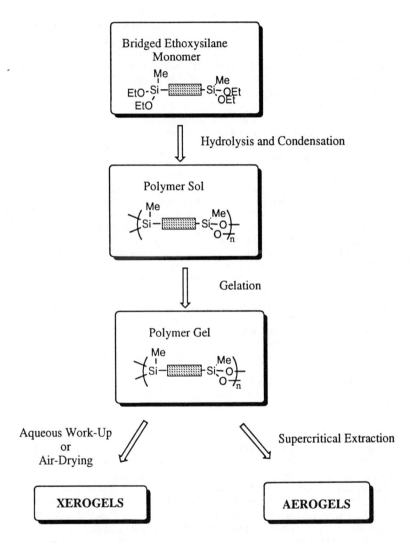

Scheme 7. Sol-gel processing of bridged diethoxymethylsilane monomers to afford xerogels or aerogels.

reasons precursor monomers to bridged polysilsesquioxanes and siloxanes have been known for some time, without any reported investigations of their polymerization to porous materials.

Solid State ^{13}C NMR Spectroscopy. We found that solid state NMR spectroscopy was probably the most useful technique for characterizing these intractable polymers. ^{13}C Cross Polarization Magic Angle Spinning (CP MAS) NMR was used to determine if the integrity of the bridging functionality had been maintained during the sol-gel process. In previous studies (7), for example, we discovered by ^{13}C and ^{29}Si CP MAS NMR that acetylene bridging groups in these materials were hydrolytically labile. However, the alkylene and arylene bridging groups have, as a rule, been demonstrated to be completely stable to the conditions we use in the polymerizations and subsequent processing. The ^{13}C CP MAS NMR spectra of the hexylene-bridged polymethylsiloxane prepared under HCl catalyzed polymerization conditions is shown in Figure 1a. As expected, there are four peaks in the spectrum. The three types of bridging hexylene carbons are observed at 33.7, 23.8, and 18.1 ppm and can be assigned to the carbons γ, β, and α to the silicon, respectively. The fourth peak at 0.3 ppm represents the methyl carbons attached to the silicon atoms. Only a vanishingly small amount of residual ethoxy substituents is left in the material as evidenced by a slight peak barely rising above the baseline noise at 58 ppm; the ethoxy methyl would be expected to lie near 18 ppm in the region obscured by the bridging carbons. The hexylene-bridged polymethylsiloxane prepared with NaOH catalyst shows slightly more of the the residual ethoxy than the acid catalyzed gel. The minor contribution by the ethoxy groups in both spectra indicates that hydrolysis of the ethoxy groups from **1** is essentially quantitative.

^{13}C CP MAS NMR spectra of the phenylene-bridged siloxane gels displayed two resonances in the aromatic region. The smaller peak at 139.2 ppm was assigned to the 1, 4-substituted carbons in the bridging phenylene; a larger peak at 132.8 ppm was assigned to the remaining four aromatic (CH) carbons. It is interesting to note that the same two aromatic carbon peaks coincide in the phenylene-bridged polysilsesquioxane materials giving the appearance of a single aromatic resonance. The methyls attached to the silicon atoms appeared as a single peak at -2 ppm. As with the hexylene-bridged polysiloxanes, there was little evidence remaining of the ethoxy groups. Hydrolysis of the ethoxy groups from the diethoxymethylsilanes was slightly more difficult under basic conditions.

Solid State ^{29}Si NMR Spectroscopy. ^{29}Si CP MAS NMR spectroscopy was used to evaluate the degree of condensation, or number of siloxane bonds to each silicon atom. Three different silicons (Figure 2a) with different degrees of condensation could be present in these materials. D^0 represents silicon atoms with no siloxane linkages and would be expected to occur in either free monomer or monomers positioned on the surface. In light of the reaction and processing conditions used to prepare these materials, it is unlikely that there is any residual monomer. D^1 represents a silicon with *one* siloxane bond and one ethoxy or hydroxy substituent. D^2 silicons are fully condensed with *two* siloxane bonds. Silsesquioxanes have four potential silicons atoms (T^0-T^3) that may be present in a sol-gel processed material (Figure 2b). Typically, only three resonances, T^1, T^2 and T^3, are observed in bridged polysilsesquioxanes prepared under the conditions described here. In both siloxanes and silsesquioxanes, the ^{29}Si NMR resonances shift upfield with increasing degree of condensation. The silsesquioxane resonances are typically 30 ppm upfield from those of the analogous siloxane polymer.

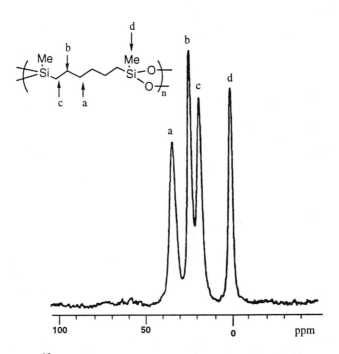

Figure 1. ^{13}C Cross Polarization Magic Angle Spinning NMR spectrum of a hexylene-bridged polysiloxane xerogel.

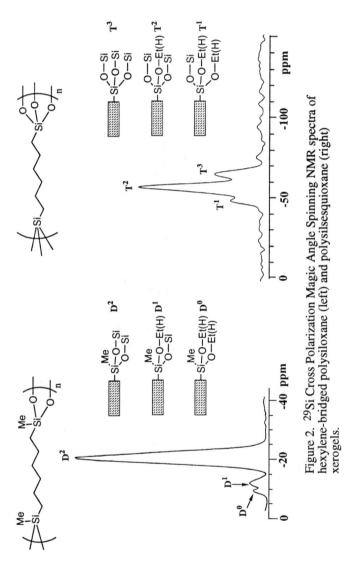

Figure 2. ^{29}Si Cross Polarization Magic Angle Spinning NMR spectra of hexylene-bridged polysiloxane (left) and polysilsesquioxane (right) xerogels.

Two resonances were observed in the ^{29}Si CP MAS NMR spectrum of a hexylene-bridged polysiloxane prepared under acidic conditions. From the relative intensities of the two peaks and the absence of the D^0 peak it is clear that these materials have a very high degree of condensation. Phenylene-bridged methylsiloxanes also exhibit two resonances at -24 and -31 ppm with no sign of the D^0 silicon.

Scanning Electron Microscopy. A scanning electron micrograph of the porous hexylene-bridged gel (**X-1-NaOH**) is shown in Figure 3. The hexylene-bridged polysiloxane has a smooth appearing surface composed of small particles (~ 50 nm in diameter). Similarly the hexylene-bridged silsesquioxane (**X-2-NaOH**, Figure 4) appears to be made of tightly packed granules 50 nm in diameter. Granular structures similar to both these materials have been observed in atomic force microscopy studies of porous phenylene-bridged polysilsesquioxane gels and is consistent with mesoporosity. Non-porous (to nitrogen sorption) bridged gels exhibit smoother, finer grained surfaces.

Thermal Analysis. Thermal stabilities of the bridged polymethylsiloxanes were determined using thermal gravimetric analyses (TGA). The samples were heated at 10 °C/min under flowing air. The hexylene-bridged polysiloxanes prepared under either acidic or basic conditions decomposed near 250 °C. Final sample weights lay near 65% at 850 °C. Phenylene-bridged polysiloxanes were stable to over 500 °C where the weight loss decreased to 65%. The greater stability of phenyl substituted siloxanes is well documented (1).

Surface Area Analysis. The gels were characterized by nitrogen sorption porosimetry in order to measure the surface areas (13), mean pore sizes, and pore size distributions. Phenylene-bridged polysilsesquioxanes have been prepared with surface areas near 600 m^2/g for acid catalyzed gels (in ethanol or tetrahydrofuran) and 800 m^2/g for base catalyzed gels. Phenylene-bridged polysiloxanes prepared under acidic (**X-3-HCl**) and basic conditions (**X-3-NaOH**) were also characterized by high surface areas (673 m^2/g and 1023 m^2/g, respectively) despite having only four siloxane linkages to each monomer repeat unit.

Gels with alkylene bridging groups continue to demonstrate a strong relationship between both spacer length and the crosslink density (silsesquioxane versus siloxane) and pore structure. Hexylene-bridged polysilsesquioxanes prepared with base catalysts were mesoporous (pore sizes between 20-500 Å) with surface areas as high as 716 m^2/g. Prepared under identical conditions, hexylene-bridged polysiloxane (**X-1-NaOH**) had a surface area one-half that of the polysilsesquioxane at 297 m^2/g. Both hexylene-bridged polysilsesquioxane and polysiloxane gels prepared using *acid* catalyst were non-porous to nitrogen sorption.

Conclusion.

Hexylene- and phenylene-bridged polysiloxanes can be prepared through sol-gel processing of bridged diethoxymethylsilane monomers **1** and **3**. The monomers polymerized rapidly to form gels at monomer concentrations as low as 0.2 M. Gelation occurred in approximately the same time that was required for the analogous silsesquioxanes. The gels could be further processed to afford either xerogels or aerogels. The importance of the bridging group on the formation of network materials capable of forming gels is clear. Silsesquioxanes *without* bridging groups

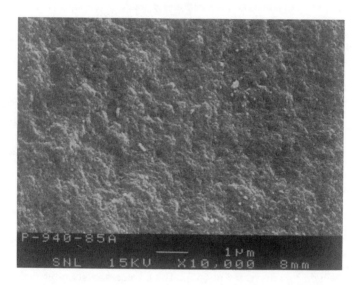

Figure 3. Scanning electron micrograph of hexylene-bridged polysiloxane
X-1-NaOH.

Figure 4. Scanning electron micrograph of hexylene-bridged
polysilsesquioxane **X-2-NaOH**.

gel only at higher monomer concentrations and harsher reaction conditions; siloxanes only polymerize to form non-crosslinked materials.

We have demonstrated that bridged polysiloxanes can be prepared and that the they resemble their polysilsesquioxane counterparts in physical appearance. Even electron microscopy cannot be used to distinguish between the two. Surface area measurements using nitrogen sorption also revealed strong similarities between the two classes of bridged polymers. Both hexylene-bridged polysiloxanes and polysilsesquioxanes formed porous gels when prepared with NaOH catalysts and non-porous gels when prepared under acidic conditions. Similarly, both acid and base catalyzed phenylene-bridged polysiloxanes were found to have surface areas as high as the analogous polysilsesquioxanes.

Bridged polysiloxanes differ from the bridged polysilsesquioxanes in an important fashion. This work demonstrates the ability to place a second kind of organic modifier on the siloxane network in order to tailor the materials' properties. For example, the presence of the methyl substituents in combination with the higher degree of condensation (and, therefore, fewer silanols) required for gelation makes the polysiloxane networks more hydrophobic materials. Other alkyl groups are presently being substituted for the methyl groups to determine any effects on the polymerization chemistry and to make even more hydrophobic networks or networks with novel nanophases. We also intend to place hydrophilic functional groups on the ends of the alkyl substituents to construct even more complex architectures.

Acknowledgements.

We would like to acknowledge Linda McLaughlin for the use of her porosimeter, Gary Zender for scanning electron micrographs, Duane Schneider for NMR spectroscopy, surface areas, and synthesis, and Brigitta Baugher for surface area measurements and infrared spectroscopy. Edward Russick performed the supercritical CO_2 extractions. This research was supported by the United States Department of Energy under Contract No. DE-AC04-94AL85000.

Literature Cited.

1.   Noll, W., *Chemistry and Technology of Silicones* ; Academic Press: London, 1968.
2.   For a review of sol-gel chemistry of silica, see: Brinker, C.J.; Scherer, G.W., *Sol-Gel Science* ; Academic Press: London, 1990.
3.   Brown, J.F., Jr.; Vogt, L.H.; Prescott, P.I. *J. Am. Chem. Soc.* **1964**, *86*, 1120
4.   For a review of oligosilsesquioxanes, see: Voronkov, M.G.; Lavrent'yev, V.I. *Top. Curr. Chem.* **1982**, *102*, 199.
5.   Shea, K. J.; Loy, D. A.; Webster, O. W. *Chem. Mater.***1989**, *1*, 572.
6.   Shea, K. J.; Webster, O.; Loy, D. A. in Better Ceramics Through Chemistry IV Mater. Res. Soc. Proc., B. J. J. Zelinski, C. J. Brinker, D. E. Clark and D. R. Ulrich, Eds., Pittsburgh, PA, **180**, 1990, p. 975.
7.   Shea, K. J.; Loy, D. A.; Webster, O. *J. Am. Chem. Soc.* **1992**, *114*, 6700.
8.   Small, J. H.; Shea, K. J.; Loy, D. A. *J. Non- Cryst. Solids*, .**1993**, *160*, 234.
9.   Oviatt, H. W., Jr.; Shea, K. J.; Small, J. H. *Chem. Mater.*, **1993**, *5*, 943.
10.   Corriu, R. J. P.; Moreau, J. J. E.; Thepot, P.; Man, M. W. C. *Chem. Mater.* **1992**, *4*, 1217.
11.   Loy, D. A.,: Jamison, G. M.; Assink, R. A.; Shea, K. J. *Poly. Sci. and Engineering Preprints* **1994**, .
12.   Imori, T.; Woo, H. G.; Walzer, J. F.; Tilley,T. D. *Chem. Mater.* **1993**, *5*, 1487.
13.   Brunauer, S.; Emmett, P.H.; Teller, E. *J. Am. Chem. Soc.* **1938**, *60*, 309.

RECEIVED September 15, 1994

HIGH-TEMPERATURE POLYMERS

Chapter 21

Structural Design of High-Performance Polymers for Sol–Gel Processing

T. D. Dang[1], J. P. Chen[2], and F. E. Arnold[3]

[1]University of Dayton Research Institute, Dayton, OH 45469
[2]Ad Tech Systems Research, Inc., Dayton, OH 45433
[3]Wright Laboratory, Materials Directorate, Wright-Patterson
Air Force Base, Dayton, OH 45433

A number of structural modifications on high performance benzazole and benzobisazole polymer systems have been carried out in our laboratory. This paper will focus on those structural variations to promote this class of materials amenable to the state-of-the-art sol-gel processing. Our research on these systems involves tailoring the structure for improved solubility properties, as well as functionalizing their backbone structure for co-reaction with tetraalkoxysilanes. A series of benzazole and benzobisazole polymers were prepared containing sulfo pendent groups. Organic ammonium salts of these polymers were found to exhibit excellent solubility in methanol. It was found that commercially available silane coupling agents containing trialkylamino and trialkoxysilyl functionality provided the required solubility in methanol leading to clear homogeneous gels.

The Air Force Office of Scientific Reseach (AFOSR) has sponsored a new initiative in ceramer technology. A three-part approach to the ceramer initiative encompasses low-temperature alkoxide chemistry, functionalized high temperature polymers, and sol-gel processing science. The objective is to investigate the appropriate polymer, ceramic and processing technologies for fabricating ceramic/polymer alloys. The uniqueness of this initiative is to seek a common processing domain between the two technologies instead of the current approaches of favoring one technology at the expense of the properties of the other technology. The initiative addresses the modification of both polymer and ceramic technologies to enable a processing temperature target of 300° - 400°C.

This class of materials would offer many new desirable properties and eliminate many drawbacks of the polymer and ceramic technologies. It is anticipated that the improvement in ceramer technology would impact many structural and nonstructural Air Force applications. A variety of high performance thermally stable polymeric structures are known, but they are intractable and virtually impossible to process. The

0097–6156/95/0585–0280$12.00/0

sol-gel ceramer technology is one logical solution to this problem. In the same manner, the optical quality of silicate glass can be combined with many high performance electrically conductive and non-linear-optically active polymers to produce low cost opto-elcetronic, electro-optic, and photonic components for telecommunication.

As part of the AFSOR initiative, the Polymer Branch, Materials Directorate has initiated a synthetic effort in structural tailoring of high-temperature polymers for sol-gel processing. The classes of high-temperature polymers selected for study are the aromatic and aromatic heterocyclics which exhibit exceptional thermal and chemical stabilities. The scope of materials utilized in our studies includes both extended chain and rigid-rod polymeric geometries. Research is currently being carried out on a number of high-temperature systems such as aromatic imidazoles, thiazoles, oxazoles, imides, and amides. Although these polymers have excellent high-temperature properties, they only exhibit solubility in high boiling aprotic or acidic solvents. Our research on these systems involves tailoring the structure for improved solubility properties as well as functionalizing their backbone structure for co-reaction with tetraalkoxysilanes. This paper reviews our research to date on structural tailoring high-temperature benzazole and benzobisazole polymer systems for sol-gel processing.

Experimental

Monomers. Polymer grade 4,4'-[2,2,2-trifluoro-1-(trifluoromethyl)ethylidene]bis[2-aminophenol] was obtained from Day Chem, Inc., Dayton, Ohio, and used as received. 2,5-Diamino-1,4-benzenedithiol dihydrochloride was prepared (1) by the bromine catalyzed cyclization of p-phenylenebisthiourea to form 2,6-diaminobenzo[1,2-d:4,5-d']bisthiazole. Hydrolysis of the cyclized product in aqueous potassium hydroxide and treatment with 10% hydrochloride acid gave monomer which was then recrystallized from concentrated hydrochloric acid containing 5% stannous chloride. 3,3'-Dimercaptobenzidine dihydrochloride was prepared and purified (2) by an analogous procedure. 1,2,4,5-Tetraaminobenzene tetrahydrochloride was prepared by the ammination of 1,3-dichloro-4,6-dinitrobenzene followed by catalytic reduction using 10% palladium on charcoal. Recrystallization from concentrated hydrochloric acid containing stannous chloride provided anhydrous polymer grade monomer. 2-Sulfoterephthalic acid was prepared by sulfonation of terephthalic acid (3) with fuming sulfuric acid in the presence of a catalytic amount of mercury. This was purified by recrystallization from acetic acid, mp 252-254°C. The acid anhydride was obtained from 2-sulfoterephthalic acid by reaction with acetic anhydride and recrystallized from toluene, mp 204-205°C.

Polymers. Poly[1,7-dipropylsulfobenzo[1,2-d:4,5-d']diimidazo-2,6-diyl[2,(2-sulfo)-p-phenylene]] (I) was prepared by the literature procedure (4) and exhibited an intrinsic viscosity of 6.1 dl/g as determined in methanesulfonic acid. The precursor polymer to (I), poly[1,7-dihydrobenzo[1,2-d:4,5-d']diimidazo-2,6-diyl[2-(2-sulfo)-p-phenylene]] (II) was prepared by the condensation of 1,2,4,5-tetraaminobenzene tetrahydrochloride and 2-sulfoterephthalic acid in polyphosphoric acid (PPA). Polymers with various intrinsic viscosities (3.0 to 15 dl/g) were synthesized to establish solubility effects as a function of molecular weight. Poly[benzo[1,2-d:5,4-d']bisthiazole-2,6-diyl[2-(2-sulfo)-p-phenylene]] (V) was prepared by the condensation of 2,5-diamino-1,4-benzenedithiol dihydrochloride with the acid anhydride of 2-sulfoterephthalic acid in PPA. In contrast to the reported (5) viscosity of 1.6 dl/g, modifications using the acid anhydride and carrying out the polycondensation in the liquid crystalline phase (12wt % polymer) provided polymer with intrinsic viscosities above 10 dl/g.

Sulfo Pendent Polymer (III) Derived From 4,4'[2,2,2-trifluoro-1-(trifluoromethyl)ethylidene]bis[2-aminophenol]. Into the bottom of a resin flask, equipped with a high torque mechanical stirrer, and a nitrogen adapter was placed 1.871 g (5 mmole) of 4,4'[2,2,2-trifluoro-1-(trifluoromethyl)ethylidene]bis[2-aminophenol], 1.35 g (5 mmole) of 2-sulfoterephthalic acid monohydrate, and 12.97 g of PPA (83% P_2O_5 content). The mixture was heated under a positive nitrogen flow at 100°C for 4h and 180°C for 48h. On cooling to room temperature, the polymer was precipitated in water, mixed with a blender, collected by suction filtration, washed with water, and dried under reduced pressure (0.05 mm Hg) at 100°C. The resulting white polymer 2.58 g (95.5%Y) exhibited an intrinsic viscosity of 1.1 dl/g as determined in methanol at 30°C.

Sulfo Pendent Polymer (IV) Derived From 3,3'-Dimercaptobenzidine dihydrochloride. Into the bottom of a resin flask, equipped with a high torque mechanical stirrer, a nitrogen adapter, a pressure regulator, and a side opening for additions, was placed 2.378 g (9 mmole) of 2-sulfoterephthalic acid monohydrate, 2.891 g (9 mmole) of 3,3'-dimercaptobenzidine dihydrochloride, and 17.09 g of PPA (77% P_2O_5). The resulting mixture was dehydrochlorinated under reduced pressure (176 mm) while heating slowly to 65°C. The reaction temperature was maintained at 65°C for 16h raised to 85°C for 4h, then cooled to 40°C and 9.9 g P_2O_5 was added to the mixture to bring the polymer concentration to 12 wt%. The mixture was heated under a nitrogen atmosphere at 100°C for 16h, 140°C for 24h, and 190°C for 24h. The polymer was precipitated in water, mixed with a blender, collected by suction filtration, washed extensively with water and dried under reduced pressure (0.05 mm Hg) at 100°C. The dark red polymer 3.78 g (99.4%Y) exhibited an intrinsic viscosity of 1.42 dl/g as determined in methanesulfonic acid at 30°C.

Coupling Agents. (N,N-Diethyl-3-aminopropyl)trimethoxysilane (b.p. 120°C/20mm) and N-[3-(triethoxysilyl)propyl]-4,5-dihydroimidazole (b.p. 134°C/2mm) were obtained from Huls America, Inc., Piscataway, NJ, and used as received.

Results and Discussion

Alcohols are the most favorable solvent for the state-of-the-art sol-gel processing since they are the by-product of the alkoxide condensation. In the area of aromatic heterocyclic polymer synthesis, alcohols are normally used as a coagulation solvent for the isolation of polymers. For this reason, to provide solubility of these materials in alcohols is most difficult. Recent work (4) in our Laboratory on the synthesis of water soluble rigid-rod polymer systems has uncovered some unusual solubility properties. Polymer I as the sodium or lithium salt is completely soluble in water and exhibits

I   II

lyotropic behavior at concentrations above 20 weight percent. The polymer is prepared from the sodium hydride induced polyanions of poly[(1,7-dihydrobenzo[1,2-d:4,5-d']diimidazole-2,6-diyl)-2-[(2-sulfo)-p-phenylene]] II with subsequent reaction with propane sultone. These polyelectrolytes are currently being investigated as ionic conductors (6) for battery applications. Most interesting, we have determined that organic ammonium salts of Polymer I exhibit excellent solubility in methanol. One obtains complete solubility on addition of triethylamine to a methanol slurry of the polymer. The polymer is maintained as a slurry to provide a maximum surface area of swollen polymer particle. Isolated polymer from the reaction mixture is dissolved in methanesulfonic acid, precipitated into water, washed extensively with water, methanol, benzene, and then freeze dried from the swollen benzene mixture. This procedure allows dried polymer with the maximum surface area for testing various solubility characteristics of the material.

The precursor Polymer II without the pendent propane sulfonic acid groups also exhibited excellent solubility in methanol as the triethylammonium salt. Polymers with different intrinsic viscosities (3.2, 10.0, and 15 dl/g) were prepared and examined to see the effect of molecular weight on solubility properties. As molecular weight increased, solubility of the polyelectrolyte in methanol decreased. Polymer with an intrinsic viscosity above 15 dl/g exhibited only a partial solubility in methanol. In all cases, the soluble polymers were extremely viscous, an inherent characteristic of a charged, expanded, polymer backbone structure. It was interesting to note that cast films from methanol on drying reverted back to the free acid Polymer II with loss of triethylamine. As the triethylammonium salt, the polymer exhibits a light yellow color, and on standing at room temperature, it changes to the parent red color of the free acid sulfo pendent polymer. This proved to be an excellent spin coating method of processing the material into very thin films. It is anticipated that this approach will be a generic method for coating all the aromatic benzobisazole rigid-rod polymer systems.

In an effort to determine if the unusual solubility property was unique only to the benzobisimidazole polymers, sulfo pendent benzoxazole and benzothiazole systems were prepared (Scheme I) and characterized. Condensation of 2-sulfoterephthalic acid with 4,4'-[2,2,2-trifluoro-1-(trifluoromethyl)ethylidene]bis[2-aminophenol] and 3,3'-dimercaptobenzidine in polyphosphoric acid provided high molecular weight polymers

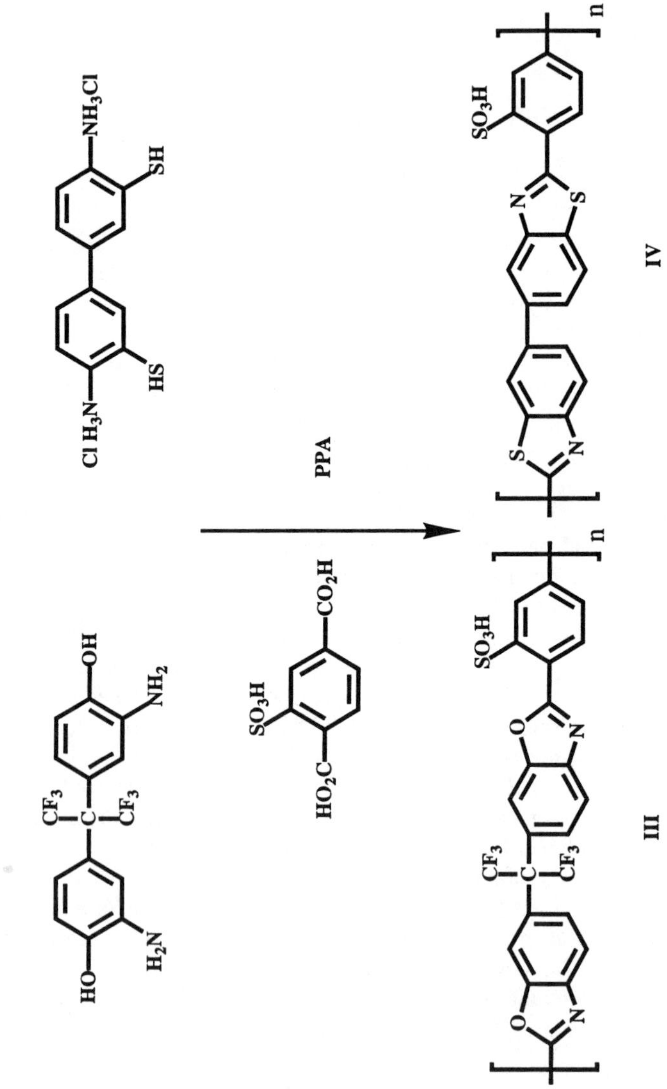

Scheme I. Synthesis of benzoxazole and benzothiazole polymer systems.

for evaluation. Surprisingly, the benzoxazole Polymer III was soluble in methanol without conversion to the ammonium salt. The triethylammonium salt of the benzothiazole Polymer IV exhibited only partial solubility in methanol; however, the tributylammonium salt readily dissolved. The benzothiazole polymer system demonstrates that the solubility is not only the effect of charges on the backbone structure but also the bulkiness of the pendent ammonium salt. For a direct comparison of an extended-rod IV versus a rigid-rod, the 2-sulfobenzobisthiazole system was prepared (Scheme II). The polymer V was obtained from the polycondensation of 2,5-diamino-1,4-benzenedithiol dihydrochloride with the acid anhydride of 2-sulfoterephthalic acid in PPA. It was found that higher intrinsic viscosities could be obtained using the anhydride instead of the free acid (5). Even with intrinsic viscosities above 10 dl/g, the polymer was completely soluble in methanol as the tributylammonium salt.

Thermogravimetric-mass spectral analysis was employed to assess the thermal stability of the organic ammonium salts. As indicated above, the triethylammonium salt disassociates at room temperature; however, the tributylammonium salt is stable at room temperature. Figure 1 shows the TGA mass-spec of the sulfo pendent benzobisimidazole rigid-rod polymer system as the tributylammonium salt. The analysis was performed *in vacuo* at a heating rate of 3°C/min. It can be seen that the initial breakdown of the tributylammonium salt occurs at 60°C maximizing at 180°C with the evolution of tributylamine, butane, and hydrogen. These temperatures are well within the processing parameters for the ceramer initiative. The TGA-mass spec. (Figure 2) of the 2-sulfo pendent benzobisimidazole polymer establishes the overall thermal limitations of a sulfo pendent material systems. The initial degradation of the sulfo group begins at 310°C with the evolution of sulfur dioxide, followed by carbon disulfide and hydrogen sulfide. Other main chain degradation products not shown are ammonia, hydrogen cyanide and nitrogen.

Having established selected high-temperature polymers which were soluble in methanol, the polymers required functionality to coreact with the tetraalkoxysilane for preparing homogeneous gels. We have attempted to utilize the 2-sulfo pendent group for also accomplishing this task. Commercially available silane coupling agents such as (N,N-diethyl-3-aminopropyl)trimethoxysilane [1] (b.p. 120°C/20mm) and N-[3-(triethoxysilyl)propyl]-4,5-dihydroimidazole [2] (b.p. 134°C/2mm) were employed for both solubilizing the polymers and.providing compatibility of the gels produced. The N,N-diethylamine or dihydroimidazole portion of the molecules provides the required solubility in methanol, and the trialkoxysilane portion allows co-reaction with the added tetraalkoxysilane.

Extensive parameters are involved in the formation of any new sol-gel material. Preliminary results on providing homogeneous gels are shown in Table I (low polymer content) and Table II (high polymer content). The low polymer content would provide a toughen ceramic, whereas, a high polymer content would favor the desirable properties of the polymer. Dilute polymer concentrations were used to resolve the viscosity problems associated with these high molecular weight materials and their polyelectrolyte behavior. At low polymer content, stoichiometric amounts of water were added to homogeneous solutions containing amine catalyst, coupling agent, and either TMOS or ethylsilicate-40. The time to gelation varied from one to five minutes after the addition of water which is typical of a base catalyzed alkoxy condensation.

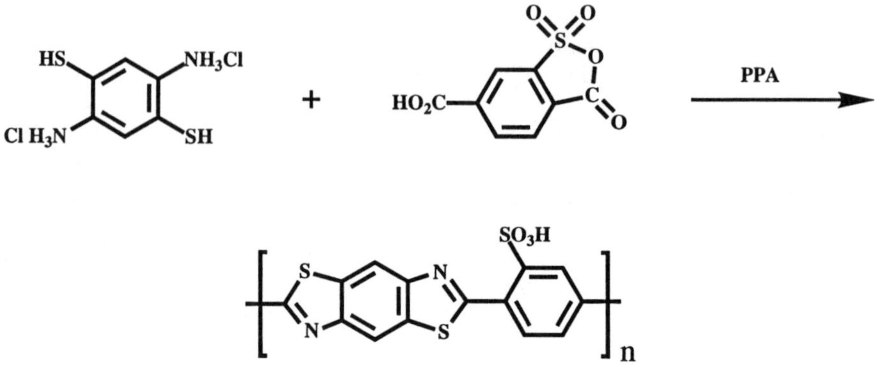

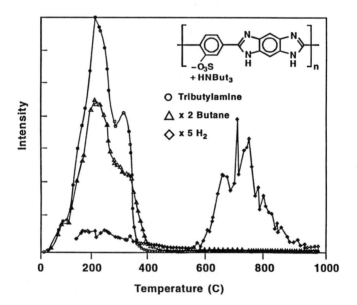

Scheme II. *Synthesis of a-sulfobenzobisthiazole rigid-rod polymer system.*

Figure 1. TGA mass-spec. of the sulfo pendent benzobisimidazole rigid-rod polymer as the tributylammonium salt.

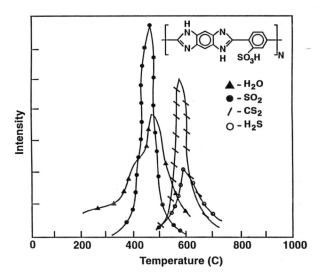

Figure 2. TGA mass-spec. of the sulfo pendent benzobisimidazole rigid-rod polymer.

Table I. Preparation of Gels (Low Polymer Content)

Polymer	$[\eta]$ [a]	Polym. (g)/ MeOH (ml)	Silane Coupling/ SO_3H	Wt % [b] Polym./ Coupling/SiO_2	H_2O/ c or d	Solution	Gel Time (min)
I	6.10	0.02/1.5	[2] 1.45	5.8/7.8/86.4 [c]	0.67	clear green	5
II	3.20	0.01/1.5	[2] 6.0	5.3/15.8/79.2 [d]	0.50	clear orange	2
III	10.00	0.01/3.0	[1] 7.0	4.8/18.7/76.5 [d]	0.50	clear yellow	3
IV	1.42	0.01/1.5	[1] 12.3	2.8/13.9/83.3 [c]	0.73	clear green	2

a. As determined in methanesulfonic acid.
b. Calculated amount, assuming hydrolysis reaction complete.
c. Ethylsilicate - 40
d. TMOS

Table II. Preparation of Gels (High Polymer Content)

Polymer	$[\eta]$ [a]	Polym. (g)/ MeOH (ml)	Silane Coupling/ SO_3H	Wt % [b] Polym./ Coupling/SiO_2	H_2O/ c or d	Solution	Gel Time
III	1.10	0.10/1.0	[1] 0.26 Et₃N 0.83	53/4.5/42.4 [c]	0.84	clear green	5 m
IV	1.42	0.15/15	[1] 1.0 Bt₃N 1.0	52/23/26 [d]	ambient moisture	clear yellow	12 hr
V	9.40	0.20/35	[1] 1.0 Bt₃N 1.0	50/25/25 [d]	ambient moisture	clear green	12 hr

a. As determined in methanesulfonic acid.
b. Calculated amount, assuming hydrolysis reaction complete.
c. Ethylsilicate - 40
d. TMOS

At high polymer content, the bulk viscosity for the extended-rod IV and rigid-rod V were too high for homogeneous mixing and required solutions of these materials to absorb water slowly from the atmosphere to gel. In all cases, clear solutions were obtained which led to clear homogeneous gels. Our current research with these polymer systems involve the formation and characterization of xerogel films at various silica content.

Summary

A variety of 2-sulfo pendent benzazole and benzobisazole polymer systems have been prepared in high molecular weight by polycondenstion in PPA. The thermal stability limitation of the polymers is approximately 300°C as evidenced by TGA-mass spec. analysis. Organic ammonium salts of the sulfo pendants provide excellent solubility in methanol, the target solvent for this study. The solubility of the polymers was a function of their molecular weight and in some cases the bulkiness of the tertiary amine utilized in the formation of the ammonium salt. Bulky silane coupling agents such as (N,N-diethyl-3-aminopropyl) trimethoxysilane or N-[3-(triethoxysilyl)propyl]-4,5-dihydroimidazole provide both the basicity to promote the solubility in methanol and alkoxy groups that prevent phase separation during the alkoxy condensation. Homogeneous gels were produced with both low and high polymer content.

References

1. Wolfe, J. F.; Loo, B. H.; Arnold, F. E. *Macromolecules* **1981**, *14*, 915.
2. Houben, W. In *Methoden der Organishen Chemie*, Miller, E., Ed,; **1955**; Vol. 9, pp 39.
3. Horn, C. F.; Vineyard, H.; Fowler, G. W. *U. S. Patent* 3088956, **1955**.
4. Dang, T. D.; Arnold, F. E. *Mat. Res. Soc. Symp. Proc.* **1993**, *305*, 49.
5. Reynolds, J. R., et. al. *Polym. Prepr.(Am. Chem. Soc., Div. Polym. Sci.)* **1993**, *34 (1)*, 1056.
6. Dang, T. D.; Bai, S. J.; Heberer, D. P.; Spry, R. J.; Arnold, F. E. *J. Polym. Phys.* **1993**, *31*, 1941.

RECEIVED August 3, 1994

Chapter 22

Preparation and Properties of High-Clarity Polyamide–Silica Hybrid Materials

Z. Ahmad[1], Shuhong Wang[2], and J. E. Mark

Department of Chemistry and Polymer Research Center,
University of Cincinnati, Cincinnati, OH 45221–0172

New hybrid materials synthesized by incorporating silica in poly(trimethylhexamethylene terephthalamide) using the sol-gel process were found to have high transparency and flexibility. They showed an increase in tensile strength at the yield point of up to 25 %, but the elongation at rupture was found to decrease sharply upon the addition of the silica. Water uptake under saturated conditions was reduced from 8 wt % for the pure polymer, to approximately 3 wt % for the hybrid material containing 15 % silica. Since there is nearly a 20 % drop in the value of the yield stress reported for the pure polymer upon long storage in water, the present class of ceramers could be particularly useful in polyamide applications requiring dimensional stability under high-humidity conditions.

The commercial use of transparent polyamides (1) has increased considerably in recent years. These materials are typically glassy and, since they are generally copolymers of irregular structure, are also non-crystalline. Important examples are Trogamid T®, Grilamid TR-55®, and Hostamid LP700®. In some applications requiring transparency, these polymers are competitive with poly(methyl methacrylate), polycarbonate, and polysulphone. Trogamid T® is poly(trimethylhexamethylene terephthalamide), and the fact that a mixture of 2,2,4- and 2,4,4-trimethylhexamethylene diamine is used as its monomer imparts the irregularity leading to the polymer's amorphous structure. Compared to the aliphatic polyamides, however, it has some excellent properties, for example, good rigidity, low thermal expansion coefficient, as well as good thermal resistance and insulating properties. On the basis of this combination of properties (1,2), it is used in many industrial applications including several in electric engineering, water and filter technology, and mechanical and instrumental engineering. The optics industry also uses it in the manufacture of correction lens and safety glasses. The presence of amide groups in Trogamid T®, however, results in some undesirable water absorption. In the molding

[1]Permanent address: Department of Chemistry, Quaid-i-Azam University, Islamabad 45320, Pakistan
[2]Current address: PPG Industries, Inc., 440 College Park Drive, Monroeville, PA 15146

compounds of the polymer, the moisture absorption is nearly 3 wt % under 50 % relative humidity. Under saturation, the value of the yield stress of Trogamid T5000® has been reported (3) to decrease by 20 % by this water absorption.

The present study reports a new sol-gel preparation (4-6) of transparent hybrid materials consisting of various proportions of this polymer and silica. Other polymers, such as poly(dimethylsiloxane) (7-9) and polyimide (10), have been extensively studied in this manner. The silica was produced *in situ* by the hydrolysis of tetramethoxysilane in the polymer solution, followed by a poly-functional condensation polymerization to give a silica network. Diethylamine was used as catalyst, and the most important reaction is

$$Si(OCH_3)_4 + 2 H_2O \rightarrow SiO_2 + 4 CH_3OH \tag{1}$$

The polyamide-silica hybrid materials generated in this way were studied with respect to their physical properties. Of particular interest was the question of whether silica could reduce the extent of water absorption in these materials, thereby improving their tensile strengths and dielectric properties.

Experimental

Materials. The pure sample of poly(trimethylhexamethylene terephthalamide) was donated by Hüls America Inc. It was dried in vacuum oven for 12 hours at 50 °C before use. Anhydrous dimethylformamide (DMF) was chosen as solvent, and was used as received from Aldrich Chemicals. The diethylamine catalyst (Aldrich Chemicals) and the tetramethoxysilane (TMOS) (Hüls America Inc.) were both described as 99 % pure and were also used as received.

Preparation and Characterization of Amide-Silica Hybrid Films. A homogeneous solution of the polyamide in DMF was made by placing 10 g of polymer into 65 g of the solvent, and stirring them in a sealed flask at 35 °C for five days. A known amount of polymer solution was taken and a chosen amount of TMOS (in DMF) was added to it. The mixture was stirred for 30 min, and then a stoichiometric amount of water containing 5 % of diethylamine as catalyst in DMF was added to begin the hydrolysis and condensation of TMOS into a silica network. The resulting homogeneous mixture was stirred for an additional six hours at room temperature, and then cast in Petri dishes, which were held at 50 °C to remove solvent and to complete the reaction. The films thus obtained were further dried under vacuum for 48 hours and then kept under vacuum until characterized. The resulting films had 5 to 30 wt % silica.

The tensile properties of these hybrid films were measured using an Instron Universal Testing Instrument Model-1122 at room temperature (23 °C), using an extension ratio of 0.2 in/min. The sample dimension was approximately 40x3x0.1 mm, and the length between the clamps was 25 mm.

Water absorption by the films under saturated conditions was measured by the method described in ASTM-D570-81.

Results and Discussison

The films of the hybrid material containing up to 30 % silica content were all transparent, as can be seen from Figure 1. Figure 2 describes the stress-strain results obtained on these films. Those containing more than 20 %, however, could not be tested in this way for their mechanical strength because they were too brittle. The value of the tensile strength at the yield point for the pure material was 61.0 MPa, which is close to the value 67.7 MPa quoted by the Hüls Laboratory for a sample

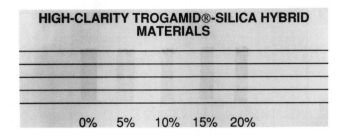

Figure 1. Vertically-mounted strips of some of the polyamide-silica hybrid films, demonstrating their transparency.

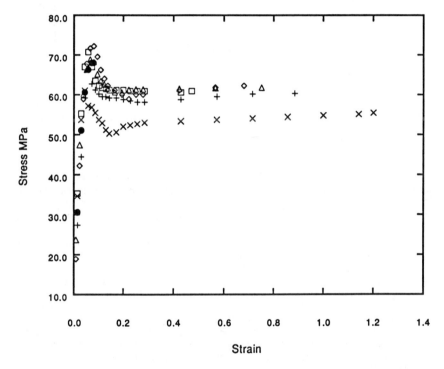

Figure 2. Stress-strain curves for polyamide-silica hybrid materials with various silica contents: (x) 0 , (+) 5 , (Δ) 7.5 , (◊) 10 , (☐) 15, and (●) 20 wt %.

drawn from a polymer melt and measured under test specification ASTM D-638. The small discrepancy between the two values is probably mainly due to the difference in sample history and the methods by which the values were measured. In comparison with the pure polyamide, the value of the yield stress was found to be larger and to increase with silica content. Figure 3 shows the variation of the tensile yield strength and tensile break strength with the amount of silica present. The tensile strength at yield increases first, but then decreases sharply; in fact, the films with 20 % silica

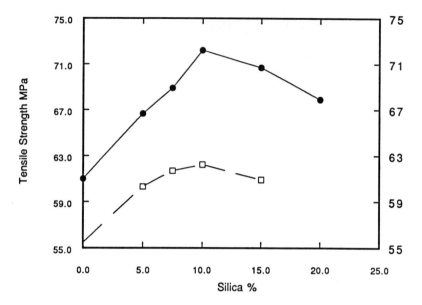

Figure 3. Variation of tensile yield strength (●) and tensile break strength (□) for the hybrid materials.

ruptured just beyond the yield point. The values of yield stress measured at room temperature were higher than those at rupture for all the samples. The tensile modulus also increased with addition of silica. Figure 4 shows the length at rupture as a function of the silica content. It is seen to decrease rapidly with increase in silica content, as expected.

The relative rate of water absorption under saturated conditions was measured by initially drying the films in a vacuum oven at 50 °C for 12 hours, immersing the films in distilled water, and then measuring the weight change until it became constant. Figure 5 shows the extent of water absorption in the hybrid films. It is evident that there is a considerable decrease in the water absorption in the films containing 5 - 10 % silica, with a slight additional decrease for higher silica contents. Thus, the change in the water absorption is not proportional to the weight of silica present. It may be that the *in-situ* generated silica becomes linked with the amide chains through secondary bond forces and that this engagement of the polar amide groups decreases the extent of water absorption. Similar results have been found in aramid-silica hybrid materials (11).

This decrease in absorption is important since water can influence many physical properties such as electric insulating ability, dielectric loss, dimensional stability, and appearance. The nearly 20 % loss in yield stress reported (3) for Trogamid T5000® resulting from water absorption under saturated conditions during long storage is shown in Figure 6. The present results show that this loss can be reduced significantly by incorporating small amounts of silica in the polyamide matrix. This new class of ceramers could thus be extremely useful in polyamide applications requiring dimensional stability under high-humidity conditions.

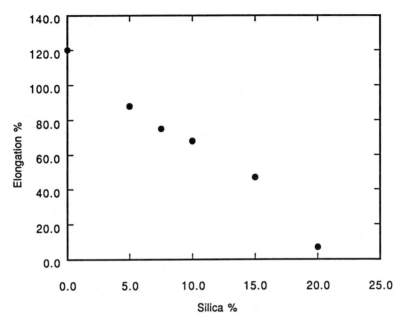

Figure 4. Percent elongation at rupture as a function of silica content for the polyamide-silica hybrid materials.

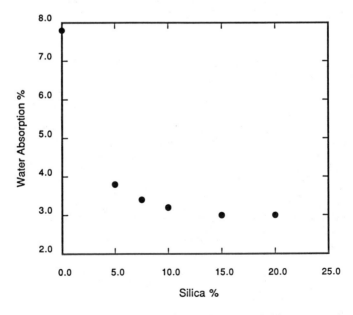

Figure 5. Extent of water absorption as a function of silica content for the hybrid materials.

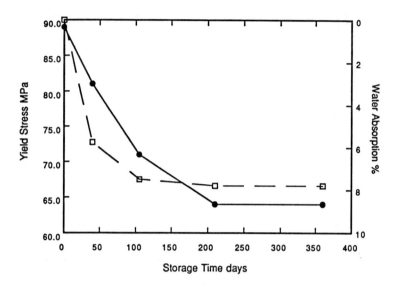

Figure 6.   Variation in the yield stress (●) of Trogamid T5000® and the extent of water absorption (□) during long storage.

Acknowledgement

It is a pleasure to acknowledge the financial support provided by the Air Force Office of Scientific Research through grant F49620-92-J-0322.

References

1.   Brydson, J. A. Plastic Materials; Butterworth: London, 1986.
2.   Michael, D. *Kunstoffe* **1980** 70, 629.
3.   Trogamid-T® property data taken from *Computer -Aided Material. Preselection by Uniform Standards:* Hüls Aktiengesellschaft, Germany, 1990.
4.   Mark, J. E. *CHEMTECH* **1989** 19, 230.
5.   Brinker, C. J.; Scherer, G. W. *Sol-Gel Science. The Physics and Chemistry of Sol-Gel Processing;* Academic Press: San Diego, 1990.
6.   Wang, S.; Ahmad, Z.; Mark, J. E. *Polym. Bulletin* **1993** 31, 323.
7.   Mark, J. E.; Erman, B. *Rubberlike Elasticity. A Molecular Primer*, Wiley-Interscience: New York, 1988.
8.   Mark, J. E. *Kautschuk+Gummi Kunstofe* **1989** 42, 191.
9.   Xu, P.;Wang, S; Mark, J. E. In *Better Ceramics Through Chemistry, Part IV,* Zelinski, B. J. J.; Brinker, C. J.; Clark D. E.; Ulrich, D. R., Eds; Materials Research Society: Pittsburgh, 1990.
10. Wang, S.; Ahmad, Z; Mark, J. E. *J. Macromol. Sci., Macromol. Reports* **1994** 31, 411.
11. Ahmad, Z; Wang, S.; Mark, J. E. In *Better Ceramics Through Chemistry, Part VI,* Sanchez, C.; Mecartney, M.L.; A. Cheetham, A., Eds. Materials Research Society: Pittsburgh, 1994.

RECEIVED August 3, 1994

Chapter 23

Preparation and Mechanical Properties of Polybenzoxazole–Silica Hybrid Materials

J. P. Chen[1], Z. Ahmad[2,4], Shuhong Wang[2,5], J. E. Mark[2], and F. E. Arnold[3]

[1]Systran Corporation, Dayton, OH 45432
[2]Department of Chemistry and Polymer Research Center, University of Cincinnati, Cincinnati, OH 45221–0172
[3]Wright Laboratory, Materials Directorate, Wright-Patterson Air Force Base, Dayton, OH 45433

High molecular weight benzoxazole copolymers prepared from 4,4'-[2,2,2-trifluoro-1-(trifluoro-methyl)ethylidene]bis[2-aminophenol], 4,4'-oxybis-(benzoic acid) and 5-hydroxyisophthalic acid or 5-phosphonoisophthalic acid exhibited solubility in tetrahydrofuran. These thermo-oxidative stable copolymers with the Tg of ca. 330°C can be incorporated with alkoxysilane by the sol-gel method. Hydroxypolybenzoxazoles could be reacted with an isocynatosilane coupling agent and then hybrid consequently with silica. The resulting hybrid films containing one third of silica were transparent. The mechanical properties of the hybrid materials were highly dependent on the nature of the organic polymers. In general, the tensile modulus of the hybrid materials increased with addition of silica, but the elongation at break decreased at higher silica contents.

Organic and inorganic hybrid materials prepared through sol-gel processing have the potential to possess the desired properties of both organic and inorganic components, such as high tensile modulus, scratch resistance, thermal and dimensional stability from inorganic network or toughness, flexibility, and lightweight from the organic portion. A particular interest in this area is to investigate a new hybrid material which would possess excellent mechanical properties for use in structural applications. Choosing a tough, processable, high temperature polymer and developing a well-controlled sol-gel process which can produce the hybrid material with a uniform structure are likely to be the two basic requirements in this investigation. Among the high temperature polymers (*1,2*), polybenzoxazoles show excellent thermal and chemical stabilities and good tensile strength. One of the benzoxazole polymers, poly(phenoxyphenyl-6F-benzoxazole) (PP-6FAP) prepared from 4,4'-oxybis(benzoic acid) (PP) and 4,4'-[2,2,2-trifluoro-1-

[4]Permanent address: Department of Chemistry, Quaid-i-Azam University, Islamabad 45320, Pakistan
[5]Current address: PPG Industries, Inc., 440 College Park Drive, Monroeville, PA 15146

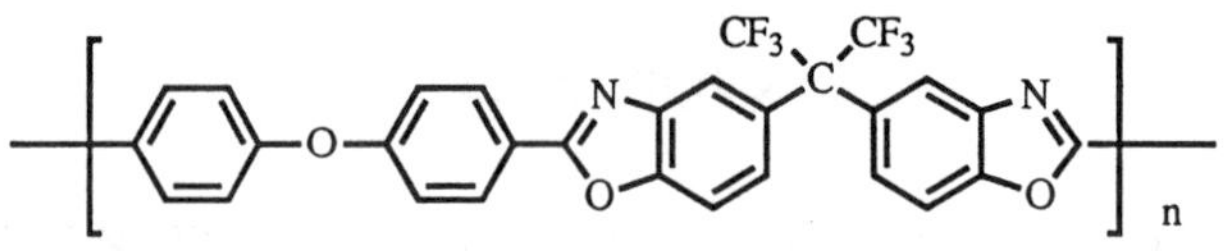

PP-6FAP

Scheme I. High temperature benzoxazole polymer (PP-6FAP), Tg at 300°, is soluble in tetrahydrofuran.

(trifluoromethyl)ethylidene]bis[2-aminophenol] (6FAP) (Scheme I), is soluble in tetrahydrofuran (3). The good solubility in a water miscible organic solvent of this polymer leads to the possibility of making hybrid materials from the benzoxazole polymers and metal alkoxides.

Based on this polymer, we have successfully prepared a series of copolymers with hydroxy or phosphonic acid pendants (Scheme II). We anticipate that the polar pendants will increase the interaction and compatibility between organic polymers and oligomers of the metal oxides, and also to bridge the organic and inorganic phases through covalent bonding after the heat treatment of dried gel or by applying a silane coupling agent. The synthesis of the functionalized copolymers, the preparation of hybrid gels from the copolymers and tetramethoxysilane, and their mechanical properties are reviewed in this paper.

Experimental

Monomer Compounds. 5-Hydroxyisophthalic acid (HIPA) and 4,4'-oxybis(benzoic acid) (PP) were received from Aldrich. 4,4'-[2,2,2-Trifluoro-1-(trifluoromethyl)ethylidene]bis[2-aminophenol] (6FAP) was obtained from Day Chem Inc. (Dayton, Ohio). 5-Phosphono-isophthalic acid (PA) was prepared from 5-bromo-1,3-xylene in three steps. According to Tavs' method (4), treatment of bromoxylene with triethyl phosphite in the presence of $NiBr_2$ produced oil-like diethyl xylenylphosphonate in good yields. The phosphonate diester was converted to acid, mp 197-9°, with ethyl bromide as by-product by refluxing the diester with 48% hydrobromic acid overnight (5). Oxidation of xylenyl phosphonic acid in aqueous potassium permanganate gave phosphono-isophathalic acid as white needle crystals from HCl aqueous solution, mp 360°, m/e 246 $(M)^+$.

Functionalized Copolymers Syntheses. Hydroxy functionalized copolymers were prepared from 5-hydroxyisophthalic acid (HIPA), 4,4'-oxybis(benzoic acid) (PP), and 4,4'-[2,2,2-trifluoro-1-(trifluoromethyl)ethylidene]bis[2-aminophenol] (6FAP). Polycondensations were carried out in polyphosphoric acid with P_2O_5 content at 83-84% at 180° for 64 h. In the work-up step, the reaction mixture was cooled to 90° and diluted with 85% phosphoric acid and water to reduce the P_2O_5 content to ca. 67%. The homogeneous solutions were stirred at 70° overnight to hydrolyze the phosphate ester, a side reaction product from polymer hydroxy pendant and polyphosphoric acid, then poured into water to coagulate the polymer.

Phosphonic acid functionalized copolymers were prepared by using various amounts of 5-phosphonoisophthalic acid (PA), 4,4'-oxybis(benzoic acid) (PP), and 4,4'-[2,2,2-trifluoro-1-(trifluoromethyl)ethylidene]bis[2-aminophenol] (6FAP). Polycondensations were carried out by the same method as the preparation of copolymers with the hydroxy pendants (Scheme II).

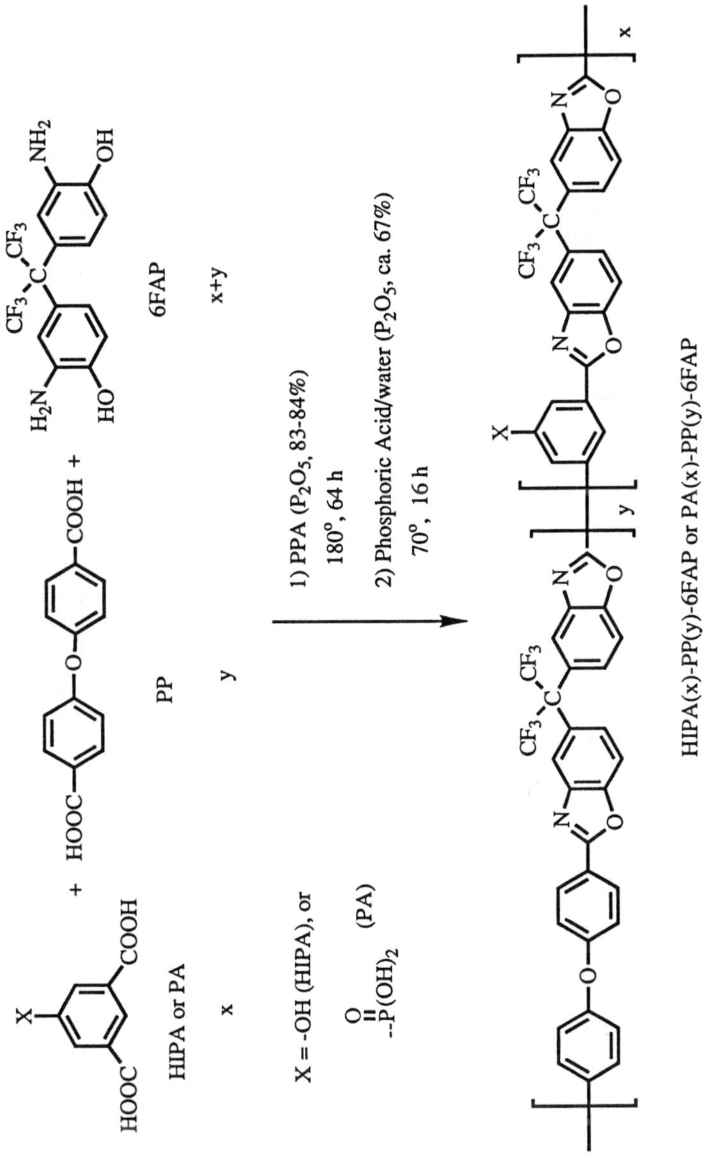

Scheme II. High molecular weight hydroxy- and phosphono-benzoxazole copolymers with Tg at 320-330° were prepared in polyphosphoric acid.

Sol-Gel Synthesis of Polymer-Silica Hybrid Materials.   A general procedure for preparing polymer-silica hybrid materials was to dissolve the polymer in anhydrous THF at the concentration of ca. 7% weight to volume. A measured amount of water and tetramethyl orthosilicate (TMOS, Aldrich) at the mole ratio 3.0-3.5 to 1 and a catalytic amount of triethylamine were added. A viscous, homogeneous solution was obtained after stirring for 2 h. The solution was transferred to a Petri dish and dried slowly over a period of 16 to 24 h at room temperature. The resulting film was further dried under vacuum for 48 h at 80-100°.

In case of the use of a silane coupling agent, hydroxypolybezoxazole reacted with isocynatopropyltriethoxysilane (Huls America) (6) in anhydrous THF and triethylamine as catalyst at 35° overnight. The resulting polymer and coupling agent product solution was treated with water and TMOS as described above.

The tensile properties of the unfilled polymers and the polymer-silica hybrid films were measured using an Instron Universal Testing Instrument (Model 1122) at the drawing rate of 0.2 in/min at room temperature.

Results and Discussion

Copolymers Synthesis and Properties.   A series of high molecular weight functionalized benzoxazole copolymers were prepared from 4,4'-[2,2,2-trifluoro-1-(trifluoro-methyl)ethylidene]bis[2-aminophenol], 4,4'-oxybis(benzoic acid) and 5-hydroxyisophthalic acid or 5-phosphonoisophthalic acid. Due to the low nucleophilicity of hexafluoroisopropylbis(aminophenol) (6FAP) (7), the polycondensations were carried out in polyphosphoric acid with a P_2O_5 content at 84% at 180° for 64 h. From previous work (8) and model compound studies, it was found that phosphate ester was formed during the polymerization by the side reaction of the pendent hydroxy groups and polyphosphoric acid. The phosphate ester can be hydrolyzed in situ without cleaving the polymer chain by diluting the polyphosphoric acid with phosphoric acid and water to hydrolytic conditions and stirring at 70° overnight. After several trials, most copolymers exhibited intrinsic viscosities above 1.0 dl/g, as measured in methansulfonic acid (Table I.) The solubility of hydroxypolybenzoxazoles in THF remains the same as PP-6FAP, but the solubility of phosphonopolybenzoxzaoles decreases while increasing the amount of phosphonic acid moiety; presumably, the strong hydrogen bonding of the phosphonic acid between polymer chains retards the solubility. The hydrogen bonding can be interrupted by adding a small amount of water to the THF. It was shown in the case of the polymer (PA(17)-PP(83)-6FAP) which was only swollen in THF but was dissolved in THF including 3-5 % of water.

Table I. Intrinsic viscosities and solubility in THF of the resulting copolymers

X	x/y	copolymers [i.v. (dl/g)]	Soluble in THF
OH	25/75	A: HIPA(25)-PP(75)-6FAP [0.9 - 1.19]	Yes
OH	40/60	B: HIPA(40)-PP(60)-6FAP [0.95 - 1.34]	Yes
PO(OH)$_2$	10/90	C: PA(10)-PP(90)-6FAP [1.0-1.46]	Yes
PO(OH)$_2$	17/83	D: PA(17)-PP(83)-6FAP [1.16]	Swollen[*]
PO(OH)$_2$	25/75	E: PA(25)-PP(75)-6FAP [0.90]	NO

* Clear solution can be obtained by adding 3 - 5 % of water.

The functionalized copolymers exhibited Tg's at 320-330° which are slightly higher than that of homopolymer PP-6FAP. Tough transparent films could be obtained by casting from the THF solutions of those copolymers. Infrared spectra of hydroxy or phophono polybenzoxazoles are similar to that of PP-6FAP except for the presence of a strong broad absorption at 3100-3200 cm^{-1} and a weak to medium broad absorption at 2300-3400 cm^{-1} which correspond to the hydroxy (-OH) and phosphonic acid (-POH), respectively. Thermal stability was investigated by using TGA-MS. The major degradation fragments were at 525-550° and 650-660° corresponding to the decomposition of 6F-isopropyl, phenyl ether, and benzoxazole.

Sol-Gel Process of Copolymers and Tetramethyl Orthosilicate. The extraordinary shrinkage accompanied with cracking is the most difficult problem for producing sol-gel glasses. A wet gel with weak interparticle interaction and uneven micropores is most likely to have cracking. The problem can be minimized to some extent by using very time consuming processes such as extensive aging to increase the strength of the interparticle networks, the enlarging of the pore size, and the very slow evaporation rate to reduce the capillary stresses. Alternatively, a sol-gel process consisting of a tough organic polymer and inorganic particles has the advantage of excluding the cracking problem because of the formation of tough organic polymer matrix interaction during the drying process. On the other hand, phase separation between organic polymer and inorganic particles might be the critical point of the sol-gel processing for hybrid materials.

At the early stage of the sol-gel process, the incomplete hydrolysis and condensation gives the low molecular weight silica particle containing organic alkoxy and polar hydroxy groups which are compatible with organic polymer in sol-gel solvent; but the growth of silica particle leads to macro phase separation of the mixture. The growth and fractal structure of silica particle are highly influenced by the hydrolysis and condensation reactions. It is known that the rate of these reactions can be affected by the pH of the solution, the catalysts, the amount of water, and even the silicon-containing starting materials. After preliminary trials, we found that the use of tetramethyl orthosilicate as the starting compound and triethylamine as catalyst gave better results in homogeneity compared with the use of ethyl silicate - 40 and HCl catalyst. A sol-gel solution was prepared as described above. The solvent of the clear sol-gel solution was removed slowly under atmosphere over a period of 16 to 24 h. The resulting film was further dried under vaccum at 80-100° for 48 h to complete the hydrolysis and to increase the condensation of silica. By using this approach, we have successfully prepared the transparent, uniform, and thick hydroxypolybenzoxazole-silica hybrid films with 33 wt % of silica (Figure. 1). In attempts to verify the formation of covalent bondings between functionalized copolymers and silica, it was found that the hybrid films redissolved in THF after soaking in the solvent overnight, which showed no significant bonding between polymer and silica.

The crosslinking between polymer and silica can be achieved by utilizing a coupling agent. Hydroxy functional copolymer was dissolved in anhydrous THF; a stoichiometric amount of isocyanatopropyltriethoxysilane (*6*) and a catalytic amount of triethylamine were added (Scheme III). The solution was heated at 35° overnight. Infrared spectrum of the product showed the absorption of -C=O of carbamate at 1760 cm^{-1} indicating the

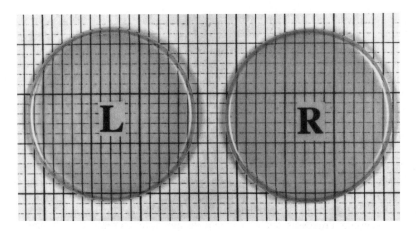

Figure 1. The transparent thick hybrid films, 1.7 inch in diameter, including 33.3% of silica and 66.6% of HIPA(40)-PP(60)-6FAP, shows left: 0.45g, without silane coupling agent; right: 0.50g, with silane coupling agent.

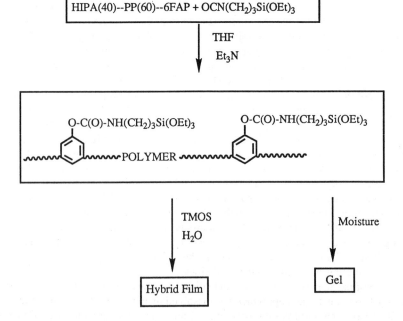

Scheme III. The preparation of hybrid materials from hydroxybenzo-xazolecopolymer, silane coupling agent, and TMOS.

reaction of functionalized copolymer and silane coupling agent. Before gelation, TMOS and water were added, and the mixture was heated at 40° with stirring for 2 h, followed by the procedure mentioned above for hybrid film formation. The resulting hybrid films including functionalized copolymer, silane coupling agent, and silica (Figure 1) were completely insoluble in THF. The results agree well with the occurrence of crosslinking of polymer and silica.

Homopolymer, PP-6FAP, and TMOS were not miscible entirely in THF. Compared to PP-6FAP, the phosphonobenzoxazolecopolymer, PA(10)-PP(90)-6FAP, showed limited compatibility with TMOS. The sol-gel solution including the copolymer and 10% silica was clear at the beginning but turned cloudy before the solvent was removed. The resulting films were opaque indicating the occurrence of macro phase separation.

Mechanical Properties of Copolymers and the Hybrid Films. Four unfilled copolymers (see Table I), a (HIPA(25)-PP(75)-6FAP, i.v.: 0.9 dl/g), b (HIPA(25)-PP(75)-6FAP, i.v.: 1.19 dl/g), c (HIPA(40)-PP(60)-6FAP, i.v.: 0.95 dl/g), and d (PA(10)-PP(90)-6FAP, i.v.: 1.11 dl/g), and the transparent hybrid films with various amount of silica from TMOS were prepared for stress-strain measurement. Figure 2 shows the stress-strain curves for the unfilled polymers. Polymers a and b have the same structure and composition, but polymer b film is much tougher than the film from polymer a. It indicates that the mechanical properties of polymers are highly influenced by the polymer molecular weight. Since both the composition and the structure of c and d are different from a or b, it is not clear that the lack in strength of c and d is caused by the lower molecular weights, or the amount of meta structure segments, or the structure of functional groups.

Figures 3 and 4 show the mechanical strength of hybrid films derived from polymers a and b. The modulus and tensile strength increased with addition of silica in both cases, but the elongation at break decreased at the higher silica contents. This is a typical mechanical behavior of polymer-silica hybrid composites which possess well-dispersed micro size silica particles in the polymer matrix. It was also consistent with the interaction between polymer and silica through the secondary bonding and large surface area of the micro silica filler. In case of polymer c (Figure 5), there is a considerable increase in mechanical strength with increasing silica contents. Although the strength of unfilled polymer c is lower than that of polymers a and b, the polymer c and silica (20%) hybrid film has the strength as good as those from polymers a and b. It is interested to note that the elongation at break even increased with addition of silica in case of polymer c. This may be due to the strong interaction including the slightly crosslinking between polymer and silica resulting from the number of hydroxy groups in polymer c.

In conclusion, this study presents the results of preparing high Tg, THF soluble, thermally stable functionalized copolymers to incorporate with TMOS. Hydroxypolybenzoxazoles can chemically hybrid with a silane coupling agent and TMOS through sol-gel processing. Phosphonopolybenzoxazole was miscible with TMOS in THF but failed to form homogeneous gels due to the occurrence of phase separation. The mechanical behavior of the hybrid materials showed a reinforcement in strength, indicating a strong interaction between the polymers and silica through the homogeneous dispersion of the micro silica particles in the polymer matrix.

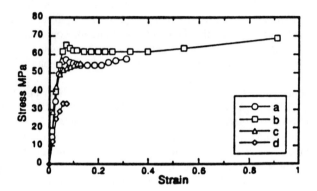

Figure 2. Stress-strain curves for unfilled copolymers a, b, c, and d.

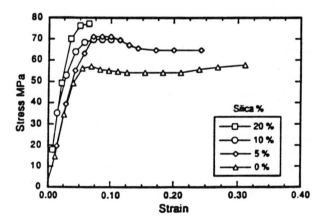

Figure 3. Stress-strain curves of hybrid films prepared from polymer a and TMOS.

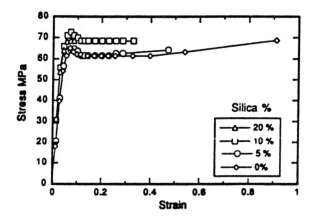

Figure 4. Stress-strain curves of hybrid films prepared from polymer b and TMOS.

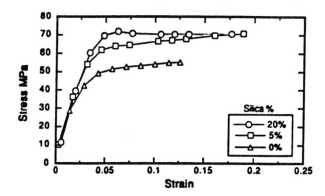

Figure 5. Stress-strain curves of hybrid films prepared from polymer c and TMOS.

Acknowledgment

The authors are pleased to acknowledge the financial support provided by the Air Force Office of Scientific Research.

References

(1) Morikawa, Y.; Iyoku, M.; Kakimoto, M.; Imai, Y. *Polym. J.* **1992**, *24*, 107
(2) Ahmad, Z.; Wang, S.; Mark, J. E. *Polym. Prepr. (Am. Chem. Soc., Div. Polym. Sci.)* **1993**, *34(2)*, 745
(3) Unroe, M. R. et. al. *22nd Science of Advanced Materials and Process Engineering Conference Proceedings,* **1990**, *22*, 186
(4) Tavs, P. *Chem. Ber.* **1970**, *103*, 2429
(5) Nagarajar, K. et. al. *Can. J. Chem.* **1987**, *65*, 1731
(6) Noell J. L.; Wilkes G. L.; Mohanty, D. K.; MacGrath, J. E. *J. Appl. Polym. Sci.* **1990**, *40*, 1177
(7) Maruyama, Y.; Oishi, Y.; Kakimoto, M.; Imai, Y.; *Macromolecules,* **1988**, *21*, 2307
(8) Arnold, F. E.; Chen, J. P. *Polym. Prepr. (Am. Chem. Soc., Div. Polym. Sci.)* **1991**, *32(2)*, 209

RECEIVED August 3, 1994

ELECTRICAL, OPTICAL, AND INTERFACIAL PROPERTIES

Chapter 24

Morphological Studies of Conductive Polymers Deposited onto High-T_c Superconductors

Steven G. Haupt, Rung-Kuang Lo, Jianai Zhao, and John T. McDevitt[1]

Department of Chemistry and Biochemistry, University of Texas at Austin, Austin, TX 78712

Methods for preparing and characterizing conductive polymer/high-T_C superconductor bilayers are reported. A number of procedures, such as cyclic voltammetry, chronoamperometry, and spray coating, have been utilized to deposit conductive polymers directly onto thin film samples of $YBa_2Cu_3O_{7-\delta}$. These structures have been fabricated in order to examine interesting electron and energy transfer phenomena between conductive polymers and superconductors as well as to explore prototype electronic devices and optical sensors. This paper focuses on an analysis of the growth of conductive polymer structures onto high-T_C superconductors. Scanning electron microscopy and electrochemical techniques are utilized to explore the hybrid structures.

A number of research groups, including our own, have initiated studies designed to explore the synthesis and characterization of conductive polymer/high-T_C structures (1-5). Conductive polymers possess many unique properties that complement those of high-T_C superconductors. Thus, conductive polymers may find utility for the construction of high-T_C composites and electronic devices.

In fact, the successful construction of a prototype superconductor/conductive polymer device based on a polypyrrole layer coated onto a $YBa_2Cu_3O_{7-\delta}$ thin film microbridge has been reported recently (1). In this device, the oxidation state of the polymer layer can be used to control the superconducting properties of the underlying high-T_C structure. Accordingly, when the polymer layer is oxidized to its conductive state, the transition temperature (T_C) and critical current (J_C) of the underlying superconductor film are suppressed. Upon reduction of the conductive polymer layer back to its non-conductive form, T_C and J_C are found to return to values close to those acquired for the underivatized $YBa_2Cu_3O_{7-\delta}$ film. Reversible modulations in the values of the transition temperatures up to 50 K are noted for these structures. Thus, the principle of a molecular/superconductor switch for controlling superconductivity has been demonstrated. Moreover, we have also explored the use

---

[1]Corresponding author

0097–6156/95/0585–0308$12.00/0

of conductive polymers as semiconducting contact elements in conjunction with high-temperature superconductors (6) and as coatings to modify the optical response of superconducting microbridges (7).

In order to form well-behaved junctions between high-T_C superconductors and conductive polymers, it is necessary to avoid chemical damage to the superconductor and the conductive polymer materials. Unfortunately, both the cuprate compounds as well as the conductive polymer systems tend to degrade chemically upon exposure to the atmosphere. It is this environmental reactivity that makes fabrication of the hybrid structures challenging.

In addition to chemical compatibility issues, the morphology of the conductive polymer layer is important for a variety of reasons. Successful construction of polymer/high-T_C circuits and devices will require that methods be developed to control the deposition of the polymer onto selected regions of circuits. Moreover, the ability to generate smooth polymer layers of variable thickness will be important for future studies in which molecule/superconductor energy and electron transfer phenomena are modeled from theoretical points of view. Recently, we have begun to explore how the morphology of dye layers influence the optical response of dye/superconductor sensors (8). Interestingly, dyes with high extinction coefficients and smooth morphologies serve as well-behaved sensitizing layers. On the other hand, dyes with similar optical characteristics but which lack smooth morphologies behave rather poorly under similar conditions (8). This paper focuses on the preparation of conductive polymer/superconductor structures as well as on an analysis of polymer morphology. Electrochemical and scanning electron microscopy measurements are utilized to study these hybrid systems.

DEVICE FABRICATION PROCEDURES

While the growth of conductive polymer layers onto bulk high-T_C ceramic pellets has been demonstrated previously (2-5), such bulk superconductor structures do not lend themselves to the construction of useful electronic devices and circuit elements. Rather, the high-T_C thin film geometry is preferred for the construction of such systems. For this reason, superconductor thin films were utilized in the present study for the analysis of conductive polymer growth. Consequently, thin films of YBa$_2$Cu$_3$O$_{7-\delta}$ (~200 to 5000 Å in thickness) were deposited onto single crystalline MgO (100) substrates using the pulsed laser ablation method (9). After establishing a base vacuum of ~10^{-6} torr, 150 mtorr of oxygen was introduced into the deposition chamber and the films were deposited with a substrate temperature of ~730°C using a KrF excimer laser operating at 248 nm to irradiate a ceramic target of YBa$_2$Cu$_3$O$_{7-\delta}$. Following their deposition, the thin films were cooled to ~450°C and 1 torr of oxygen gas was introduced for the annealing step which typically lasted about 15 minutes. The resulting superconductor films possess smooth structures as illustrated in Figure 1a. As we have demonstrated previously (1), for some devices it is necessary to create points of weak electrical contact between the superconducting grains. Superconductivity is easily disrupted in such weak links making these structures more sensitive to the influence of light, magnetism and heat. Rougher films such as the one illustrated in Figure 1b are better suited for such applications.

A number of different polymeric compounds and synthetic techniques are now available to grow conductive polymer materials. However, not all polymers and polymer deposition methods are suitable for the preparation of polymer/superconductor structures. Many of the conductive polymer systems lose their high electrical conductivities when exposed to the atmosphere (10). Furthermore, the conditions utilized to grow some conducting polymers can cause

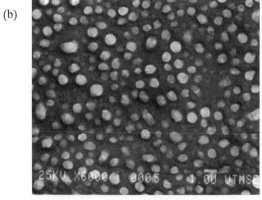

Figure 1. Scanning electron micrographs showing the surface of $YBa_2Cu_3O_{7-\delta}$ films that were deposited onto MgO (100) substrates using the laser ablation method. (a) Image of a ~1500 Å thick c-axis film and (b) ~800 Å thick polycrystalline specimen.

degradation of the superconductor surface. For example, the preparation of polyaniline films is normally accomplished using an acidic solvent. Although the use of such corrosive fluids is not problematic for the deposition of polyaniline films onto noble metal surfaces such as Pt or Au, growth of polyaniline onto bulk $YBa_2Cu_3O_{7-\delta}$ has been reported to result in the loss of up to 20% of the mass of the superconductor pellet (3).

We have, to date, utilized two different procedures for depositing conductive polymers onto superconductors thin film structures. The first involves the use of electrochemical techniques to polymerize the materials directly onto the surface of a $YBa_2Cu_3O_{7-\delta}$ thin film working electrode. Electrochemical procedures provide convenient and versatile methods for forming such polymers. For example, pyrrole can be electrochemically polymerized at relatively low potentials (1.0 V vs. SCE) and is reasonably stable in its conductive form. Thus, this material is an ideal candidate for the fabrication of polymer/superconductor structures.

The second approach for the preparation of polymer/superconductor systems involves the use of a solution processable polymer, such as a poly(3-hexylthiophene). Although this same material can be polymerized using electrochemical techniques, the use of high potentials that are required to polymerize this material causes chemical damage to the superconductor surface. Formation of the polymer chemically prior to deposition onto the superconductor provides a useful strategy for avoiding damage to the high-T_C element (11). After polymerization, poly(3-hexylthiophene) can be dissolved in a dry solvent that is compatible with the superconductor and the material can be deposited onto the high-T_C thin surface using a commercial artist's airbrush.

Once a polymer film is deposited, electrochemical techniques can be used to cycle the polymer layers between their neutral and oxidized forms. To dope polypyrrole, the polymer/superconductor assembly is immersed into a solution of $0.1M$ Et_4NBF_4 in acetonitrile and the electrode potential is raised above 0.5 V vs. SCE. While electrochemical polymerization of polythiophene from the monomer cannot be accomplished within the available stable potential window of the superconductor, electrochemical doping of the pre-formed poly(3-hexylthiophene) material can be achieved readily at low potentials (0.7 V vs. SCE).

MORPHOLOGY OF POLYMERS ON $YBa_2Cu_3O_{7-\delta}$

Polypyrrole. The two different methods utilized to deposit the conductive polymer materials onto $YBa_2Cu_3O_{7-\delta}$ thin film assemblies lead to drastically different polymer morphologies. The morphological differences noted here can, to a large extent, be attributed to differences in the deposition mechanism. When polypyrrole is grown electrochemically on the surface of superconductor, the morphology is determined by the local conductivity properties of the superconductor electrode surface. The typical polypyrrole film morphology is very globular in nature when grown on $YBa_2Cu_3O_{7-\delta}$ and is similar in some details to that of films grown on platinum under similar conditions. However, polypyrrole films deposited onto $YBa_2Cu_3O_{7-\delta}$ are less uniform than those films grown on platinum electrodes.

Figure 2a shows a low magnification image of a polypyrrole coated $YBa_2Cu_3O_{7-\delta}$ microbridge. While the polymer film is relatively smooth, areas of thicker polymer can be seen. Figure 2b is a high magnification image that clearly shows the globular morphology of the polypyrrole.

Figure 2. (a) Low and (b) high magnification scanning electron micrographs of polypyrrole film that was grown electrochemically onto the surface of a 100 μm wide $YBa_2Cu_3O_{7-\delta}$ microbridge supported on a single crystal MgO (100) substrate.

Images we have obtained from atomic force microscopy of polypyrrole during the initial stages of deposition reveal that the polypyrrole structures are more dispersed on the superconductor than on platinum or gold (12). These isolated areas that are observed during the early stages of film growth at $YBa_2Cu_3O_{7-\delta}$ surface are probably responsible for the somewhat rougher film morphology that is obtained at longer deposition times as compared to platinum. Nucleation of the polymer onto c-axis oriented films of $YBa_2Cu_3O_{7-\delta}$ occurs more slowly and in fewer locations when compared to growth on platinum film electrodes. Studies are currently in progress to evaluate the differences in growth behavior of polypyrrole on a-axis, a,c-axis and c-axis oriented $YBa_2Cu_3O_{7-\delta}$ thin films.

Unlike chemically robust noble metals which possess isotropic structures, the high-T_C compounds exhibit high chemical reactivities and anisotropic electrical properties which complicate their electrochemical behavior. Parasitic corrosion reactions which occur when high-T_C materials are exposed to water, CO_2, CO and acids have been shown to inhibit electron transfer reactions which occur at cuprate surfaces (13). Surface degradation reactions can result in local regions on the film that display large variability in their electroactivity properties. As a result, the actual electrochemically active area of the high-T_C specimen may vary considerably across the surface and this lack of uniformity may play an important role in determining the morphology of the superconductor localized polypyrrole films.

Careful analysis of the current densities for the polypyrrole growth supports the idea that the growth of the polymer is influenced by the superconductor surface. We have found that the growth rate of polypyrrole is often much slower at the superconductor than that obtained at a platinum electrode. The retarded growth observed for the superconductor surface is likely due to surface degradation of the cuprate film in certain regions. In previous electrochemical studies in which encapsulated pellets of superconductor were used as the working electrodes, procedures were developed in which the surface of the electrode was polished periodically to remove the insulating degradation phases from the electrode surface. Such surface treatments were shown to dramatically improve the performance of the high-T_C electrodes (14). Unfortunately, these treatments are not suitable for use with the thin films of $YBa_2Cu_3O_{7-\delta}$.

However, an immersion of the $YBa_2Cu_3O_{7-\delta}$ film in a 1% solution of Br_2 in ethanol can be used to remove the surface degradation and restore a chemically pristine surface (15). Chronoamperometry experiments have shown that etching by bromine treatment yields a more active surface of the high-T_C film and appears to be an effective surface treatment procedure for the preparation of high-T_C thin film electrodes.

Poly(3-hexylthiophene). In contrast to the electrochemically polymerized systems, the morphology of the poly(3-hexylthiophene) is, for the most part, independent of the quality of the surface of the superconductor film. Consequently, the polymer films can be deposited easily onto insulating materials as spray coated films. This property has been exploited to create contacting layers that can electrically bridge superconducting elements over relatively large insulating surfaces. As seen in Figure 3a, the small scale features of these spray coated films appear to be dominated by the micro droplets that are employed in the fabrication procedure. Interestingly, the features shown in Figure 3b appear to be more uniform than the electrochemically polymerized pyrrole. Polymer films with smooth morphologies are obtained by spray coating both onto superconductor as well as insulator substrates.

(a)

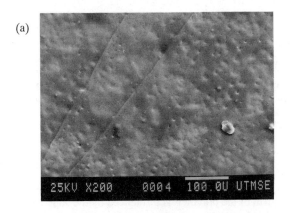

(b)

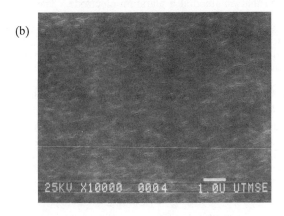

Figure 3. (a) Low and (b) high magnification scanning electron micrographs of a poly(3-hexylthiophene) film that was deposited onto a $YBa_2Cu_3O_{7-\delta}/MgO$ (100) thin film assembly using a spray coating procedure.

We have found that the overriding factor responsible for the quality of the spray coated films is the size of the micro droplets in the spray. Aspirators with large apertures that produce a relatively coarse mist create the roughest films. The smoothest films that we have produced have been fabricated by using a commercial artist's airbrush. While spin coating can create smoother and more uniform films, the spray coated films are much thicker and have a higher current carrying capacity. Another factor that influences the morphology of the spray coated films is the volatility of the solvent. More volatile solvents tend to produce films having smaller feature sizes.

CONCLUSION

In summary, we have developed a number of techniques that can be used to create conductive polymer/high-T_C superconductor structures. The polymer films that are electrochemically polymerized directly onto superconductor thin film tend to have a less uniform morphology when compared to those that have been spray coated. However, the electrochemical deposition procedure provides a convenient method for exploring issues related to the local surface conductive properties of high-T_C films. Further work in this area is warranted because knowledge of the surface chemical properties of the high-T_C system is important for the development of future high-T_C electronic devices.

ACKNOWLEDGMENTS

This work was supported by the National Science Foundation, the Texas Advanced Technology Program and the Welch Foundation. Princeton Applied Research, Texaco and Exxon Corporations are thanked for generous gifts.

REFERENCES

1. Haupt, S.G.; Riley, D.R.; Jones, C.T.; Zhao, J.; McDevitt, J.T. *J. Am. Chem. Soc.* **1993**, *115*, 1196-1198.
2. Kaneto, K.; Yoshino, K. *Jpn. J. Appl. Phys.* **1987**, *26(11)*, L1842-L1844.
3. Izakovich, E.N.; Geskin, V.M.; Stepanov, S.V. *Synth. Metals* **1992**, *46*, 71-77.
4. Osteryoung, J.G.; Magee, L.J.; Carlin, R.T. *J. Electrochem. Soc.* **1988**, *135 (10)*, 2653-2654.
5. McDevitt, J.T.; McCarley, R.L.; Dalton, E.F.; Gollmar, R.; Murray, R.W.; Collman, J.P.; Yee, G.T.; Little, W.A. in *Chemistry of High-Temperature Superconductors II* , Nelson, D.L.; George, T.F., Eds.; ACS Symposium Series 377, American Chemical Society Press: Washington, D.C., 1988, Chapter 17.
6. Haupt, S.G.; Riley, D.R.; Zhao, J.; McDevitt, J.T., *J. Phys. Chem.* **1993**, 97, 7796-7799.
7. Jurbergs, D.; Haupt, S.G.; Lo, R.; Zhao, J.; McDevitt, J.T. *J. Mol. Cryst. Liq Cryst.*, accepted.
8. Jurbergs, D.; Lo, R.; Zhao, J.; McDevitt, J.T., *Bolometric and Nonbolometric Superconductor Detectors*, SPIE Symp. Series, Los Angeles CA, 1994, accepted.
9. Dijkkamp, D.; Venkatesan, T.; Wu, X. D.; Shaheen, S. A.; Jisrawi, N.; Min-Lee, Y. H.; McLean, W. L.; Croft, M. *Appl. Phys. Lett.* **1987**, *51*, 619-621.

10.   *Handbook of Conducting Polymers*, Skothheim, T.A., Editor; Marcel Dekker, Inc: New York, 1986.
11.   Sugimoto, R.; Takeda, S.; Gu, H.; Yoshino, K. *Chem. Express* **1986**, *1*, 635-638.
12.   Haupt, S.G.; Riley, D.R.; Grassi, J.; Lo, R.; Zhao, J.; Zhou, J.P. and McDevitt, J.T., *J. Am. Chem Soc.*, submitted.
13.   Riley, D. R.; McDevitt, J. T. *J. Electroanal. Chem.* **1990**, *295*, 373-384.
14.   McDevitt, J.T.; Riley, D.R.; Haupt, S.G. *Anal. Chem.* **1993**, *65*, 535A-545A.
15.   Vasquez, R.P.; Hunt, B.D.; Foote, M.C. *Appl. Phys. Lett.* **1988**, *53*, 2692.

RECEIVED August 3, 1994

Chapter 25

Novel Organic–Inorganic Composite Materials for Photonics

Paras N. Prasad, Frank V. Bright, Upvan Narang, Run Wang, Richard A. Dunbar, Jeffrey D. Jordan, and Raz Gvishi

Photonics Research Laboratory and Department of Chemistry, State University of New York at Buffalo, Buffalo, NY 14214

A polymeric composite structure offers one the opportunity to systematically optimize individual properties independently in order to produce useful materials for photonics. This contribution summarizes a portion of our work on the local microenvironment and photonic properties in a new class of inorganic:organic hybrid composites prepared by sol-gel processing. The ability to form sol-gel processed hybrid materials into films, monoliths, and fibers *and* to control the microstructure for producing optically transparent bulk materials with desired porosity make sol-gel-derived materials suitable for development of chemical biosensors. However, in order to achieve this final goal one must understand how the evolving sol-gel matrix affects the "chemical recognition element" entrapped within the sol-gel network. Toward this end, we have used steady-state and time-resolved fluorescence spectroscopy to determine how the microenvironment around a dopant (organic fluorophore) entrapped within a sol-gel matrix evolves with time. We have successfully used this new information to develop new chemical sensing platforms (i.e., thin films) based on artificial recognition element, enzymes, and intact antibodies. The sol-gel matrix clearly provides a convenient avenue to entrap organic dopants. This in turn provides an attractive vehicle for fabricating nonlinear optical glasses. With this goal in mind, we have prepared a series of inorganic glass:polymer composites for third-order nonlinear optical applications. The nonlinear optical response has been investigated using femtosecond degenerate four-wave mixing, optical Kerr gating, and transient absorption. Stable electric field-induced alignment has been achieved for second-order nonlinear optical effects such as second harmonic generation and electro-optic modulation.

Sol-gel processing lends itself to conveniently prepare novel inorganic:organic composite materials *(1-4)*. This approach involves the hydrolysis of a metal

0097–6156/95/0585–0317$12.00/0

alkoxide, followed by a cascade of condensation and polycondensation reactions *(1-4)*. During this process, one starts with a solution phase which is transformed first into a gel and finally into a solid glass upon removal of the solvents. The low temperature conditions associated with the sol-gel process allow one to encapsulate organic species within a sol-gel-derived inorganic matrix without decomposing the organic moiety. Sol-gel-processed materials have been used in areas ranging from the development of novel solid-state dye lasers to chemical biosensors *(1-14)*. These sol-gel-processed materials have also been used to fabricate non-linear optically active composites *(5)* for applications in optical telecommunications.

Sol-gel-derived inorganic:organic hybrid materials have been prepared by various procedures. As examples, one can mix the sol-gels with the organic materials while the sol is in the solution phase, impregnate or dope the organic species into the pores of a sol-gel-processed composite material, hydrolyze a precursor material with an organic moiety covalently bonded to it (ORMOSIL), or mix an ormosil with an unmodified sol-gel solution.

In a majority of the sol-gel-derived inorganic:organic composites, the chemical recognition, activity, or overall composite function arises from the organic dopant. Thus, it becomes very important to understand how the local microenvironment around the dopant, within a sol-gel-derived material, is affected by the growing/evolving sol-gel network. Toward this end, we have encapsulated rhodamine 6G (R6G) and 6-propionyl-2-(dimethylamino)naphthalene (PRODAN) within tetramethyorthosiliane (TMOS)-derived sol-gel glasses *(15-17)*. R6G and PRODAN are fluorescent molecules with unique properties. R6G is an isotropic rotor and its emission spectrum is essentially insensitive to its local environment *(16)*. As a result, we can use static and time-resolved anisotropy techniques to study the rotational freedom/mobility of this dopant within a sol-gel matrix. The excited-state decay kinetics and emission profile of PRODAN are highly dependent on the local microenvironment. Thus, PRODAN reports on the local sol-gel microenvironment around the probe. In addition, PRODAN has very good nonlinear optical properties.

Sol-gel-processed materials have emerged as platforms for chemical biosensors. To date, there have been several encouraging reports on the encapsulation of organics, artificial receptors, enzymes, and antibodies within sol-gel-derived materials *(9-14)*. For example, our group reported the first successful encapsulation of an artificial chemical recognition element within a sol-gel matrix *(7)*. Soon after that work we also demonstrated the first antigen-antibody function within a sol-gel matrix *(12)*.

Photonics represents a multidisciplinary frontier of science and technology that has captured the interest of scientist and engineers around the world. Much of the current interest arises because of the applicability of photonics to current and future information and image processing technologies. Photonics is analogous to electronics in that it describes the technology in which photons (not electrons) are used to acquire, store, transmit, and process information. Nonlinear optical processes are critical to the ultimate processing of photonics information. Typical examples of key nonlinear optical phenomena include the ability to alter the frequency (or color) of light and to amplify one source of light with another *(5,18-21)*.

At the molecular level, the nonlinear optical response can be described in terms of the dipole induced by the applied electric field which can be expanded by the following power series *(19)*:

$$(\mu - \mu_0) - \alpha \cdot E + \beta : EE + \gamma : EEE. \tag{1}$$

In the above expression, μ and μ_0 are the total and permanent dipole moment; E is the applied external electric field, the coefficients α, β, and γ are linear, second, and third order molecular polarizabilities, respectively. The terms involving β and γ describe the nonlinear optical response of the molecule when E is derived from an applied optical field.

To describe bulk optical nonlinear response, one can use expansion of the bulk polarization, P, as follows *(19)*:

$$P - \chi^{(1)} \cdot E + \chi^{(2)} : EE + \chi^{(3)} : EEE + ... \tag{2}$$

This expression is the bulk analog of equation 1. The terms $\chi^{(n)}$ are the nth order bulk susceptibilities. These can be recovered from the corresponding molecular polarizabilities (α, β, and γ) by using orientationally-averaged site sums with Lorentz local field correction factors, relating the applied field to the local field at a molecular site *(18)*.

Microenvironment Within Sol-Gels

One of the critical aspect of using sol-gels for the development of chemical sensors and non-linear optical materials is information on how the evolving sol-gel matrix or network affects the chemical recognition element or nonlinearly active material. In order to address this issue we set out to investigate changes in microviscosity, rotational mobility, and local environment around model dopants during all stages of the sol-gel aging cycle. The model dopants used in these studies were the luminescent probes R6G *(15,16)* and PRODAN *(17)*.

The first experiments used the steady-state and time-resolved anisotropy to determine the dynamics of R6G doped within a TMOS-derived sol-gel matrix *(16)*. Fluorescence anisotropy experiments provide insight into the mobility of fluorescent centers within a given microdomain. Our results showed that the average anisotropy ($<r>$) after the completion of sol-gel aging cycle was statistically smaller than the limiting anisotropy for R6G (r_0) recovered in a vitrified solvent (e.g., 1,2-propanediol). These results demonstrated that the R6G molecules (radius $\approx$ 5 Å) are partially mobile within the porous sol-gel network. Time-resolved anisotropy results revealed that the observed anisotropy resulted from two distinct reorientational motions throughout the entire sol-gel aging cycle. One of the rotational correlation times remains constant throughout the sol-gel aging cycle and the other varied with aging time. The rotational correlation time data was transformed into microviscosity data (Figure 1), and these results were interpreted in terms of a two-microdomain model. One of the domains exhibits a low viscosity and remains "constant" throughout the sol-gel aging cycle. The second domain, in contrast, is "variable"; it is low in viscosity initially but substantially more viscous as the sol-gel ages. The relative ratio of the "constant" to "variable" domain decreases significantly as the sol-gel ages. These results demonstrate the heterogeneity of the sol-gel matrix, the complexity of the microencapsulation process, and show that there is a subpopulation of dopants that are mobile within the sol-gel network.

The excited-state decay kinetics of PRODAN is strongly dependent on the physicochemical properties of its local environment (i.e., the cybotactic region). Thus, PRODAN doped within a sol-gel can provide information on the evolution of the network throughout the sol-gel aging cycle (17). The PRODAN fluorescence over the aging cycle indicated that solvent expulsion from the sol-gel matrix is a step-wise process, in which the removal of ethanol is followed by water. The excited-state intensity decay profiles of PRODAN in sol-gels is best described by a unimodal Gaussian distribution throughout aging cycle. This result suggests that PRODAN experiences an ensemble of microenvironments.

Figure 2 depicts the change in average PRODAN lifetime (within a sol-gel matrix) as a function of aging time and processing pH (8, • and 4, ∇). The average PRODAN lifetime is 2.6 ns during the first week of the sol-gel aging cycle. This is intermediate to the lifetime of PRODAN in water (2.1 ns) and ethanol (3.6 ns). Upon further aging, the average lifetime drops to 2.1 ns suggesting a water-like environment. The average lifetime of PRODAN increases and levels off at 2.6 - 2.7 ns on further aging of the sol-gels. These results indicate that cybotactic region within a sol-gel is heterogeneous immediately after gelation and throughout the entire aging cycle and that expulsion of a more volatile solvent, ethanol, is followed by the expulsion of water.

Biosensors

The selectivity of antibodies has been widely used for the development of antibody-based biosensors (22-30). In these schemes, an immobilized antibody or fragment thereof serves to recognize selectively the target analyte and the binding process leads to an optical, mass, or electrochemical response related to the concentration of analyte in the sample (22-30).

There are many steps associated with the actual development of any real-world antibody-based biosensor (22-30). As examples, one must select an appropriate target analyte-antibody pair, develop a detection scheme, and "immobilize" the antibody (28) such that it retains its affinity and is stable over time. The decision on a particular antigen-antibody pair is straight forward and the choice of detection method depends on issues like required dynamic range and needed detection limits. The issue of protein immobilization is not, however, so straight forward, yet it represents one of the keys to any biosensor development (22-28).

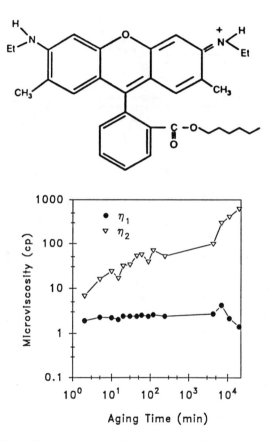

Figure 1. Effects of aging time on the recovered microviscosities associated with the two discrete R6G rotational motions in TMOS-derived sol-gel. η_1 and η_2 corresponded to R6G in the "constant" and higher viscosity, "variable" sol-gel microdomains, respectively. The chemical structure of R6G is also shown.

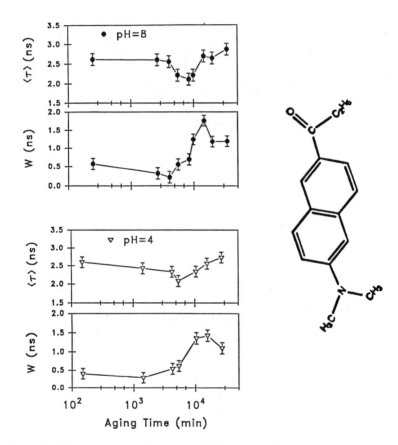

Figure 2. Effects of aging time and pH [8 (•); 4 (▽)] on the recovered mean excited-state lifetime (<τ>) and the width (W) of PRODAN in a TMOS-derived sol-gel. The error bars represent the uncertainty at the rigorous 67% confidence interval. The chemical structure of PRODAN is also shown.

The most common methods of immobilization involve non-covalent (entrapment and adsorption) or covalent attachment schemes *(29)*. Unfortunately, although many of these methods work reasonably well, they can be non-trivial, the immobilized antibodies are often unstable or loose a significant portion of their affinity with time *(29,30)*, and the final biosurface is prone to fouling in a real biological milieu. Thus, biosensor development is very much limited by the lack of a simple, generic immobilization protocol. As a result, a simple (ideally one-step) method to immobilize and stabilize active antibodies such that they can be located at sensor or transducer interfaces would offer many advantages.

Recently, ambient condition sol-gel methods have yielded a new way to immobilize (i.e., encapsulate) proteins within porous, optically transparent glasses *(9-14)*. Because of the mild conditions associated with the sol-gel glass processing, encapsulated biomolecules have been shown to retain a degree of function/activity *(9-14)*. This approach is unique compared to the conventional methods involving adsorption on glass surfaces, entrapment in polymer matrices, or impregnation of porous glass powders because encapsulation is based on actually growing of siloxane polymer chains *around* the biomolecule within an inorganic oxide network.

We recently demonstrated for the first time that an antibody (antifluorescein) could be entrapped within a sol-gel-derived glass matrix *(12)*. In this particular series of experiments the hapten (fluorescein) served simultaneously as the fluorescent probe. When fluorescein is free in solution it exhibits a strong emission at about 520 nm. When bound to antifluorescein its fluorescence is quenched by a factor of ~ 10 and there is a distinct shift in the emission spectrum. Thus, we were able to use changes in the fluorescence spectrum, intensity, and excited-state decay kinetics to probe the antifluorescein -fluorescein binding process within the sol-gel matrix (Figure 3). Our results demonstrated that: (1) we could indeed entrap an antibody within a sol-gel glass matrix; (2) analyte could diffuse into the porous sol-gel network; (3) the entrapped antibody retained its inherent function (i.e., it selectively bound its hapten); and (4) the antibody retained a remarkable fraction of its affinity ($K_f = 10^{10}$ M^{-1} native; $K_f = 10^7$ M^{-1} in sol-gel).

Nonlinear Optics

The ultimate realization of advanced photonics technologies rests on the development of multifunctional materials which are capable of simultaneously satisfy many requirements. In order to implement nonlinear optical materials to perform photonics functions (e.g., optical frequency conversion, light control by electric field or by another light beam) and fabricate photonic devices (e.g., frequency convertors, light modulators, optical switches, limiters), one needs material of high optical quality with large, stable optical nonlinearities. One also requires a means to control the linear refractive index and thickness of the final material *(21)*.

The best currently known optical medium of low optical losses (high optical quality) is silica glass. Sol-gel processing is a novel way to make optically

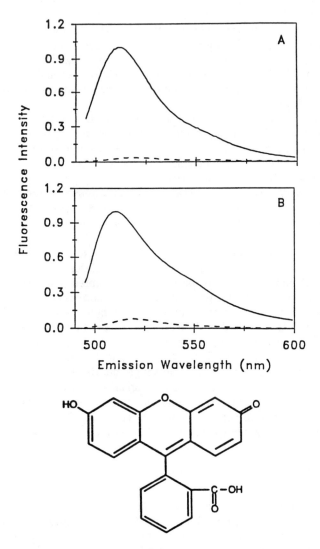

Figure 3. Emission spectra of fluorescein without (___) or with (---) antifluorescein. (Panel A) In solution. (Panel B) In a sol-gel glass matrix. The chemical structure of fluorescein is also shown.

transparent *composite* glasses which are extremely useful in the design of passive and active optical devices *(1,31-33)*. In the following section we summarize some of our more recent work on the fabrication of waveguides and electro-optic modulators based on sol-gel technology. We also present a new approach to preparing third-order optical materials.

Waveguides and electro-optic modulators. Preparing oxide glass:organic polymer composites by the sol-gel process allows one to vary the linear refractive index of the medium by controlling the composition of the precursors. This provides a simple, convenient means to control the waveguide properties of the films prepared by the sol-gel process. In this way, one can easily produce passive waveguides and channels on different substrates. As an example, we have mixed poly-vinylpyrrolidone (PVP) with tetraethoxysilane (TEOS) and cast films on glass microslides. The linear refractive index of the composite films varied linearly with added PVP (1.49 at 50% PVP to 1.54 100% PVP). The resulting films were optically transparent and of good waveguiding quality. Using this protocol we have made films with waveguide losses as small as 0.2 dB/cm *(34)*. The fact that these composites exhibit excellent optical quality suggests that the components, if they are not molecularly mixed, are restricted to domains which are much smaller than the wavelength of light.

An additional application of sol-gel:polymer composites involves fabricating electro-optic modulators where the refractive index changes with an applied electric field. The key properties of the ideal composite includes: (1) high $\chi^{(2)}$ value (r-coefficient > 30 pm/V; comparable to r-coefficient of $LiNbO_3$); (2) wide spectral window such that one can use different wavelengths for modulation (important wavelengths for optical communication are 810 nm, 1.3 μm, and 1.55 μm); (3) thermal stability up to 300 °C; and (4) high stability of the electrically aligned dopants at elevated temperatures.

The key to obtaining electro-optic activity involves orienting a nonlinearly active dopant in a noncentrosymmetric arrangement. Sol-gel-processed materials present an almost ideal host since electric field poling to induce alignment can be carried out at room temperature well before the sol-gel glass matrix becomes fully densified. Once the active dopants become oriented in the electric field subsequent heat treatment causes the pores within the film to shrink and thus "lock" the active molecules in the poled orientation.

The aforementioned scheme has been used in our laboratory to prepare a series of second-order active chromophore-doped composites poled under various conditions. We then investigated the second harmonic intensity of these new materials after poling (with and without the applied field) as a function of time *(35,36)*. We obtained fairly high and stable second order activity at room temperature. However, when we heated the samples to 200 °C the second order intensity dropped significantly.

More recently, we used a new chromophore, N,N'-diethoxyamin-4'-nitrostilbene, which has two terminal -OH groups chemically bonded to silica. With this chromophore, we fabricated thin films based on the sol-gel process. The resulting films were very hard, of good optical quality, and one could easily cut the edge of the films and effect "end-fire" coupling to couple light in to the planar waveguiding film. These films exhibit high thermal stability, the recovered r-coefficient is over 30 pm/V at 633 nm, and the films were actually poled at 280 °C for a short period of time *(37)*. These materials show tremendous promise and after further optimization of the processing steps may even meet the thermal requirements.

Third-order nonlinear optical materials. Important third-order nonlinear optical effects include third harmonic generation, degenerate four wave mixing (DFWM), and intensity-dependent refractive index devices. The key factor associated with these third-order processes is a large $\chi^{(3)}$ value. Conjugated polymers have emerged as an important class of third-order nonlinear optical materials *(18,20)*. Unfortunately, films formed from these polymers often exhibit poor optical quality. Composites, based on sol-gel processed inorganic oxide/conjugated polymer (e.g., poly-p-phenyline-vinylene (PPV): silica; 1:1 by weight), have proven to be more suitable as third-order nonlinear optical materials *(5,38)*.

We previously reported studies of the third-order nonlinearity of PPV-silica films using femtosecond time-resolved transient absorption, optical Kerr gate (OKG), and DFWM measurements *(38)*. Although we demonstrated third-order nonlinearity in these materials, the values were not sufficiently large. One means to improve the low nonlinearity is to simply increase the interaction length. Toward this end, we introduced a third-order-active chromophore into bulk silica glass by using the method developed by Gvishi, et al. *(39)*.

Briefly, the chromophores were dissolved in a methylmethacrylate (MMA) monomer solution. Highly porous silica xerogels were prepared by the sol-gel process and dried finally at 500°C. The bulk xerogel was then immersed in a chromophore-MMA monomer solution, which was simultaneously radical catalyzed. The MMA-chromophore solution that diffused into the xerogel pores was polymerized therein. These poly(methylmethacrylate) (PMMA):silica glass composites were cleaned and polished until good optical transmission was achieved.

The chromophores used in these experiments are new oligomers of polyphenyl (didecyl- and didecyloxy-para-poly-phenyl heptamer) that were synthesized by Reinhardt *(40)*. These particular oligomers have a very strong two-photon absorption between 600 - 700 nm which may prove useful as part of an optical power limiter. That is, on increasing the intensity there is an increase in absorption. Thus, the transmission through the bulk will decrease as the incident photo flux increases in the wavelength range of 600 - 700 nm.

The third-order nonlinear optical coefficients of the heptamer in the PMMA:silica composite glass were also studied by DFWM and OKG measurements. The apparatus used in our DFWM experiments is the same as that described previously *(41,42)*. The laser source delivers amplified and nearly transform-limited 400 fs pulses at 602 nm at a repetition rate of 30 Hz (up to 0.4 mJ/pulse). In a DFWM experiment, the time response of the third-order optical nonlinearity is obtained by measuring the decay of the signal as a function of the time delay of the probe beam with respect to the interfering beams. Figure 4 illustrates the temporal behavior of the DFWM signal for the heptamer in a composite PMMA:silica glass.

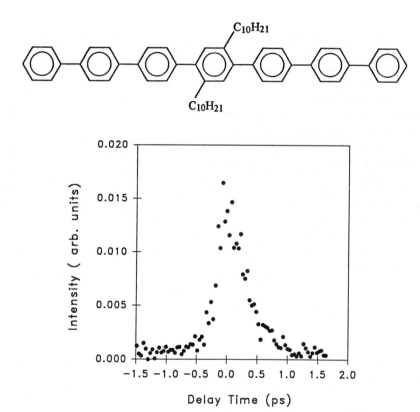

Figure 4. The temporal DFWM profile for $\approx 10^{-2}$ M heptamer in a PMMA:silica composite glass (1 mm thickness). The chemical structure of heptamer is also shown.

An effective third-order susceptibility was calculated from the intensity of the signal at time delay equal zero for different concentrations of heptamer using bare fused quartz as a reference *(40)*:

$$\chi^{(3)} = (\frac{n}{n_{ref}})^2 \ (\frac{I}{I_{ref}})^{1/2} \ (\frac{l_{ref}}{l}) \ \chi^{(3)}(ref) \ F \qquad (3)$$

where I stands for DFWM signal intensity, n is the refractive index of the medium, l is the interaction length, and the subscript ref refers to a reference sample. $\chi^{(3)}$ (ref) is the third-order susceptibility of the reference, fused quartz, taken as 0.3×10^{-14} esu *(43)*. F is the correction factor taking into account the sample absorption. Since there is no one-photon absorption for this dye at 602 nm only two-photon absorption correction is necessary. We obtained a $\chi^{(3)}$ value of approximately 1×10^{-12} esu.

In the OKG experiments, a strong pump beam creates an optically-induced birefringence in the nonlinear medium. This birefringence is detected by passing a linear polarized probe beam through the medium and monitoring the transverse component through a polarizing analyzer. The OKG experiments were performed using the 60 fs pulses from a colliding pulsed mode-locked dye laser. Femtosecond pump-probe experiments show a very fast response followed by a response associated with an intermediate state. In contrast, when one does the same study in a PMMA:silica composite, the relaxation pathways are altered, and we no longer see the intermediate state build up. These preliminary results suggest that one can control the dynamics of the resonant processes by controlling the processing condition. Additional studies on this phenomenon are currently in progress.

Summary

We have shown how sol-gel-derived materials can serve as a novel platform for a new generation of chemical and biosensors and demonstrated that the tunability of the sol-gel matrix can be used to develop new types of nonlinear optical materials and composites.

Acknowledgments

A portion of the results presented here have been obtained in collaboration with Laser Photonics Technology. The research in our laboratories has been supported by the Air Force Office of Scientific Research and the National Science Foundation. We also thank Dr. Charles Lee for his encouragement.

Literature Cited

1.   Brinker, C. J.; Scherer, G. W. *Sol-Gel Science, The Physics and Chemistry of Sol-Gel Processing*, Academic Press, Inc.: San Diego, 1990; pp. 1-18.
2.   *Chemical Processing of Advanced Materials*; Hench, L. L.; West, J. K., Eds.; Wiley: New York, 1992.

3. Hench, L. L.; West, J. K. *Chem. Rev.* **1990**, 90, 33.
4. Paul, A. *Chemistry of Glasses*; 2nd Ed.; Chapman and Hall: New York, 1990; p. 51.
5. Burzynski, R.; Prasad, P. N. *Photonics and Nonlinear Optics with Sol-Gel Processed Inorganic Glass:Organic Polymer Composite*, Klein, L. C., Ed.; Kluwer Academic: Boston, 1994, Chapter 19.
6. Wolfbeis, O. S.; Rodriguez, N. V.; Werner, T. *Mikrochim. Acta* **1992**, 108, 133.
7. Narang, U.; Dunbar, R. A.; Bright, F. V.; Prasad, P. N. *Appl. Spectrosc.* **1993**, 47, 1700.
8. MacCraith, B. D.; Ruddy, V.; Potter, C.; O'Kelly, B.; McGilp, J. F. *Electron. Lett.* **1991**, 27, 1247.
9. Ellerby, L.; Nishida, C.; Nishida, F.; Yamanaka, S. A.; Dunn, B.; Valentine, J. S.; Zink, J. I. *Science* **1992**, 225, 1113.
10. Shtelzer, S.; Rappoport, S.; Avnir, D.; Ottolenghi, M.; Braun, S. *Biotech. Appl. Biochem.* **1992**, 15, 227.
11. Braun, S.; Rappoport, S.; Zusman, R.; Avnir, D.; Ottolenghi, M. *Mater. Lett.* **1990**, 10, 1.
12. Wang, R.; Narang, U.; Bright, F. V.; Prasad, P. N. *Anal. Chem.* **1993**, 65, 2671.
13. Tatsu, Y.; Yamashita, K.; Yamaguchi, M.; Yamamura, S.; Yamamoto, H.; Yoshikawa, S. *Chem. Lett.* **1992**, 1615.
14. Wu, S.; Ellerby, L. M.; Cohan, J. S.; Dunn, B.; El-Sayed, M. A.; Valentine, J. S.; Zink, J. I. *Chem. Mater.* **1993**, 5, 115.
15. Narang, U.; Bright, F. V.; Prasad, P. N. *Appl. Spectrosc.* **1993**, 47, 229.
16. Narang, U.; Wang, R.; Prasad, P. N.; Bright, F. V. *J. Phys. Chem.* **1994**, 98, 17.
17. Narang, U.; Jordan J. D.; Prasad, P. N.; Bright, F. V. *J. Phys. Chem.* **1994**, in press.
18. Prasad, P. N.; Williams, D. J. *Introduction to Nonlinear Optical Effects in Molecules and Polymers*, Wiley: New York, 1991.
19. Shen, Y. R. *Principles of Nonlinear Optics*, Wiley: New York, 1985.
20. Chemla, D. S.; Zyss, J. *Nonlinear Optical Properties of Organic Molecules and Crystals*, Academic Press: New York, 1987.
21. Prasad, P. N. *Thin Solid Films* **1987**, 152, 275.
22. Janata, J. *Principles of Chemical Sensors*, Plenum Press: New York, 1989.
23. *Fiber Optic Chemical Sensors and Biosensors*, Vol. I and II, Wolfbeis, O. S., Ed., CRC Press: Boca Raton, FL, 1991.
24. Wise, D. L.; Wingard, C. B., Jr. *Biosensors with Fiber Optics*, Humana Press: Clifton, NJ, 1991.
25. Tromberg, B. J.; Sepaniak, M. J.; Vo-Dinh, T.; Griffin, G. D. *Anal. Chem.* **1987**, 59, 1226.
26. Bright, F. V.; Betts, T. A.; Litwiler, K. S. *Anal. Chem.* **1990**, 62, 1065.
27. Bright, F. V.; Litwiler, K. S.; Vargo, T. G.; Gardella, J. A., Jr. *Anal. Chim. Acta* **1992**, 262, 323.
28. Thompson, M.; Krull, U. J. *Anal. Chem.* **1991**, 63, 393A.
29. Taylor, R. F. *Protein Immobilization: Fundamentals and Applications*, Marcel Dekker, Inc.: New York, 1991; Chapter 8.

30. Weetall, H. H. *Immobilized Enzymes, Antigens, Antibodies, and Peptides: Preparation and Characterization*, Marcel Dekker, Inc.: New York, 1975; Chapters 6 and 8.

31. *Sol-Gel Technology for Thin Film, Fibers, Preforms, Electronics, and Specialty Shapes*; Klein, L., Ed.; Noyes: New Jersey, 1988.

32. Hench, L. L.; West, J. K. *Chem. Rev.* **1990**, 90, 61.

33. Livage, J.; Henry, M.; Sanches, C. *Prog. Solid. State Chem.* 1988, 18, 259.

34. Yoshida, M.; Prasad, P. N., manuscript in preparation.

35. Zhang, Y.; Prasad, P. N.; Burzynski, R. *Chemical Processing of Advanced Materials*; Hench, L. L.; West, J. K., Eds.; Wiley: New York, 1992, p. 825.

36. Zhang, Y,; Prasad, P. N.; Burzynski, R. *Chem. Mat.* **1992**, 4,

37. Burzynski, R.; Prasad, P. N., unpublished results.

38. Pang, Y.; Samoc, M.; Prasad, P. N. *J. Chem. Phys.* **1991**, 94, 5282.

39. Gvishi, R.; Reisfeld, R.; Burshtein, Z. *SPIE Proc.* **1993**, 1972.

40. Zhao, M.; Cui, Y.; Samoc, M.; Prasad, P. N.; Unroe, M. R.; Reinhardt, B. A. *J. Chem. Phys.* **1991**, 95, 3991.

41. Zhao, M.; Singh, Y.; Prasad, P. N. *J. Chem. Phys.* **1988**, 89, 5535.

42. Zhao, M.; Samoc, B.; Singh, B.; Prasad, P. N. *J. Phys. Chem.* **1989**, 93, 7916.

43. Thomazeau, I; Etchepare, J.; Grillon, G.; Migus, A. *Optics Letters* **1985**, 10, 223.

RECEIVED December 12, 1994

Chapter 26

Inorganic—Organic Hybrid Coatings for Metal and Glass Surfaces

H. Schmidt[1], R. Kasemann[1], T. Burkhart[1], G. Wagner[1], E. Arpac[2], and E. Geiter[1]

[1]Institut für Neue Materialien, Im Stadtwald, Gebäude 43, D—66123 Saarbrücken, Germany
[2]University of Antalya, Guzeloluk Mahallesi 1846, 07200 Antalya, Turkey

Inorganic-organic composite materials have been tailored by sol-gel synthesis with different properties for different applications. A composite from SiO_2, diphenyl silanes and methylvinyl silanes was modified by an amino silane and crosslinked by diepoxides, leading to a sealing strength of more than 10 N/cm when used as a hot melt for sealing copper to polyimide. An ormocer type of coating was modified by nanoscaled γ-Al_2O_3 or boehmite, leading to a low-curing transparent coating material with very high scratch resistances and a ZrO_2 nanocomposite modified with silanes with perfluorinated groupings having high transparency combined with low surface free energy with 18 mJ/m² was synthesized.

The combination of properties related to ceramic materials and polymeric materials is an attractive conception and allows the tailoring of material properties in a wide range. For this reason, composite materials have been developed and applied for many purposes. In general, ceramic-polymer composite materials are prepared by mechanical mixing. With a few exceptions, ceramic filler components are produced by mechanical processing which restricts the particle size to the μm range. The exception is fumed silica which is available with smaller particle sizes (50 nm), but, in general, is rather agglomerated. If particles in the lower nano range are of interest, other methods have to be considered. The synthesis of inorganic components, starting from molecular precursors by a growth process and stopped at the desired particle size, seems to be an adequate route. In the sol-gel synthesis, as a rule, molecular precursors are used which react in a growth process to colloidal systems ranging from one to several 100 nm [1 - 5]. If these reactions are carried out in the presence of organic groupings or organic grouping-carrying molecules (e.g. organofunctional silanes), a variety of different types of composites can be synthesized (ormosils, ormocers, polycerams, ceramers) [6 - 9]. Using these principles, materials with

0097—6156/95/0585—0331$12.00/0

special optical [*10*], mechanical [*11*] or protective properties on various surfaces can be tailored. For example, the mechanical properties can be influenced by the type of crosslinking (three-dimensional versus two-dimensional) or the phase dimensions of the inorganic and the organic phase. Depending on the dimension of the inorganic phase, one can distinguish between so-called molecular-type and nano-type of composites, however, this differentiation is hardly done in the literature. In the present paper three examples for material development are given using inorganic-organic sol-gel processing with the objective of tailoring very special, but different surface properties: a high surface free energy system with adhesive properties based on a molecular type of composite, a low surface free energy system and a system with a very high abrasion resistance. Furthermore, in this paper, the material development aspect using sol-gel synthesis principles is emphasized.

Results

Materials with Tailored Mechanical Properties for Adhesion of Metal Foils to Polyimide

In order to achieve a good adhesion, adhesive materials have to show two basic properties: a good wetting behavior to the surface and mechanical properties leading to stress dissipation under peeling stress. If such materials have to be used in a hot-melt process, wetting is of extremely high importance since no solvents can be used which, in general, provide a good wetting behavior. If, in addition to this, high temperature stability and good sealing properties to surfaces with very different chemical properties (e.g. copper to polyimide for electronics) are required, the material basis becomes very narrow, and silicones are the most interesting materials at present. As already shown in [*11*], a hot-seal system has been developed by sol-gel techniques from diphenyl silanes (($(RO)_2Si(C_6H_5)_2$ (I)), methylvinyl silanes ($CH_3(CH_2=CH)Si(OR)_2$ (II)) and tetraethylorthosilicates ($Si(OR)_4$ (III)). As starting compound, the corresponding dichlorides can be used also. This system, however, developed for sealing aluminum foils to glass and having excellent properties with respect to moisture resistance, shows a relatively low peel strength of about 3 N/cm in the peel test. The sealing medium shows a typical brittle behavior under stress, since the strain zone cannot be dissipated by the system, as schematically shown in Figure 1. The experimental details are given elsewhere [*12*].

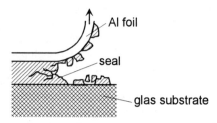

Fig. 1: Schematics of a peeling test with a too brittle sealing agent [after *11*].

The brittleness of the system can be explained by a high degree of three-dimensonal crosslinking based on a condensation process taking place at 150 °C through benzene elimination,

$$\equiv Si\text{-}C_6H_5 + HO\text{-}Si\equiv \rightarrow \equiv Si\text{-}O\text{-}Si\equiv + C_6H_6, \qquad (1)$$

which has been proved by IR spectroscopy as well as by analysis of volatiles during the curing process, forming stiff "units". Nevertheless, the resin-like systems show a thermoplastic behavior, which is attributed to the low molecular weight "units" (3000 - 5000 a.m.u., as proved by molecular weight measurements), stabilized at low temperatures by hydrogen bounds of $\equiv$SiOH groupings still present in the material. The most interesting property of the system is its excellent wettability to various surfaces (glass, metals, polymers) and a good hot-melt behavior. The wetting behavior is attributed to the polarity of the system caused by a relatively high content of SiOH groupings to be controlled by IR. The wettability decreases drastically if the ratio $E_{3620\,cm^{-1}}$ (SiOH) : $E_{3070\,cm^{-1}}$ (phenyl CH) ≤ 0.19.

In the present investigations, in order to reduce the brittleness of the system, short organic groupings have been introduced into the basic composition using amino silanes (e.g. $H_2NCH_2CH_2CH_2(CH_3)Si(OR)_2$ crosslinked by commercially available Bisphenol/A-diepoxides (Araldit$^{®}$). The experimental details have been presented elsewhere [*13*]. The objective of this investigation was to maintain the good wetting and adhesion properties as well as the thermoplastic behavior and to improve the stress dissipation power of the system by adding "elasticity". For this reason, in a sol-gel synthesis process the starting components (I, II, III, equation 2) have been hydrolysed under acidic conditions, and a condensation reaction was carried out to build up the inorganic backbone. In figure 2, the principles of the synthesis are shown. This synthesis is the result of a systematic development using different silanes and epoxides as well as an optimized reaction route. The use of amino grouping-containing silanes, for example, was found out to be a suitable means to improve the adherence to polyimide surfaces. However, the amino silane has to be bound to the inorganic -backbone during the first synthesis step (hydrolysis and condensation of the phenylsilane, the methyl vinyl silane and the TEOS) in order to be efficient.

$$(C_6H_5)_2Si(Cl_2)_2 + Si(OR)_4 + CH_2=CH(CH_3)SiCl_2 + H_2N(CH_2)_3Si(CH_3)(OR)_2$$

$$\xrightarrow[\text{condensation}]{\text{hydrolysis}} \left[\begin{array}{c} C_6H_5 \\ | \\ Si\text{-}O \\ | \\ C_6H_5 \end{array}\right]_m \left[\begin{array}{c} O \\ | \\ Si\text{-}O \\ | \\ O \end{array}\right]_n \left[\begin{array}{c} CH_3 \\ | \\ Si\text{-}O \\ | \\ CH=CH_2 \end{array}\right]_o \left[\begin{array}{c} CH_2\text{\textasciitilde}NH_2 \\ | \\ Si\text{-}O \\ | \\ CH_3 \end{array}\right]_p \qquad (2)$$

$$m : n : o : p = 15.4 : 1.4 : 39.3 : 14.6$$

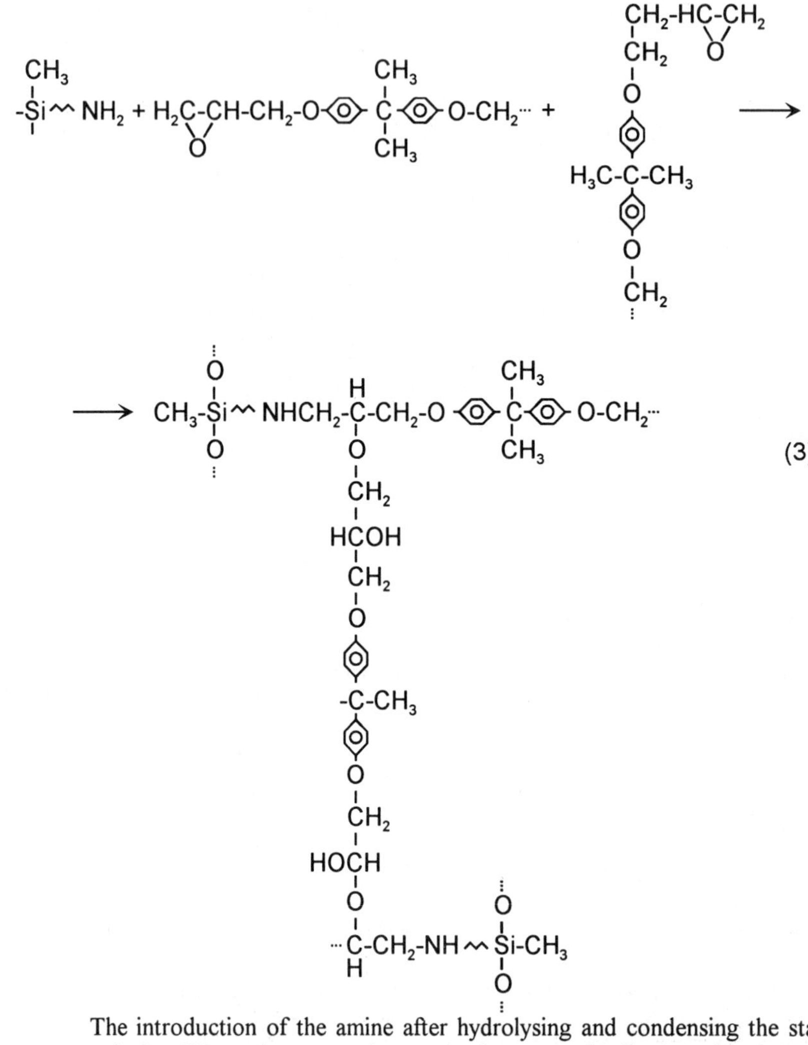

$$
\begin{array}{l}
\mathrm{CH_3} \\
\mathrm{-\overset{|}{\underset{|}{Si}}\!\sim\! NH_2} + \mathrm{H_2C\text{-}CH\text{-}CH_2\text{-}O\text{-}\langle\!\circ\!\rangle\text{-}\overset{\overset{\textstyle CH_3}{|}}{\underset{\underset{\textstyle CH_3}{|}}{C}}\text{-}\langle\!\circ\!\rangle\text{-}O\text{-}CH_2\cdots} + \quad\quad\quad\longrightarrow
\end{array}
$$

(Upper right fragment):

$$
\begin{array}{c}
\mathrm{CH_2\text{-}HC\text{-}CH_2} \\
\mathrm{\underset{|}{CH_2}\ \ \ O} \\
\mathrm{O} \\
\langle\!\circ\!\rangle \\
\mathrm{H_3C\text{-}C\text{-}CH_3} \\
\langle\!\circ\!\rangle \\
\mathrm{O} \\
\mathrm{CH_2} \\
\vdots
\end{array}
$$

$$
\longrightarrow\quad
\mathrm{CH_3\text{-}\overset{\overset{\textstyle O}{\vdots}}{\underset{\underset{\textstyle O}{\vdots}}{Si}}\!\sim\! NHCH_2\text{-}\overset{\overset{\textstyle H}{|}}{\underset{\underset{\textstyle\ }{|}}{C}}\text{-}CH_2\text{-}O\text{-}\langle\!\circ\!\rangle\text{-}\overset{\overset{\textstyle CH_3}{|}}{\underset{\underset{\textstyle CH_3}{|}}{C}}\text{-}\langle\!\circ\!\rangle\text{-}O\text{-}CH_2\cdots}
\qquad (3)
$$

(Lower chain, descending from the C–H carbon):

$$
\begin{array}{c}
\mathrm{CH_2} \\
\mathrm{HCOH} \\
\mathrm{CH_2} \\
\mathrm{O} \\
\langle\!\circ\!\rangle \\
\mathrm{-C\text{-}CH_3} \\
\langle\!\circ\!\rangle \\
\mathrm{O} \\
\mathrm{CH_2} \\
\mathrm{HOCH} \\
\mathrm{O} \qquad\qquad \overset{\vdots}{O} \\
\mathrm{\cdots C\text{-}CH_2\text{-}NH\!\sim\! Si\text{-}CH_3} \\
\mathrm{H} \qquad\qquad \overset{}{\underset{\vdots}{O}}
\end{array}
$$

The introduction of the amine after hydrolysing and condensing the starting compounds I - III to the uncured state leads to good adhesion, but low stress dissipation, since the epoxide then is not distributed uniformly within the inorganic network. The best results have been obtained by a molar ratio of amine to diepoxide of 1:1.

The process has been investigated by ^{29}Si-NMR [12]. In Figure 3 the ^{29}Si-NMR spectrum is shown after the polycondensation reaction before the diepoxide addition, and one can see that oligomers of a variety of types of species could be identified, based on comparison with the single components.

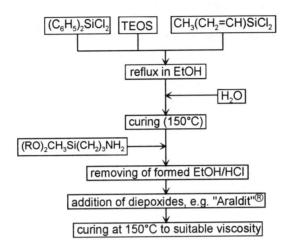

Fig. 2: Flow chart of the synthesis of the hot melt sealing agent.

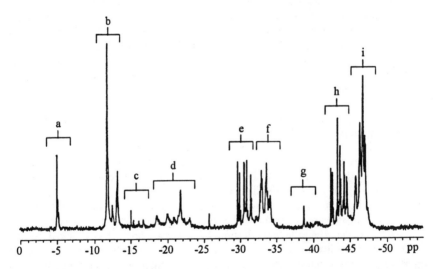

Fig. 3: ^{29}Si-NMR of the hydrolysed adhesive sol-gel systems from diphenyl silane, methylvinyl silane, amino silane and SiO$_2$ (four-component system without the diepoxide), a: monomeric amino silane; b: oligomeric and dimeric unit of the condensed amino silane; c: co-oligomers of the amino silane with the diphenyl silane; d: co-oligomers of the amino silane with the methylvinyl silane; e - h: oligomers and co-oligomers of the methylvinyl silane and SiO$_2$ with the amino silane and the diphenyl silane; i: oligomers of the phenyl silane.

In Figure 4, the solid state NMR spectrum of the system is shown after curing, which shows basically the same basic type of peaks, but monomers can no longer be detected. The system is still soluble in organic solvents.

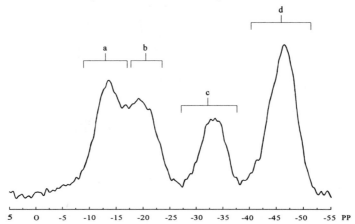

Fig. 4: ^{29}Si-NMR (solid state) of the cured four-component system. a: T_1 units of diamines; b: Si resonances of the amine with phenyl silanes (co-oligomers) and methyl vinyl silanes (co-oligomers); T_2 oligomers of the diamine; c: Si resonances of the methyl vinyl groupings with phenyl silane (co-oligomers) and with the diamine (co-oligomers); oligomeric diphenyl silanes; d: Si resonances of the phenyl silane group with the methyl silane (co-oligomers); phenyl silane oligomers.

The results show that during the condensation and curing process the monomers disappear. Similar to the epoxide-free system, after addition of the epoxide component the wettability in the uncured state with a viscosity of about 220 Pa · s (controlled by curing at 150 °C) as well as the temperature stability up to 140 °C after curing the seal at 100 °C for 5 hrs remain unchanged, but the peel strength increases substantially due to the higher deformability and stress dissipation of the seal in the peeling experiment. In Figure 5, the peel strength of compositions without amines and diepoxide are compared to the amine/diepoxide containing systems.

As one can see, the modification by organic short chains linked to the inorganic backbone leads to a strong increase of the peel strength without decrease of important properties like hot-melt processing and wetting. The peel strength values exceed the values of commercial siloxane seals (≈ 4 N/cm) used for polyimide-to-copper sealing remarkably. SEM analysis after peeling shows the typical surface patterns of a cohesion rupture, that means, both sides, the polyimide as well as the Cu surface, are covered by the sealant.

The results show clearly that by tailoring the structure of the sol-gel-derived inorganic-organic composite through organic crosslinking, the material properties can be tailored to suit the desired application. A structural model was developed, based on IR and NMR analysis. In these composites, no phase separation can be detected. The model is shown in Figure 6.

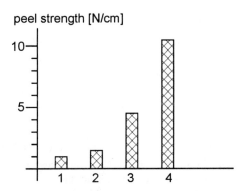

Fig. 5: Peel strength of different compositions. Methylvinyl silane (1); diphenyl silane (2); SiO_2 (3); amine (4); diepoxide (Araldit®) (5) [mole %].
1: (1) : (2) : (3) : (4) = 45 : 42.5 : 2.5 : 0
2: (1) : (2) : (3) : (4) = 60 : 37.5 : 2.5 : 0
3: (1) : (2) : (3) : (4) = 70 : 27.5 : 2.5 : 0
4: (1) : (2) : (3) : (4) : (5) = 39.3 : 15.4 : 1.4 : 14.6 : 14.6; peeling rate: 5 mm/min; sealant thickness: 28 μm.

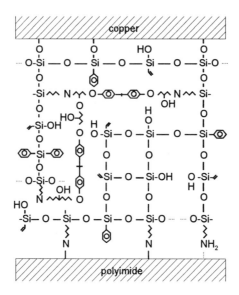

Fig. 6: Structural model of the epoxy crosslinked thermoplastic sealant.

It should be mentioned that the system is based on units crosslinked only by ≡Si-O-Si≡ bonds, additionally connected by epoxy bridging. The thermoplastic behavior again is attributed to the ≡SiOH bridging, getting weaker at elevated temperatures. The sealing agent can be applied to Cu foils by roll coating at 100 °C and then sealed to polyimide. After curing the seal is stable up to 180 °C.

Surface-Mechanical and Surface-Chemical Properties in Inorganic-Organic Nanocomposites

While hybrid sol-gel systems using SiO_2 or ≡Si-O-Si≡ bonds as inorganic network formers only seem to be preparable as molecular (or nearly molecular) types of composites, systems based on more reactive types of inorganic components with increased phase dimensions should be more difficult to be synthesized. In a molecular composite, the inorganic part looses its identity with respect to inorganic material properties, but mainly acts as a structure-modifying agent. In various cases, however, it seems to be desirable to preserve (or to use) properties related to the solid state properties of materials (ceramics or metals), e.g. passive or active optical properties, hardness or even catalytic properties. In case of optical transparency to be obtained, the particle size should be roughly less than 1/20th of the wavelength of the used light to reduce Rayleigh scattering, which means for the visible region particle sizes below 20 nm. Particle sizes in this range can be obtained by sol-gel processing very easily during the growth reaction, starting from alkoxides. For utilizing these colloids, e.g. for nanocomposite fabrication, however, it is necessary to control their surface properties with respect to particle-to-particle interaction (aggregation or agglomeration), which occurs spontaneously if the colloids are not covered by a "protecting" layer such as electric charges or non-reactive groupings. Electric charge stabilization, in general, leads to gelation with pH change or up-concentration, resulting in low-density porous gels. If nanocomposites with high optical performance have to be synthesized, gelation of the inorganic part has to be prevented at any rate, since gelation would lead to uncontrolled agglomeration with particle sizes causing undesired light scattering. As shown elsewhere [13, 14], transparent composites can be produced from methacrylic acid-modified Zr alkoxides reacted together with other methacrylates. An unsolved question of this reaction was related to the structure of the ZrO_2 nanoparticles synthesized from Zr alkoxide/methacrylic acid complexes [15]. Several investigations have been carried out for a better understanding of the process. A synthesis scheme is shown in Figure 7.

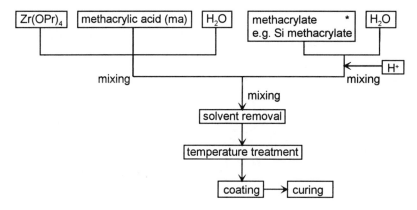

Fig. 7: Schematics of the formation of ZrO_2 (ma) composites with functional silanes.
*Other functional silanes can be added to the reaction at this stage.

As shown in [16], coatings on glass surfaces obtained from materials fabricated according to the scheme in Figure 7, the hydrolytic stability of the ma to ZrO_2 bond is very stable in the solid material. The reaction of ma with $Zr(OPr)_4$ can be monitored by IR from the shift of the   C=O frequency from 1718 to 1551 and by ^{13}C NMR. The IR as well as the NMR shift remain unchanged after hydrolysis of the $Zr(ma)(OPr)_3$. However, with excess water, precipitation takes place. The analysis of the Zr ma bond in the precipitate shows that it is still unchanged. The ZrO_2 precipitation limit, however, is rather low, and it is difficult to synthesize composites with ZrO_2 contents above 5 mole %. The reason for this can be attributed to the hydrolysis and condensation of Zr (ma) alkoxides, leading to large agglomerates and gelation before being incorporated into the inorganic-organic network as a consequence of high local H_2O concentrations, as schematically shown in Figure 8.

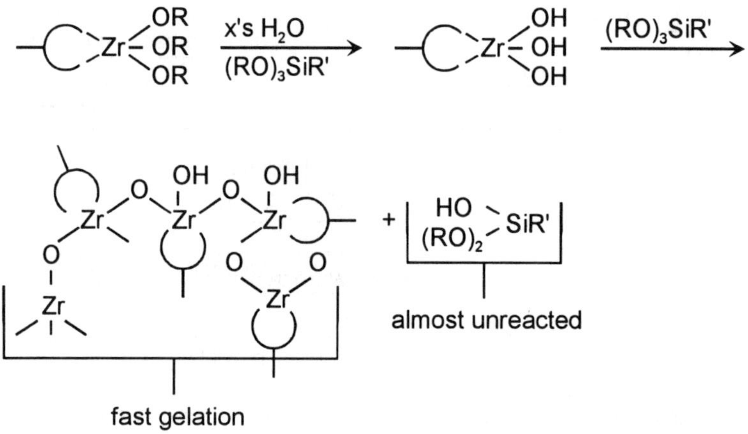

Fig. 8: Model for the precipitation of Zr (ma) alkoxides with excess water addition.

In order to avoid precipitation, the controlled chemical condensation process (CCC [*17*]) was used. The basic idea is to introduce water in form of ≡SiOH groupings using the high reactivity of the ≡ZrOR grouping.

$$\equiv SiOH + ROZr\equiv \;\rightarrow\; \equiv SiOZr\equiv + HOR \tag{4}$$

This would require the formation of an (at least intermediate) ≡SiOZr≡ bond, as it has been proved by Babonneau [*18*] recently in similar systems. In addition to this, a catalytic effect for the condensation of silanols can be expected from ≡Zr(OR) groupings. As shown in [*19*], the prehydrolysis of (RO)₃SiR' to the maximum ≡SiOH content possible for the chosen system leads to a maximum Zr(ma)(OPr)₃ to be added without precipitation. After refluxing the system, excess water for gelation can be added without causing phase separation.

For optimizing the silane system for Zr(ma)(OPr)₃ addition, H₂O content and ≡SiOH content can be monitored by Karl-Fischer titration and ²⁹Si NMR spectroscopy. In Figure 9, it is shown how the different species develop with time and how ZrOR addition affects ≡Si-O-Si≡ formation.

One can clearly see the development of the different species with time. If methacryloxy silane is used without solvent and prehydrolysed with the amount of water necessary for the complete hydrolysis of the ≡SiOR groupings with HCl as catalyst, up to 50 mole % of a Zr(ma)(OPr)₃ complex can be added and after refluxing and water addition, no precipitation takes place. After radical polymerization, the solid material does not show any peaks of free methacrylic acid C=O frequency, leading to the conclusion that ma is still stably bond to Zr⁴⁺. The data were confirmed by ¹³C NMR spectroscopy. Since in the IR spectrum the C=C double bonds are reduced drastically and degrees of conversion above 90 % (95 % at 120 °C) can be obtained, one has to conclude that the ma bond to Zr⁴⁺ participates in the polymerization process [*20*].

$$\geqslant\!Zr\!\!\bigcirc\!\!-\!\underset{\underset{CH_3}{|}}{C}\!=\!CH_2 \quad CH_2\!=\!\underset{\underset{CH_3}{|}}{C}\!\wedge\!Si\!\leqslant \;\overset{cat.}{\longrightarrow}\; \geqslant\!Zr\!\!\bigcirc\!\!-\!\underset{\underset{CH_3}{|}}{\overset{H}{\underset{|}{C}}}\!-\!\overset{|}{\underset{H}{C}}\!-\!C\!\wedge\!Si\!\leqslant \tag{5}$$

The question arises, however, what type of structure is formed by the Zr(ma) complex. For this reason, a diffraction analysis was carried out with bulk materials [*20, 21*]. With X-ray, no crystalline phase could be detected, but with electron diffraction it could be shown that crystalline ZrO₂ is present, mainly in the form of the monoclinic modification with crystallite sizes below 2 nm, which is rather surprising since the preparation temperature never exceeds 120 °C (Figure 10a). One has to conclude that the crystallites are "covered" with ma bond to Zr⁴⁺ ions, forming a chelate type of interface to the ZrO₂ surface. From this, a structural model can be developed (Figure 10b).

The methacryloxy silane can be partially substituted by other silanes. If silanes with perfluorinated side chains are used [*19*], the fluorinated units diffuse preferentially to the surface of the wet film leading to a low surface free energy layer. This type of coatings adheres very well to polar surfaces despite of the fluorinated groupings due to the polar $\equiv$SiOH groupings. The fluorinated groupings preferentially turn to the air interface, whereas the ZrO_2 nanoparticles remain distributed homogeneously in the layer (EDX profiling). In Figure 11, the process is shown schematically.

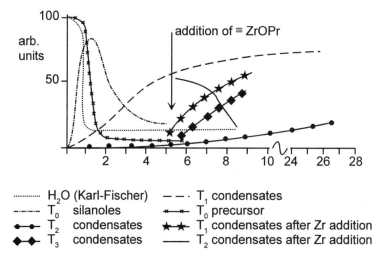

·········· H_2O (Karl-Fischer)	— — · T_1 condensates
—·—·— T_0 silanoles	—×—×— T_0 precursor
—•—•— T_2 condensates	★—★—★ T_1 condensates after Zr addition
◆—◆—◆ T_3 condensates	—— T_2 condensates after Zr addition

Fig. 9: Hydrolysis and condensation of methacryloxy silane monitored by ^{29}Si NMR spectroscopy and Karl-Fischer titration with and without $\equiv$ZrOR.

Fig. 10a: Electron diffraction pattern of the ZrO_2 containing nano-composite .

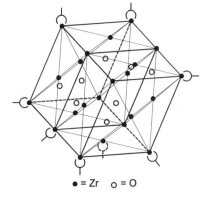

• = Zr ○ = O

Fig. 10b: Model of the ZrO_2 ma based on the monoclinic ZrO_2 unit cell.

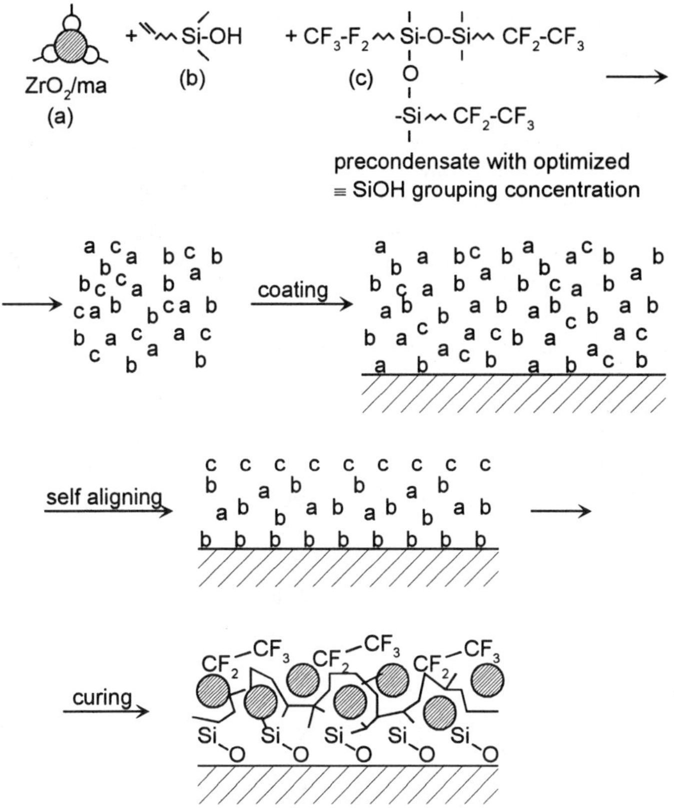

Fig. 11: Synthesis and processing of a perfluorinated grouping-containing low surface free energy nanocomposite coatings.

Due to the selfaligning effect, the concentration of silanes with a perfluorinated side chain can be kept low (only > 0.6 mole %) to obtain surface free energies of about 18 mJ/m². The coatings are highly transparent and abrasion-resistant (2 % haze after taber abrader standard test) and can be easily cured by UV. They can be applied to all types of substrates. Since the processing and properties are mainly depending on the interface of the nanoparticles (e.g. avoiding of agglomeration and gelation during processing, refractive index 1.50 to 1.55 depending on ZrO₂ content, abrasion resistance, transparency), one can speak of so-called interfacial phase determined nanocomposites. This can be considered as a new class of materials with interesting properties.

Another type of interfacial phase-determined materials has been developed based on an epoxide and boehmite or γ-Al₂O₃ with particle sizes in the range of 15 nm. Epoxy silane-based systems have been developed for plastic eye glass lens protection and have been successfully on the market for many years [*10*]. As shown in [*21*], polycondensates based on (RO)₃Si-(CH₂)₃-O-ĈHCH₂Ö with TEOS or Al[iprop]₃, in general, do not show the formation of nanoparticles to be detected by photon correlation spectroscopy and can be considered more or less as molecular types of composites after polymerization of the epoxide by methyl imidazole, for example. If boehmite powder suspended in ethanol is added to a prehydrolysed mixture of the epoxide TEOS and Al[iprop]₃ after having carried out a ligand exchange reaction of iprop against butoxyethanol in order to slow down the Al alkoxide reactivity according to [*22, 23*],

$$Al(iprop.)_3 + 3CH_3(CH_2)O(CH_2)_2OH \rightarrow Al(O(CH_2)_2O(CH_2)_3CH_3)_3 + 3iprop. \qquad (6)$$

it could be shown that the boehmite catalyses the epoxy polymerization to form a polyethylene oxide network, probably preferentially around the active particles. At the same time, as shown by NMR spectroscopy (Figure 12), =Al-O-Si≡ bonds are formed by heterocondensation between =AlOH and ≡SiOR.

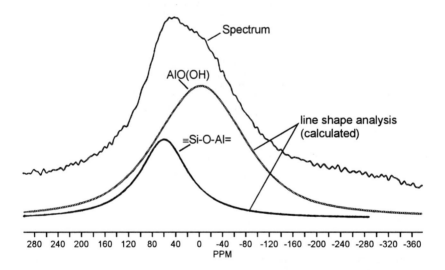

Fig. 12: NMR spectrum of the boehmite/epoxy silane reaction mixture and deconvo-
luted peaks [*23*].

The liquid systems can be used for coating of glass or plastics and cured between 90 and 120 °C without using any additional catalysts or addition of methyl imidazole. Due to the small particle size, they are transparent. In Figure 13, the schematics of the synthesis of the composite system is shown.

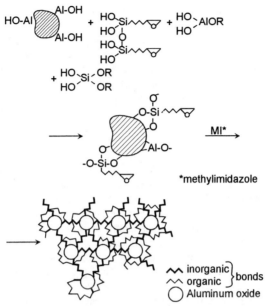

Fig. 13: Reaction scheme of the formation of a aluminum oxide particle-containing nano composite.

After curing, the scratch resistance of these systems (determined by a Vickers diamond scratched over the surface) is surprisingly high compared to polymers or ormocer coatings. In Figure 14, the load of the diamond causing the first visible scratch is shown and compared to polymers and ormocer coatings.

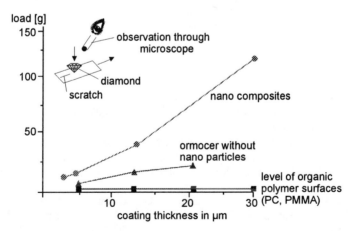

Fig. 14: Comparison of the scratch resistance of various types of coatings; ●: nano-composite; ▲: ormocers and polysiloxanes; ■: organic polymer films.

The surprisingly high scratch resistance is attributed to the change of the matrix structure at the particle's surface, probably leading to more ordered structures at the particle surface (which, for example, already have been established at polymer/metal interfaces). This could explain the surface hardness far above what has been expected from similar composition where alumina is introduced by the sol-gel process in a molecular form. The coating system has been adapted to CR39 eye glass lenses and is industrially used.

Conclusion

The three examples show that special material properties can be obtained by sol-gel based synthesis techniques and control of reaction condition and composition. For this reason, it seems to be of special interest to tailor the dimension of organic as well as of inorganic units. In the case of the hot melt composite, it was of importance to establish organic chains of well-defined length in order to obtain sufficient stress dissipation behavior to increase the overall seal strength considerably. In the case of ZrO_2-containing composites, the formation of nanoparticles with controlled sizes and well developed crystallinity can be obtained, allowing to tailor the refractive index and mechanical properties. The mechanical properties and (which is of high importance) the processing properties depend strongly on the nanoparticle surface properties and the interface to be developed during the composite synthesis. The "interfacial phase" can play an important role if its concentration is high enough. Nanocomposite systems with high surface areas open up an interesting way to new interfacial phase-determined materials.

Acknowledgments

The authors want to express their thank to the Minister of Research and Culture of Saarland and several industrial companies for their financial support.

Literature Cited

[1]   Aegerter, M. (ed.),
        J. NonCryst. Solids 1990, 121, North Holland Publishers.
[2]   Equivias, L. (ed.),
        J. Non-Cryst. Solids 1992, 147&148, North Holland Publishers.
[3]   Zelinski, B. J. J., Brinker, C. J., Clark, D. E., Ulrich, D. R., Eds.;
        Mat. Res. Soc. Symp. Proc 180; Materials Research Society: Pittsburgh/PA, 1990.
[4]   Hampden-Smith, M. J., Klemperer, W. G., Brinker, C. J., Eds.,
        Mat. Res. Soc. Symp. Proc 271; Materials Research Society: New York/NY, 1992.

[5] Schmidt, H.,
J. Sol-Gel Science and Technology **1994**, *1*, pp. 217 - 213.

[6] Mackenzie, J. D., Chung, Y. J., Hu, Y.,
J. Non-Cryst. Solids **1992**, *147&148*, pp. 271.

[7] Schmidt, H.
In *Chemical Processing of Advanced Materials*; Hench, L. L., West, J. K., Eds., John Wiley & Sons: New York/NY, 1992, pp. 727.

[8] Motakef, S., Boulton, J. M., Teowee, G. T., Uhlmann, D. R., Zelinski, B. J. J.,
Sol-Gel Optics II **1992**, *SPIE Vol. 1758*, pp. 432.

[9] Wilkes, G. L., Brennan, A. B., Huang, H.-H., Rodrigues, D., Wang, B.,
Mat. Res. Soc. Symp. Proc. 171; Materials Research Society: 1990, pp. 15.

[10] Schmidt, H., Seiferling, B., Philipp, G., Deichmann, K.
In *Ultrastructure Processing of Advanced Ceramics;* Mackenzie, J. D., Ulrich, D. R., Eds., John Wiley & Sons: New York/NY, 1988, pp. 651.

[11] Schmidt, H., Scholze, H., Tünker, G.,
J. Non-Cryst. Solids **1986**, *80*, pp. 557.

[12] Schmidt, H., Tünker, G., Scholze, H.,
Verfahren zum Verbinden von Substraten durch Heißsiegeln; European Patent No. 81 10 21 21.1; 0 03 6648; 03-20-1981.

[13] Schmidt, H., Krug, H., Kasemann, R., Tiefensee, F.,
SPIE Proc. **1991**, *1590*, pp. 36 - 49.

[14] Schmidt, H.,
Proc. 1993 PAC RIM Meeting; Honolulu, November 1993; American Ceramic Society, in press.

[15] Popall, M., Meyer, H., Schmidt, H., Schulz, J.,
Mat. Res. Soc. Symp. Proc. 180; Materials Research Society: 1990, pp. 995 - 1001.

[16] Gerhard, V., private communication.

[17] Schmidt, H., Seiferling, B.,
Mat. Res. Soc. Symp. Proc. 73; Materials Research Society: 1986, pp. 739.

[18] Babonneau, F.,
Proc. MRS Spring Meeting; San Francisco, April 94; *Better Ceramics Through Chemistry* VI, in press.

[19] Kasemann, R., Schmidt, H., Brück, S.,
Bol. Soc. Esp. Ceram. Vid **1992**, *31-C, 7*, pp. 75.

[20] Tiefensee, F.,
Ph. D. Thesis, University of Saarland: Saarbrücken, 1994.

[21] Krajewski, T., private communication.

[22] Jones, R. W.,
Fundamental Principles of Sol-Gel Technology; The Institute of Metals, North American Pupblication Center: Brookfield/USA, 1989.

[23] Kasemann, R., Geiter, E., Schmidt, H.,
private communication, to be published in detail later.

RECEIVED October 11, 1994

Chapter 27

Surface Modification of Carbon Fibers for Advanced Composite Materials

Yuechuan Wang[1] and Roderic P. Quirk[2]

Maurice Morton Institute of Polymer Science, University of Akron, Akron, OH 44325

Carbon fibers were treated with a strong oxidizing agent, 2 % $KClO_3$ in concentrated H_2SO_4, to introduce polar phenol and/or hydroxyl groups on the fiber surfaces. A convenient benzoate labeling and analysis method for surface hydroxyl and phenol groups was developed to evaluate this treatment. Both ESCA and the benzoate labeling analyses indicated that oxidation for only 3-10 minutes at room temperature increased the oxygen content to a maximum of 2-3 times the original amount. The concentration of phenol and/or hydroxyl groups on the surfaces of treated fibers corresponded to approximately 2-3 phenol and/or hydroxyl groups/ nm^2. The contact angle of water (pH=7) on highly ordered pyrolytic graphite after oxidation for 30 sec. was 53-58°, compared with 87° for the original surface. There was no obvious damage to the surface morphology of the treated fibers as indicated by SEM. The tensile breaking strength of single filaments of AS-4 carbon fibers was not affected by oxidation treatments for up to 1.25 hours; a decrease of only 3 % was observed after oxidation for 10 hours. Treatment of oxidized carbon fibers with trimellitic anhydride chloride formed anhydride-labeled surfaces at a concentration of approximately 1-2 anhydride groups / nm^2.

[1]Visiting scientist from Changchun Institute of Applied Chemistry, Chinese Academy of Science, Changchun, China
[2]Corresponding author

0097–6156/95/0585–0348$12.00/0

Carbon fibers, with their low density, high strength and stiffness, are useful materials in combination with matrix resins for the fabrication of high performance composite materials (*1-3*). It has been generally recognized that the mechanical properties of these composites are strongly dependent upon the interfacial bonding between reinforcing fibers and the resin matrix (*4,5*). Effective bonding promotes the efficient transfer of the applied forces among the fibers and the resin matrix (*5*). In the case of the composites of carbon fibers with polyimides, effective bonding also increases the thermal stabilities of the composites; strong interfacial bonding prevents oxygen from penetrating into the interface, where carbon fibers are more easily oxidized than the polyimide matrix (*6*).

The interaction between the fibers and matrices depends on both physical and chemical interactions. The chemical interaction component relies on the chemical bonding interactions of functional groups in the matrix resin with complementary functional groups on the fiber surface. Physical interactions depend on the mechanical interlocking and wetting of the matrix on fiber surfaces. Processes which increase surface functional groups and surface roughness or eliminate surface defects on carbon fibers can improve interfacial bonding (*4,5*). Native carbon fibers have relatively smooth surfaces, and only limited amounts of surface functional groups, which are mainly located on the edges of the exposed graphitized layers, and at the imperfection in the graphite structure, like vacancies, dislocations, and steps in the outer planes. Such surfaces do not promote the wetting of resin matrices on the fibers. Surface treatment of carbon fibers is often necessary in order to achieve the desired properties for advanced composites.

A variety of methods have been investigated for the surface treatment of carbon fibers for composite materials (*2,3,7*). Treatment methods include using plasmas of various gases, high temperature treatment with gases, chemical treatment with oxidizing agents or anodic treatment in basic or acid media. Anodic oxidation is generally recognized as the best method and has been used commercially (*8,9*). However, there are controversial reports on whether fiber surfaces are damaged by anodic treatment using an alkaline electrolyte, for example with NaOH (*8,10-12*). Furthermore the effects of anodic treatment are sensitive to the pH value of the medium, and sodium ions are difficult to wash out from the fiber surface with distilled water (*11*). Recent work using scanning tunneling microscopy found that almost any of these treatments of carbon fibers caused physical damage such as pitting and increased surface roughness (*13*); however, most of these treatments improve the mechanical properties of the corresponding carbon fiber-reinforced composites (*7,8*).

Herein we report a rapid and convenient solution oxidization method to introduce phenol and/or hydroxyl functional groups on the surfaces of carbon fibers to improve the interfacial bonding to matrix resins for use in advanced composite materials.

Experimental

Oxidation of fibers. About 0.4g of carbon fibers (AS-4, Hercules 3K unsized, or T650, Amoco 12K, unsized) cut into 1 inch lengths was added to a yellow - red, 2 % solution of 1g of $KClO_3$ (Fisher) in 50 mL of concentrated sulfuric acid at room temperature. The contents were stirred intermittently. The reaction was stopped after various times by removing the fibers and washing them with 40 mL portions of distilled water until no absorption at 358 nm was detected by UV-visible spectroscopic analysis of the wash water. The fibers were dried at 120 - 150 °C under high vacuum (less than 10^{-4} mm Hg) for 12 hours.

Oxidation of highly ordered pyrolytic graphite (HOPG). A thin sheet (10 mm x 7 mm) of highly ordered pyrolytic graphite (HOPG, Union Carbide) was peeled from the surface of a plate with adhesive tape, which was then attached to a glass slide. The graphite sheet was oxidized by adding one drop of the 2 % solution of $KClO_3/H_2SO_4$ at room temperature. After 30-60 seconds, the plate was rinsed with distilled water for 1 min. and dried under vacuum at room temperature.

Benzoyl labeling of the oxidized fibers. The oxidized and vacuum dried fibers (0.4g) were introduced into a 100 mL dry flask containing 50 mL of THF (distilled after drying over CaH_2) under dry nitrogen pressure, followed by addition of 1 mL of triethylamine (Aldrich, distilled after drying over KOH) and 1 mL of benzoyl chloride (Aldrich, used as received) under nitrogen. The benzoylation of the fibers was effected for 6-8 hours with occasional mild stirring. The decanted fibers were then washed twice with 40 mL of acetone and eight times with 40 mL of water until no residual absorption (A < 0.002) at 230 nm was detected by UV-visible spectroscopic analysis. The fibers were used directly for the hydrolysis reaction or dried under vacuum.

Hydrolysis of benzoyl-labeled fibers. The benzoyl-labeled fibers were mixed with 100 - 150 mL of water and 3-5 mL of concentrated HCl in a 200 mL beaker. The hydrolysis of the fibers was effected at 65 °C for approximately 8 hr as determined by a constant UV-visible absorption reading for the solution at 230 nm.

Anhydride labeling and analysis of carbon fibers. Oxidized and dried fibers (about 0.4 g) were reacted with 0.1 g of trimellitic anhydride chloride (Aldrich) and 0.1 mL triethylamine in 50 mL of THF at room temperature for about 6 hr. with intermittent shaking. The decanted fibers were then washed with acetone six times (6x 40 mL), and then washed three times with distilled water until no absorption at 240 nm was detected by UV-visible spectroscopic analysis. These anhydride-labeled fibers were hydrolyzed with a mixture of 100 mL of water and 3 mL of concentrated HCl at 60 °C overnight.

Determination of the absorption coefficient of 1,2,4-benzenetricarboxylic acid. The triacid was synthesized in-situ by the reaction of 19.0 mg of trimellitic anhydride acid chloride with distilled water in a 100 mL volumetric flask at room temperature for 2.5 days. The solution was diluted by removing 1 mL of solution with a 1/100 ml pipette and diluting to 10 mL, followed by removal of 0.5 mL of this solution and diluting to 10 ml using 10 mL volumetric flasks. The absorption coefficient at 288 nm was determined with the first solution. The stronger absorption bands near 240 nm and 210 nm were determined with the second solution.

Measurements. UV-visible absorption spectra were recorded on an HP 8452 Diode Array Spectrophotometer with a 10 cm quartz cell for quantitative analyses. Contact angles for water at various pH on graphite were determined with a Rame-Hart 100-07-00 contact angle goniometer equipped with an environmental chamber, which was saturated by filling the wells in the sample chamber with distilled water. Scanning Electron Micrographs (SEM) were obtained on a JEOL 3 U microscope at 25 kV. Electron spectroscopy for chemical analysis (ESCA) was performed on a Perkin-Elmer ESCA 5400 electron spectrometer with a magnesium Kα X-ray source operated at 400 W. For ESCA many fibers were coaxially mounted and analyzed simultaneously. Quantification was based on computer-calculated peak areas. The breaking strengths of individual carbon fibers were measured on a laboratory built micro-instron with 0.2g force sensitivity. The fiber was attached to a strip of adhesive tape cut as shown below. The narrow adhesive strip was cut

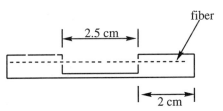

after mounting the sample on the instron. Ten samples were measured to calculate the average breaking strength of a given sample of treated fibers.

Results and Discussion.

Oxidation of carbon fibers. The solution oxidation of carbon fibers by strong oxidative agents has been used for a long time (*2,3,14*). The goal of this research was to develop a mild, efficient and rapid oxidation method to produce polar, reactive functional groups on the surface of

carbon fibers for use in promoting interfacial bonding in composites. McCarthy and coworkers (*15*) have reported that graphite fibrils could be oxidized using a dilute solution of potassium chlorate in concentrated sulfuric acid. They reported that hydroxyl and phenol groups accounted for most of the increased oxygen content of the oxidized fibrils, based on results of analysis using a variety of labeling methods which were specific for each different type of functional group; this conclusion is also in accord with other data in the literature regarding the nature of the functional groups on oxidized carbon fiber surfaces (*8,16*). Therefore, it was of interest to determine the usefulness of this oxidation method to generate phenol and hydroxyl groups on the surface of carbon fibers for use in composites.

    The oxidation of carbon fibers was effected at room temperature with a 2% solution of potassium chlorate for various periods of time as illustrated in Scheme 1 where only the formation of phenol-type oxygen functionality is indicated for simplicity.

Benzoate labeling and analysis. In order to quantitatively determine the extent of oxidation of the fibers to form phenol and/or hydroxyl groups on the surface, it was necessary to develop an indirect method based on (a) labeling the functional groups by converting them to a suitable derivative, (b) removing the label by hydrolysis and (c)

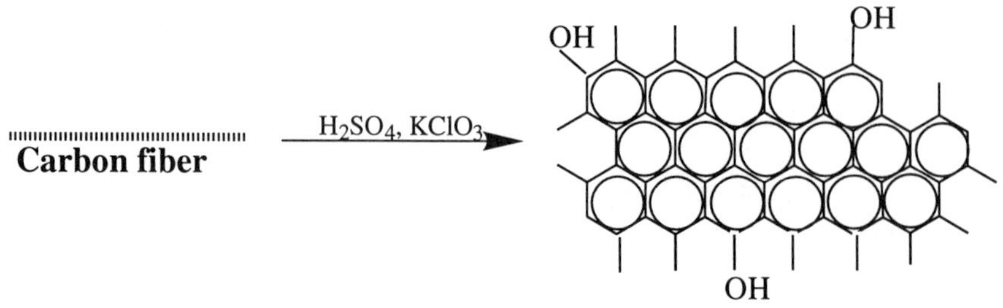

Carbon fiber $\xrightarrow{\text{H}_2\text{SO}_4,\ \text{KClO}_3}$

Scheme 1

analyzing the amount of label in the hydrolysis solution using a sensitive UV-visible spectroscopic method. For this purpose, the preparation of benzoate derivatives was utilized since this is a recommended procedure for analysis of both alcohols and phenols (*17*). Samples of fibers were reacted with benzoyl chloride in the presence of triethylamine to convert the alcohol and phenol groups on the surface to the corresponding benzoate derivatives as shown in equation 1. After 8 hours, the fibers were washed repeatedly with water to remove unreacted benzoyl

$$\text{OH} \quad \xrightarrow{\text{PhCOCl / Et}_3\text{N, THF}} \quad \xrightarrow{\text{1.Wash, 2, H}^+/\text{H}_2\text{O}} \quad \text{OH} + \text{PhCOOH} \quad (1)$$

chloride. This washing step to remove unreacted benzoyl chloride is a critically important step in this procedure because benzoyl chloride could be physically adsorbed on the surface. Therefore, the water washing process was monitored by UV-visible spectroscopy until the absorbance at 230nm was less than 0.002. Then the benzoate-labeled fibers were treated with dilute, aqueous hydrochloric acid to hydrolyze the benzoate esters and form benzoic acid. The resulting aqueous solution was analyzed for the presence of benzoic acid by UV-visible spectroscopy (λ_{max} = 230 nm, ε = 1.16 x 10^4 L/mol-cm) (*18*). From the reproducibility of these analyses, the error limits on the analytical results are estimated to be approximately ± 20%.

The results of analysis of surface hydroxyl and phenol groups using this benzoate-labeling method are shown in Figure 1. These results indicate that this chlorate oxidation procedure is both rapid and mild. Within 3 minutes at room temperature, 1.5 x 10^{-6} moles of phenol and/or hydroxyl groups per gram of fiber are formed, and the number of these phenol and/or hydroxyl groups does not increase significantly with increasing oxidation time up to a period of two hours. From control experiments, it was established that the untreated fibers contained approximately 0.5 x 10^{-6} mol/g of hydroxyl/phenol functional groups. Thus, using this mild oxidation procedure it was possible to increase the concentration of surface hydroxyl/phenol groups by a factor of approximately 2-3 within a few minutes of contact with the chlorate oxidizing solution. Preliminary experiments indicate that this procedure utilizing a bath of this oxidant is adaptable to a fiber-spinning process.

The surface concentration of phenol and hydroxyl groups was estimated from the surface area of the fibers. The diameter and density of AS-4 fibers are 7 mm and 1.8 g/cm^3, respectively; thus, the specific surface area was calculated to be 0.317 m^2/g; the actual measured

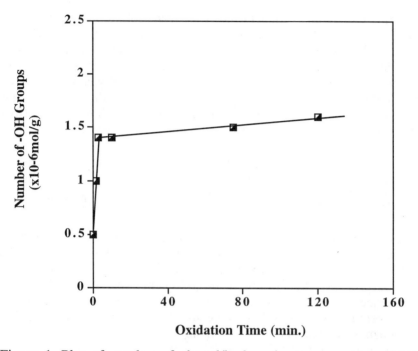

Figure 1. Plot of number of phenol/hydroxyl groups versus oxidation time determined using the benzoate labeling method.

specific surface areas of carbon fibers are reported to be in the range of 0.33-0.56 m^2/g depending on surface treatment (*13*). Therefore, the concentration of 1.5 x 10^{-6} mol of phenol and hydroxyl groups per gram of fiber corresponds to approximately 2.8 hydroxyl/phenol groups per nm^2 of surface area. For structurally perfect graphitic carbon, it can be calculated that an area of 1 nm^2 contains approximately 38 carbon atoms (*19*); therefore, the chlorate oxidation procedure results in the formation of approximately 1 hydroxyl/phenol group for every 13 carbon atoms.

Trimellitic anyhydride labeling and analysis. In order to promote chemical reactions between functional groups on the carbon fibers and a variety of matrix resins, it was of interest to attach other reactive functional groups to the surface of the oxidized carbon fibers. For this purpose, trimellitic anhydride chloride was chosen to react with the oxidized fibers and to introduce the phthalic anhydride-type of functional group on the surface. The reaction is shown in equation 2; it was envisioned that the acid chloride would react with surface hydroxyl and phenol groups to generate the corresponding anhydride-

$$\text{—OH} \quad \xrightarrow{\quad\text{ClCO}-\text{(trimellitic anhydride chloride)}\quad} \quad \text{—OC-(trimellitic anhydride ester)} \qquad (2)$$

functionalized surface. The procedure for analysis of the efficiency of anhydride labeling was via an indirect labeling and hydrolysis method analogous to the benzoate labeling procedure. Thus, samples of oxidized fibers were treated with trimellitic anhydride chloride in the presence of triethylamine to convert phenol and hydroxyl groups on the surface to the corresponding trimellitic anhydride ester groups. After six hours the fibers were washed with acetone and water to remove unreacted acid chloride and its hydrolysis products. To analyze the extent of anhydride labeling of the oxidized fibers, the anhydride-labeled fibers were treated with dilute, aqueous hydrochloric acid to remove the anhydride groups. The resulting acidic aqueous solution was analyzed for the presence of 1,2,4-benzenetricarboxylic acid by UV-visible spectroscopy at 249 nm (in water, λ_{max} = 210 nm, 249 nm and 290 nm, ε = 26,900, 10,400 and 1,900 L/mol-cm, respectively; in dilute aq. HCl (pH < 4), the absorption at 210 nm is shifted to 202 nm, ε = 35,900 L/mol-cm, and the ε_{249} increased to 11,200 L/mol-cm). For a sample of carbon fibers which had been oxidized for 30 minutes, the results of anhydride labeling and analysis indicated that the concentration of reactive hydroxyl and phenol groups was approximately 1×10^{-6} mol/g compared with a value of 1.5×10^{-6} mol/g based on benzoyl labeling. These results are in reasonable agreement and confirm the expected reactivity of the surface hydroxyl and phenol groups with acid chlorides and the ability to produce reactive anhydride groups on the surface for subsequent reactions with a variety of other functional groups in matrix resins.

ESCA analysis of oxidized carbon fibers. The effect of chlorate oxidation on carbon fibers was also investigated by ESCA and the results are shown in Table 1. After oxidation with the chlorate solution for only 10 minutes, the oxygen content increased by a factor of approximately two compared to the untreated fibers. Furthermore, in accord with the titration data (see Figure 1), the oxygen content does not further increase significantly upon prolonged oxidative treatment for up to 2 hours. The atomic compositions which correspond to the ESCA data are $C_{100}O_{8.5}$ for the untreated fibers and $C_{100}O_{18}$ for the fibers

Table 1. ESCA Analysis of Oxidized Fibers

Ox.time (min.)	atom %			
	C	O	Na	N
0	86.46	9.75	1.39	2.4
10	77.4	19	0.48	2.23
120	76.6	18.45	0.21	2.34

which were chlorate oxidized for 10 minutes, i.e. the increase in surface oxygen content corresponds to a factor of two. These results can be compared to the hydroxyl and phenol content as determined by the chemical labeling methods. Thus, it was determined that the amount of surface hydroxyl and phenol groups increases by a factor of approximately 2-3 as determined by benzoate labeling and by a factor of 1-2 as determined by trimellitic anhydride labeling. Considering the difficulty in measuring the concentrations of surface functional groups, all of these results are in reasonably good agreement.

SEM analysis of oxidized carbon fibers. In addition to the requirement of rapid, efficient introduction of surface functional groups on carbon fibers for composites, a useful chemical oxidation procedure must not seriously damage the fibers. In order to evaluate the effect of chlorate oxidation on the physical nature of the carbon fibers, the surface morphologies of treated fibers were compared with untreated fibers using scanning electron microscopy (SEM). The results are shown in Figure 2, which shows micrographs (x10,000) for T-650 carbon fibers before (A) and after (B) chlorate oxidation (30 minutes). No evidence for surface damage was observed. However, the oxidized fiber surfaces appeared to be smoother than the untreated fibers; the striations along the length of the untreated fibers are much less prominent in the micrograph of the treated fibers. The micrographs for untreated AS-4 fibers exhibited smooth, non-structured surfaces and they remained smooth after 10 hours of oxidation at a magnification of x20,000.

Single filament breaking strength of oxidized carbon fibers. The most important criterion to evaluate the effectiveness of an oxidation procedure is that the process does not significantly reduce the tensile strength of the fibers. Single filament breaking strengths of oxidized fibers were measured and compared with untreated fibers and

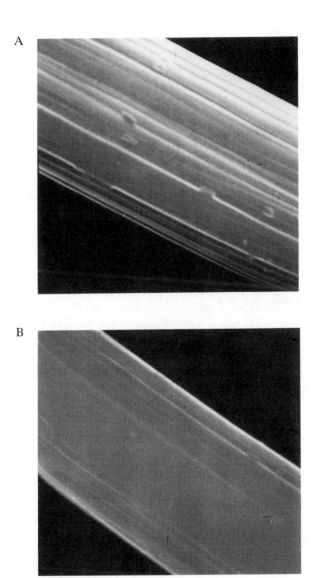

Figure 2. Scanning electron micrograph (x 10,000) for T-650 carbon fibers before (A) and after (B) oxidation with 2% KClO3 in H2SO4 for 30 min.

Table 2. Breaking Strength of Oxidized Fibers

Ox. Time	Breaking Strength (GPa)
0	3.8
1.25 h	3.8
10h	3.7

the average results for at least ten samples are shown in Table 2. After a prolonged oxidation for 1.25 hours, there was no observable change in the single filament breaking strengths. Even after an oxidation time of 10 hours, only a decrease of 3% in tensile breaking strength was observable. These tensile breaking strengths are in good agreement with the value of 3.6 GPa for AS-4 single filaments which was reported by Drzal (5).

Oxidation of highly ordered pyrolytic graphite (20). The oxidation of the surface of highly ordered pyrolytic graphite (HOPG) proceeded very rapidly; when a few drops of the $KClO_3/H_2SO_4$ solution were dropped onto the surface of HOPG, the red color discharged immediately. If another drop of oxidant solution was added after the first one was washed away, the interface at the liquid and graphite became colorful, rainbow reflective; the original mirror-flat graphite surface with a metallic luster changed into a metal-gray color. The contact angle of water (pH 7) on the original HOPG surface was 87±1⁰; the contact angle of water on the surface of the oxidized HOPG sheet was 53-57⁰. SEM analysis of the untreated and oxidized HOPG surfaces are shown in Figure 3. In contrast to the smooth original surfaces, after oxidation the surfaces exhibited domain-like structures surrounded by oxidized edges.

Chemical Safety of $KClO_3/H_2SO_4$

The reaction of $KClO_3$ and H_2SO_4 forms perchloric acid and the active oxidizing agent, chlorine dioxide, as shown in equation 3(21).

$$3\,KClO_3 + 3\,H_2SO_4 \longrightarrow HClO_4 + 3\,KHSO_4 + H_2O + 2\,ClO_2 \qquad (3)$$

The chemical reactivity of perchloric acid depends upon the temperature and concentration. At room temperature and at low concentrations, its oxidation ability is very limited (22). On the other

A

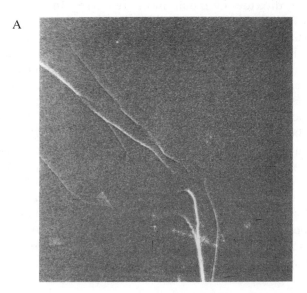

B

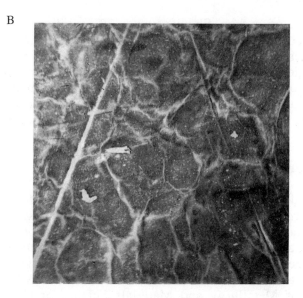

Figure 3. Scanning electron micrograph of highly ordered pyrolytic graphite before (x 100) (A) and after (x 160) (B) oxidation with 2% $KClO_3$ in H_2SO_4.

hand, chlorine dioxide is much more reactive. In fact, when its concentration is over 10%, or when chlorine dioxide comes in contact with easy oxidized organic chemicals and reducing agents, it is explosive (21,22). The chosen concentration was about 2 g/L in this work, which is well below the amount specified by the safety requirement. We have processed over 80 meters of T650 carbon fibers safely using this method. However, it is necessary to only use fibers which are free of organic chemicals, and they should be unsized. Chlorine dioxide is not stable; it can be decomposed by heat and sunlight. The $KClO_3/H_2SO_4$ solution has an orange-red color, and this color could be used as a build-in indicator: when the solution stands, especially under sunlight, for several hours, or when it reacts with carbon fibers, the color will fade. The patented method using the same chemicals was carried at high temperature for a long time, possibly because the strong oxidizing agent, ClO_2, was destroyed at the high temperature used (14). Under the conditions described in this patent, therefore, it is probable that only the $HClO_4$ was available to oxidize the fibers.

Conclusions

The oxidation of carbon fibers using a 2% solution of potassium chlorate in concentrated sulfuric acid reacts rapidly to double the oxygen content on the surface. Analysis using labeling with both benzoyl chloride and trimellitic anhydride chloride is consistent with the formation of primarily hydroxyl and phenol groups on the surface. Both SEM analysis and single fiber tensile breaking strength measurements show that this oxidation procedure does not damage the fibers.

Acknowledgment

The financial support from the NSF Center for Molecular and Microstructure of Composites (CMMC) at Case Western Reserve University and the University of Akron is greatly appreciated. The authors are grateful to Professor Galiatsatos and his research group for the help in measuring the single filament breaking strengths of fibers. The authors are also grateful to Professor Gary M. Michal in the Department of Metallurgy and Materials Science at Case Western Reserve University for his help in obtaining the ESCA data.

Literature Cited

1. "Carbon Fibers. Properties and Application", *J. Physics, D, Applied Physics, Carbon fiber conference issue*, 1987, 20(3), pp 245-322.

2. Riggs, J. P. in *Encyclopedia of Polymer Science & Engineering;* Kroschwitz, J. , Ed.; John Wiley & Sons: New York, 1985; Vol. 2; pp 640.
3. Ehrburger, P. ; Donnet, J.B. in *Handbook of Composites, V1 , Strong Fibers;* Watt , W.; Perov, B. V., Ed.; Elsevier Science Publishers B.V: Netherlands, 1985; pp 577.
4. Drzal, L. T. in *Controlled Interphases in Composite Materials;* Ishida, H., Ed.; Elsevier: New York, 1990; pp 309-320.
5. Drzal, L. T. *Adv. Polym. Sci.*, **1986**, *75*, pp. 1.
6. Scola, D. A.; Vontell, J. H. *Polym. Eng. Sci.*, **1991**, *31*, pp 6.
7. Jang, B. Z. *Composites Sci. Technol.*, **1992**, 44, pp 333-349.
8. Fitzer, E. ; Rensch, H.-P. in *Controlled Interphases in Composite Materials;* Ishda, H., Ed.; Elsevier, New York, 1990; pp 241.
9. Harvery, J.; Kozlowski, C; Sherwood, P. M. A. *J. Mater. Sci.*, **1987**, *22*, pp. 1.
10. Nakahara, M.; Shimizu, K. *J. Mater. Sci.,* **1992**, *27*, pp 1207.
11. Kozlowski, C. and Sherwood, P. M. A. *J. Chem. Soc., Faraday Trans. 1,* **1985**, *81*, pp 2745.
12. Chun, Byong-Wa; Davis, C. R.; Quan He; Gustafson, R R. *Carbon,* **1992**, *30*, pp 177.
13. Hoffman, W. P.; Hurley, W. C.; Owens, T.W.; Phan, H.T. *J. Mat. Sci.*, **1991**, *26*, pp 4545.
14. Goan, J. C.; Joo, L. A. *U. S. Pat. 3,746,560* (July 17,1973).
15. Bening, R. C.; Ivatury, S. R.; McCarthy, T. J. *Polymer Prepr. (Am.Chem.Soc., Div.Polym.Chem.)* **1990**, *31*(1), pp 420.
16. Ehrburger, P.; Herqte, J. J.; Donnet, J. B. in *Petroleum Derived Carbons;* Deviney, M. L. ; O'Grady, T., Eds.; *ACS-Symposium Series*, **1976**, *21*, pp. 324.
17. Knapp, D. R. *Handbook of Analytical Derivatization Reactions;* Wiley: New York, 1979, pp 55.
18. Jaffe, H. H.; Orchin, M. *Theory and Applications of Ultraviolet Spectroscopy;* Wiley: New York, 1962.
19. Hugh O. Pierson, *Handbook of Carbon, Graphite, Diamond and Fullerenes*, Noyes Publications, Park Ridge, 1993, Chapt. 3, p 43.
20. Chang, H. P.;Bard, A. J. *J.Am.Chem.Soc*, **1991**, *113*, 5588.
21. Durrant, P-J. and Durrant, B.; *Introduction to Advanced Inorganic Chemistry;* 2nd ed., John Wiley & Sons: New York, 1970; pp 937.
22. Downs, A. J.; Adams, C. J. in *Comprehensive Inorganic Chemistry;* Pergamon Press, Great Brittain; 1973; Vol. 2, pp 1366, 1442.

RECEIVED October 26, 1994

Author Index

Ahmad, Z., 291,297
Arnold, F. E., 280,297
Arpac, E., 331
Assink, Roger A., 237,264
Beaucage, G., 97,237
Belfiore, Laurence A., 192
Bergstrom, D. F., 237
Black, E. P., 97,237
Bossé, Francis, 192
Bourgeat-Lami, E., 112
Brennan, A. B., 142
Briat, S., 112
Bright, Frank V., 317
Bulte, J. W. M., 19
Burdon, Jeremy, 6
Burkhart, T., 331
Burns, G. T., 237
Calvert, Paul, 6
Cerveau, G., 210
Chen, J. P., 280,297
Chorro, C., 210
Clément, R., 29
Corriu, R., 210
Dang, T. D., 280
Das, Pronab, 192
Deng, Q., 66
Dickson, D. P. E., 19
Douglas, T., 19
Dunbar, Richard A., 317
Ellsworth, Mark W., 86
Espiard, P., 112
Ford, Warren T., 181
Frankel, R. B., 19
Gauthier, C., 112
Geiter, E., 331
Giwa-Agbomeirele, P. A., 237
Guyot, A., 112
Gvishi, Raz, 317
Haupt, Steven G., 308
Jamison, Gregory M., 248,264
Jethmalani, Jagdish M., 181

Jin, Danliang, 125
Jordan, Jeffrey D., 317
Kamigaito, Osami, 55
Kasemann, R., 331
Kurauchi, Toshio, 55
Lemmon, John P., 43
Lepeytre, C., 210
Lére-Porte, J. P., 210
Lerner, Michael M., 43
Lo, Rung-Kuang, 308
Loy, Douglas A., 248,264
Mackenzie, J. D., 226
Man, M. Wong Chi, 210
Mann, S., 19
Mark, J. E., 1,291,297
Mauritz, K. A., 66
McDevitt, John T., 308
Miller, T. M., 142
Moore, R. B., 66
Moreau, J., 210
Moskowitz, B. M., 19
Murray, James K., Jr., 125
Myers, Sharon, 264
Narang, Upvan, 317
Novak, Bruce M., 86
Okada, Akane, 55
Pankhurst, Q. A., 19
Perez, J., 112
Prasad, Paras N., 317
Quirk, Roderic P., 348
Schaefer, D. W., 97,237
Schmidt, H., 331
Sharp, Kenneth G., 163
Shea, Kenneth J., 248,264
Small, James H., 248
Sunkara, Hari Babu, 181
Szmania, Jeffrey, 6
Thepot, P., 210
Ulibarri, T. A., 97,237
Usuki, Arimitsu, 55
Verrier, Celine, 86

Vigier, G., 112
Vinocur, R. B., 142
Wagner, G., 331
Wang, Bin, 125
Wang, Run, 317
Wang, Shuhong, 291,297
Wang, Wei, 125

Wang, Yuechuan, 348
Wei, Gu, 125
Wei, Yen, 125
Wu, Jinghe, 43
Yang, Dachuan, 125
Yeh, Jui-Ming, 125
Zhao, Jianai, 308

Affiliation Index

Ad Tech Systems Research, Inc., 280
California Polytechnic State
  University, 19
Colorado State University, 192
Dow Corning Corporation, 237
Drexel University, 125
DuPont, 163
Institut für Neue Materialien, 331
Institut National des Sciences
  Appliquées, 112
Laboratoire de Chimie et Procédés
  de Polymérisation, 112
National Institutes of Health, 19
Oklahoma State University, 181
Oregon State University, 43
Sandia National Laboratories,
  97,237,248,264
State University of New York
  at Buffalo, 317
Systran Corporation, 297

Toyota Central Research and
  Development Laboratories, Inc., 55
Université Montpellier II, 210
Université Paris Sud, 29
University of Akron, 348
University of Antalya, 331
University of Arizona, 6
University of Bath, 19
University of California—Irvine, 248,264
University of California—Los Angeles, 226
University of Cincinnati, 1,291,297
University of Dayton Research Institute,
  280
University of Florida, 142
University of Liverpool, 19
University of Massachusetts, 86
University of Minnesota, 19
University of Southern Mississippi, 66
University of Texas at Austin, 308
Wright-Patterson Air Force Base, 280,297

Subject Index

A

Abrasion resistance, polymer-filled
  glass, 174,177f
Advanced composites, surface
  modification of carbon fibers, 348–360
Alkoxysilyl-containing polymer precursors,
  synthesis using free radical
  copolymerization method, 128–129
Alkylene-bridged bissiliconates,
  synthesis, 251

Aluminum oxide particle containing
  nanocomposite, synthesis,
  343,345
Anionic pentacoordinate siliconate,
  synthesis, 250
Antibody-based biosensor, development,
  320,323,324f
Aromatic groups bound to silica,
  complexation, 214–216
Arylene-bridged bissiliconates,
  synthesis, 251

B

Bandwidths of Bragg diffracted peaks, calculation, 182–183

Benzazole polymer systems, structural design for sol–gel processing, 280–290

Benzoate labeling and analysis, surface modification of carbon fibers for advanced composite materials, 352–354

Benzobisazole polymer systems, structural design for sol–gel processing, 280–290

Biomimicry
inorganic–protein interactions in synthesis of ferrimagnetic nanocomposite, 19–27
objective, 2
organic–inorganic hybrids with crystalline polymer matrix, 6–17

Biosensor
antibody based, development, 320,323,324*f*
chemical, use of sol–gel-derived inorganic–organic hybrid materials, 318

Bis(acetonitrile) dichloropalladium(II)–diene polymer–solvent system, isothermal ternary phase diagrams, 196–200

Bissiliconates, arylene- and alkylene-bridged, synthesis, 251

Boehmite–epoxysilane reaction mixture, synthesis, 343–344

Bone, properties compared to those of synthetic composites, 6–7

Bragg diffracted wavelength, 182

Bridged polysiloxanes, synthesis, 265–266

Bridged polysilsesquioxanes, synthesis, 265

Bulk optical nonlinear response, description, 319

C

Carbon fibers, surface modification, 349–360

Cation-transfer intercalation chemistry, MPS$_3$ layered compounds, 30–31

Ceramer technology, importance, 280–281

Ceramic(s), synthesis using sol–gel process, 1,125

Ceramic–polymer composites
challenges, 97
reaction conditions vs. properties, 331–332

Characterization
hexylene- and phenylene-bridged polysiloxane network materials, 270,272–275
magnetoferritin, 20–27
core size, 28
electron diffraction, 20,27
electrophoresis, 25–27
magnetic measurements, 25,26*f*
Mössbauer spectroscopy, 20,23–27
TEM, 20–22*f*,27
nanostructured organic–inorganic hybrid composites synthesized using simultaneous processes, 89–92*f*
poly(ethylene oxide)–layered solid nanocomposites, 43–53
vinyl polymer modified sol–gel hybrid materials, 130–131

Chemical biosensors, use of sol–gel-derived inorganic–organic hybrid materials, 318

Chemical property–surface relationship, inorganic–organic hybrid coatings for metal and glass surfaces, 338–346

Chemically bonded organic–inorganic hybrids
applications, 235
mechanical properties, 234*t*,235
structure, 231,233–235
synthesis using sol–gel process, 231,233–235

Clay mineral, candidate for polymer additive, 55

Clay–nylon 6 hybrid, 55–62

Clay–polymer hybrids
nylon 6–clay hybrid, 55–62
rubber–clay hybrid, 61,63–65

Clay–rubber hybrid, 61,63–65

Coatings for metal and glass surfaces, 331–346

Colloidal crystals of silica, solidification, 181–190
Colloidal 3-(trimethoxysilyl)propyl methacrylate–silica particles, solidification, 184
Composite(s)
advanced, surface modification of carbon fibers, 348–360
factors affecting properties, 86–87
reasons for development, 331
use of sol–gel process for synthesis, 86
See also Sol-gel-derived silica–siloxane composites
Composite polymer(s), synthesis using emulsion polymerization process, 112
Composite polymer colloid nucleated by functionalized silica, 112–123
emulsion polymerization of ethyl acrylate at silica surface, 115–119
experimental objective, 112–114
grafting process, 119–122*f*
inorganic material characteristics, 114–115,117*f*
mechanical properties, 121,123*f*,124
Composite powders, observation by electron microscopy, 116–117
Condensation of phenyltriethoxysilane with catechol, synthesis of anionic pentacoordinate siliconate, 250
Conductive polymer–high transition temperature superconductor bilayers, morphology, 309–315
Coordination compounds, MPS$_3$ materials in destruction–reconstruction intercalation mechanism, 31–32
Correlated systems, scattered, 102–107
Creep, polymer-filled glass, 170
Cross-linked siloxanes, structure, 264
Crystal(s) of silica, colloidal, solidification, 181–190
Crystalline ceramic oxides and oxide glasses, synthesis using sol–gel process, 226–227
Crystalline polymer matrix, organic–inorganic hybrid composites, 6–17

Cubic oligosilsesquioxanes, formation, 265

D

Depolymerization–polymerization of silica, synthesis of dianionic hexacoordinate siliconate, 249–250
Design of high-performance polymers for sol–gel processing, 282–290
Destruction–reconstruction intercalation mechanism involving heterogeneous equilibria, 31–32
Dianionic hexacoordinate siliconate, synthesis, 249–250
Diene polymer(s), metal–olefin coordination, 193–206
Diene polymer–bis(acetonitrile) dichloropalladium(II)–solvent system, isothermal ternary phase diagrams, 196–200
Differential scanning calorimetry, characterization of grafting process for polymers, 121,122*f*
Diyne polymerization, organizational requirements, 211–212,214
Dry gels, description, 200
Dynamic mechanical spectrometry, characterization of grafting process for polymers, 121,122*f*

E

Elastic moduli, solid films from ternary solutions, 203–206
Elastic properties, organic–inorganic hybrid composites with crystalline polymer matrix, 12,14–16*f*
Electrical properties
design of nanocomposites, 29–40
organic–inorganic hybrid composites, 3
Electron diffraction, magnetoferritin, 20,27
Electron microscopy, observation of composite powders, 116–117

Electron spectroscopy for chemical analysis of oxidized fibers, surface modification of carbon fibers for advanced composite materials, 355–356

Electrooptic modulators, nonlinear optical materials using organic–inorganic composites, 325–326

Electrophoresis, magnetoferritin, 25–27

Emulsion polymerization of ethyl acrylate at silica surface
encapsulation state, 115–116
kinetics, 118–119
observation of composite powders by electron microscopy, 116–117
polymerization results, 115–116

Emulsion polymerization process, advantages for composite polymer synthesis, 112

Entrapped organic hybrids
applications, 229–232*f*
properties, 229,231
structure, 231,232*f*
synthesis using sol–gel process, 229

Epoxy cross-linked thermoplastic sealant, structural model, 336,338*f*

Ethyl acrylate, emulsion polymerization at silica surface, 115–119

Exfoliation–adsorption method of synthesis, poly(ethylene oxide)-layered solid nanocomposites, 45

F

Ferrimagnetic model, MPS$_3$ intercalates, 34–37

Ferrimagnetic nanocomposite, synthesis, inorganic–protein interactions, 19–28

Ferritin, advantages for nanocomposite material synthesis, 19–20

Fiber-filled composites, properties, 6–7

Form factor, definition, 103–104

Formic acid, use as alternative nonaqueous medium for sol–gel chemistry, 163–164

Fractal approaches, description, 97

Free radical copolymerization method, synthesis of alkoxysilyl-containing polymer precursors, 128–129

Functionalized silica, nucleation of composite polymer colloid, 112–124

G

γ-radiation, effect on properties of organic–inorganic hybrid interpenetrating networks, 146–151

Gelation in polymeric systems, 193

Glass, polymer filled, *See* Polymer-filled glass

Glass fibers, properties, 7

Glass surfaces, inorganic–organic hybrid coatings, 331–346

Grafting process for polymers
characterization, 121,122*f*
kinetics, 120–121
polymer extraction, 119–120

Group-transfer polymerization, synthesis of polyacrylate precursors, 128

H

Hexylene- and phenylene-bridged polysiloxane network materials, 264–278
experimental procedure, 267
future studies, 277
monomer synthesis, 269–270
polymerization procedures, 268–269
scanning electron microscopy, 275,276*f*
sol–gel polymerizations, 270
solid-state ^{13}C-NMR spectroscopy, 272,273*f*
solid-state ^{29}Si-NMR spectroscopy, 272,274*f*,275
surface area analysis, 275
synthetic procedures, 267–268
thermal analysis, 275

High-clarity polyamide–silica hybrid
  materials, 291–296
applications, 296
elongation vs. silica content, 294–295
experimental procedure, 292
stress–strain curves, 292,293*f*
tensile break strength vs. silica
  content, 293–294
tensile yield strength vs. silica
  content, 293–294
transparency of films, 292,293*f*
water absorption, 294–296
yield stress vs. storage time, 294,296*f*
High-performance polymers for sol–gel
  processing, structural design, 280–290
High-temperature polymers, problems, 3
High transition temperature–conductive
  polymer superconductor bilayers,
  morphology, 308–315
Highly ordered pyrolytic graphite,
  oxidation, 358,359*f*
Hybrid coatings for metal and glass
  surfaces, inorganic–organic, 332–346
Hybrid interpenetrating networks,
  organic–inorganic, 145–160
Hybrid materials
  organic–inorganic, *See* Organic–
    inorganic hybrid composites
  polybenzoxazole–silica, 297–305
  vinyl polymer modified, 130–139
Hydrolytic stability, inorganic–organic
  hybrid coatings for metal and glass
  surfaces, 339–340
Hydrophobically shelled nanoparticles,
  production, 67,68*f*
Hydroxyl-terminated polydimethylsiloxane–tetraethoxysilane system,
  one- and two-step syntheses, 105–110
Hypervalent spiro polysiliconate and
  polygermylate ionomers, 248–262
alternative counterion polysiliconate
  characterization, 258–259
applications, 249
development, 249–251
experimental description, 249

Hypervalent spiro polysiliconate and
  polygermylate ionomers—*Continued*
future studies, 261
monomer synthesis, 253–254
polygermylate
  characterization, 260–261
  synthesis, 259–260
polysiliconate characterization, 254–258
preparation strategy, 252
previous studies, 252–253

I

Impregnated organic–inorganic hybrid
  composites
applications, 229
mechanical properties, 227,229,230*t*
structure, 227,229
synthesis using sol–gel process, 227
In situ grown silicon oxide nanoparticles,
  organic modification of surfaces, 66–82
Inorganic– and organic–polymer hybrid
  composites, 181
Inorganic constituent of hybrid materials,
  synthesis using sol–gel chemistry, 163
Inorganic–organic hybrid coatings for
  metal and glass surfaces
adhesion of metal oils to polyimide,
  332–338
boehmite–epoxysilane mixture, 343–345
experimental description, 331
hydrolytic stability, 339–340
metal amount vs. silane formation,
  341,342*f*
perfluorinated groupings vs. adhesion,
  342–343
scratch resistance, 345–346
structure of metal oxide, 341–342
surface–chemical properties, 338–346
surface–mechanical properties, 338–346
Inorganic–organic hybrid composites, *See*
  Organic–inorganic hybrid composites
Inorganic–organic hybrid composites
  synthesized using simultaneous
  processes, nanostructured, 89–96

Inorganic–organic hybrid interpenetrating networks, 145–160

Inorganic–organic hybrid silica materials, 210–221

Inorganic oxide–perfluorosulfonate ionomer nanocomposites, 67–81

Inorganic–polymer networks, synthesis, 164

Inorganic–protein interactions in ferrimagnetic nanocomposite synthesis, characterization of magnetoferritin, 20–28

Intercalation chemistry, development, 29

Interfacial properties, organic–inorganic hybrid composites, 3

Intermolecular interactions, solid formation control, 221–224

Interpenetrating networks
organic–inorganic hybrid, 145–160
route to polymer-filled glass, 163–179

Interphase bonding, polymer-filled glass, 170,172,173f

Ion-exchange intercalation into MPS_3 layered compounds, 29–41
cation-transfer intercalation chemistry of MPS_3 layered compounds, 30–31
experimental description, 29
ferrimagnetic model of MPS_3 intercalates, 34–37
magnetic properties of MPS_3 intercalates, 33–34,35f
$Mn_{0.86}PS_3[(dimethylamino)-N-methyl-stilbazolium cation]_{0.28}$, 40
molecular metal synthesis, 37–38
MPS_3 layered compounds, 29–30
optical properties of MPS_3 intercalates, 38–40

Iron biomineralization, mechanism, 19

Isothermal ternary phase diagrams for diene polymer–bis(acetonitrile) dichloropalladium(II)–solvent system
examples, 196,197f
palladium(II) salt concentration vs. sol–gel properties, 198,200
polymer molecular weight vs. sol–gel properties, 198,199f
solvent effects, 196–198,200t

K

$KClO_3$–H_2SO_4, chemical safety, 358,360

L

Ladder materials, synthesis and characterization, 248–261

Ladder structures, examples, 3

Langevin magnetization function, definition, 25

Layered solid–poly(ethylene oxide) nanocomposites, 43–47

Linear siloxane, structure, 264

Liquid colloidal optical rejection filters, development, 181–182

Li_xpoly(ethylene oxide)$_y MoO_3$ elemental analysis and synthesis, 53

Li_xpoly(ethylene oxide)$_y MoS_2$, thermal analysis, 48–50

Loading, sol–gel-derived silica–siloxane composites, 240,242–245

M

Macroscopic stiffness for entire network, definition, 204

Magnetic properties
design of nanocomposites, 29–40
magnetoferritin, 25,26f

Magnetoferritin
characterization, 20–27
magnetic properties vs. structure, 27–28

Mechanical properties
chemically bonded organic–inorganic hybrids, 234t,235
impregnated organic–inorganic hybrid composites, 227,229,230t
organic–inorganic hybrid composites, 7–9
polybenzoxazole–silica hybrid materials, 303,304–305f

Mechanical property–surface relationship, inorganic–organic hybrid coatings for metal and glass surfaces, 338–346

Melt processing of hybrids, feasibility, 17

Metal foil–polyimide adhesive materials
  brittleness of sealing medium, 332–333
  peel strength vs. composition, 336,337*f*
  requirements, 333
  ^{29}Si-NMR spectroscopy, 334–336
  structural model, 336,338*f*
  synthesis, 333–335
  wetting behavior, 332–333

Metal–olefin coordination in diene
    polymers, 192–207
  experimental procedure, 193–195
  future studies, 206
  isothermal ternary phase diagrams,
    196–200
  previous studies, 193
  properties of solid films obtained from
    ternary solutions, 200–206
  sol–gel phase transition, 195–196

Metal surfaces, inorganic–organic hybrid
    coatings, 331–346

$Mn_{0.86}PS_3[(dimethylamino)$-$N$-methyl
    stilbazolium cation]$_{0.28}$, magnetic
    properties, 40

Modulus, polymer-filled glass, 166–169*f*

Molecular magnets, synthesis, 33

Molecular metals, synthesis using
    ion-exchange intercalation into MPS_3
    layered compounds, 37–38

Molecular-weight dependence,
    sol–gel-derived silica–siloxane
    composites, 238–240

Monodisperse colloidal polymer particles,
    light transmission, 181

Montmorillonite, swelling behavior with
    ω-amino acid, 56–58

Morphology
  conductive polymer–high-T_c
    superconductor bilayers, 308–315
  device fabrication procedures, 309–311
  experimental description, 308
  future work, 315
  influencing factors, 315
  poly(3-hexylthiophene), 313–314
  polypyrrole, 311–313
  polymer-filled glass, 170–173

Mössbauer spectroscopy, magnetoferritin,
    20,23–25,27

MPS_3 intercalates
  ferrimagnetic model, 34–37
  magnetic properties, 33–35*f*
  optical intercalates, 38
  synthesis, 33
  use for molecular metal synthesis, 37–38

MPS_3 layered compounds
  cation-transfer intercalation chemistry,
    30–31
  destruction–reconstruction intercalation
    mechanism, 31–32
  intercalation of soluble polymers, 32–33
  ion-exchange intercalation, 29–40
  reactivity, 30
  schematic representation, 29–30

Multiple size scale structures in
    silica–siloxane composites studied by
    small-angle scattering, 97–110
  accounting for multiple structural
    levels, 98–102
  correlated systems, 102–107
  one- vs. two-step synthesis, 105–110
  unified equation for small-angle
    scattering, 98

Multiple structural levels, accounting,
    98–102

$M_x$poly(ethylene oxide)$_y$TiS$_2$
  electrical measurements, 51,53
  exfoliation in N-methylformamide,
    51,52*f*
  synthesis, 50–51
  thermal measurements, 51

N

Nanocomposite(s)
  ferrimagnetic, inorganic–protein
    interactions in synthesis, 19–28
  (perfluorosulfonate ionomer)–(inorganic
    oxide), 66–82
  poly(ethylene oxide)–layered solid,
    43–53
  polymer containing, 44–45

Nanocomposite(s)—*Continued*
with unusual magnetic, electrical, and
nonlinear optical properties, 29–40
Nanocomposite material synthesis,
advantages of ferritin, 19–20
Nanoparticles, in situ grown silicon
oxide, 66–82
Nanostructured organic–inorganic hybrid
composites synthesized using
simultaneous processes, 86–96
future work, 96
properties, 89,93–95
sol–gel process, 87
structural characterization, 89–92*f*
synthesis, 88–89
Naturally occurring siliconate materials,
examples, 218
Network materials
hexylene- and phenylene-bridged
polysiloxane, 267–277
synthesis and characterization, 248–261
Network polysilicate components,
synthesis, 164–165
Nonaqueous sol–gel chemistry,
polymer-filled glass, 164–165
Nonlinear optical materials using
organic–inorganic composites, 323–328
Nonlinear optical properties
design of nanocomposites, 29–40
MPS$_3$ intercalates, 38–40
Nonlinear optical response, 319
Nonshrinking sol–gel composites
properties, 89,93–95
structural characterization, 89–92*f*
synthesis, 88–89
Nucleating agent effect, organic–inorganic
hybrid composites with crystalline
polymer matrix, 9,10*f*,12,13*f*
Nylon 6–clay hybrid
application, 61,62*f*
bond character of organic and inorganic
surface, 60–61
properties, 59–60
swelling behavior of montmorillonite with
ω-amino acid by ε-caprolactam, 56–58
synthesis, 55–56,59,62*f*

O

Olefin–metal coordination in diene
polymers, 193–206
Oligosilsesquioxanes, cubic, 265
Optical path length difference between
fast and slow waves, definition, 195
Optical properties
MPS$_3$ intercalates
insertion of luminescent centers, 38
nonlinear optical properties, 38–40
nonlinear, 29–40
organic–inorganic hybrid composites, 3
Organic– and inorganic–polymer hybrid
composites, reasons for interest, 181
Organic–inorganic hybrid composites
biomimicry, 2
bridged structures, 3
chemically bonded materials,
231,233–235
electrical properties, 3
entrapped organic hybrids, 229–232*f*
future studies, 235–236
impregnated hybrids, 227,229,230*t*
interactions, 2
interfacial and optical properties, 3
pattern of mechanical properties, 7–9
photonics
antibody-based biosensor development,
320,323,324*f*
factors affecting microenvironment
with sol–gels, 319–322*f*
nonlinear optical material development,
323,325–328
sol–gel process for preparation, 1–2
property improvement, 297
routes to improved nanoscale
composites, 9
synthesis using simultaneous processes,
89–96
synthetic methods, 210
types, 227,228*f*
with crystalline polymer matrix
advantages, 9
amorphous TiO$_2$ identification, 12
elastic properties, 12,14–16*f*

Organic–inorganic hybrid composites—
    Continued
with crystalline polymer matrix—
    Continued
experimental procedure, 9,11
hydrolysis reaction, 17
nucleating agent, 9–13*f*
phase separation of alkoxide, 15,17
solidification behavior, 11–12
tensile properties, 15,16*f*
titania particle size, 12
Organic–inorganic hybrid interpenetrating
    networks
characterization, 145–146
dynamic mechanical tan δ response
    vs. temperature, 159,160*f*
equilibrium mass uptake, 149,151*f*
experimental procedure, 143–145
factors affecting morphology, 142–143
γ-radiation, effect on dynamic
    moduli, 146–150
mechanical properties, 156
mixing between phases vs. mechanical
    properties, 154–156
moisture-absorbing ability, 159–161
monomer uptake, 149,152*f*,153
pH sensitivity, 160–161
polymer formed vs. γ-irradiation–
    polymerization, 153
solubility parameter, 154–156
storage modulus vs. temperature,
    156,158*f*,159
synthetic examples, 142
tensile strength vs. elongation, 156,157*f*
Organic–inorganic hybrid silica materials
advantages, 210
complexation of aromatic groups bound
    to silica, 214–216
enzyme formation, 213–214
exothermic transformation during diyne
    polymerization, 213–214
experimental description, 211
molecular structure vs. texture of
    solid, 211
organizational requirements of diyne
    polymerization, 211–212,214

Organic–inorganic hybrid silica
    materials—*Continued*
polymerization of thiophene units inside
    matrix, 216–221
solid formation control by
    intermolecular interactions, 221–224
synthesis, 211–212
Organic modification, surfaces of in situ
    grown silicon oxide nanoparticles,
    66–82
Organically modified hybrid
    materials, synthesis using sol–gel
    process, 125
Organically modified silicate,
    synthesis, 227
Organically modified sol–gel materials,
    families, 125–126
Organically shelled nanoparticles,
    production, 67,68*f*
Oxidation of highly ordered pyrolytic
    graphite, surface modification of
    carbon fibers for advanced composite
    materials, 358,359*f*
Oxidized carbon fibers
ESCA, 355–356
SEM analysis, 356,357*f*
single filament breaking strength,
    356,358
Oxygen permeability, polymer-filled
    glass, 174

P

Particle-filled composites,
    properties, 6
Particle volume fraction, definition, 182
Peel strength, metal foil–polyimide
    adhesive materials, 336,337*f*
Pentacoordinate ladder polysiliconate,
    synthesis, 252–253
Pentacoordinate siliconate, anionic,
    synthesis, 250
Percolation-based mechanical response
    via metal–olefin coordination in diene
    polymers, 192–206

Perfluorosulfonate ionomer–inorganic
oxide nanocomposites
characterization methods, 70–71
diethoxydimethylsilane posttreatment
FTIR spectroscopy, 71–73
mechanical tensile studies, 74,76–77f
^{29}Si solid-state NMR spectroscopy,
74,75f
ethoxytrimethylsilane posttreatment
FTIR spectroscopy, 78,79f
mechanical tensile studies, 78,80–82f
experimental procedure, 67–70
future work, 81
previous studies, 66
properties, 69
Si–C bond stability in aqueous
environment, 67
surface modification vs. organic
molecule affinity, 67,69
Phenylene- and hexylene-bridged
polysiloxane network materials, 267–277
1,4-Phenylene-bridged
polysilsesquioxane, synthesis, 250–251
Photochemical synthesis of
polymer-modified sol–gel materials
alkoxide vs. reaction rate, 136
composition of hybrid materials, 136–137
experimental setup, 133,134f
polymer precursor amount vs. reaction
rate, 133–135
reactions, 133
Photonics
description, 318–319
organic–inorganic composites, 317–328
Polarized optical microscopy, solid films
from ternary solutions, 201–203,205f
Polyacrylate precursors, synthesis using
group-transfer polymerization, 128
Polyamide(s), transparent, See Transparent
polyamides
Polyamide–silica hybrid materials,
high-clarity, 291–296
Polybenzoxazole–silica hybrid materials
copolymer(s)
mechanical properties, 303,304f
properties and synthesis, 300–301

Polybenzoxazole–silica hybrid
materials—Continued
copolymer synthetic procedure, 298–299
experimental description, 298
mechanical properties, 303–305f
monomer synthetic procedure, 298
polymer synthesis, 297–298
sol–gel process of copolymers and
tetramethyl orthosilicate, 301–303
sol–gel synthetic procedure, 300
synthesis, 297–303
Polydimethylsiloxane–tetraethoxysilane
system, hydroxyl terminated, one- and
two-step synthesis, 105–110
Poly(ethylene oxide)–layered solid
nanocomposites
development, 43
electrical measurements, 46–47
exfoliation–adsorption method of
synthesis, 45
Li$_x$poly(ethylene oxide)$_y$MoO$_3$, 53
Li$_x$poly(ethylene oxide)$_y$MoS$_2$, 48–50
M$_x$poly(ethylene oxide)$_y$TiS$_2$, 50–53
stoichiometry, 45–47f
structures, 43–46
thermal measurements, 46–47
topotatic methods of synthesis, 44–45
Poly(3-hexylthiophene), morphology,
313–314
Polymer(s)
diene, metal–olefin coordination, 193–206
high-performance, structural design,
280–290
reinforcement by glass fiber and
inorganic materials, 55
Polymer–ceramic composites
challenges, 97
reaction conditions vs. properties, 331–332
Polymer–clay hybrids
nylon 6–clay hybrid, 55–62
rubber–clay hybrid, 61,63–65
Polymer-containing nanocomposites,
synthesis, 44–45
Polymer-filled glass, 163–179
abrasion resistance, 174,177f
creep, 170

Polymer-filled glass—*Continued*
  experimental procedure, 178–179
  extent of network development, 172,173*f*
  interphase bonding, 170,172,173*f*
  modulus, 165–169
  morphology, 170–173
  nonaqueous sol–gel chemistry, 164–165
  oxygen permeability, 174
  polysilicate phase continuity, 170,171*f*
  porosity, 174
  reaction mechanism, 174,177
  SAXS, 172,174,175*f*
  softening behavior, 170,171*f*
  stress–strain behavior, 165,168,169*f*,170
  structure, 170–173
  toughness, 168,170
  TEM, 174,176*f*
Polymer hybrid composites, organic– and
  inorganic–, reasons for interest, 181
Polymer–inorganic networks,
  synthesis, 169
Polymer–polysilicate interpenetrating
  networks, 166–169
Polymer-rich systems, effect of reaction
  conditions on sol–gel-derived
  silica–siloxane composite materials,
  237–245
Polymeric systems, gelation, 193
Polypropylene–titania composites,
  preparation and properties, 6–17
Polypyrrole, morphology, 311–313
Polysilicate components, network,
  synthesis, 164–165
Polysiloxane(s), bridged, synthesis,
  265–266
Polysiloxane network materials, hexylene
  and phenylene bridged, 267–277
Polysilsesquioxanes, bridged,
  synthesis, 265
Polystyrene–silica hybrid materials,
  sulfonation, 132
Poly(trimethylhexamethylene-
  terephthalamide), 291
Porosity
  description, 67
  polymer-filled glass, 174

Properties
  entrapped organic hybrids, 229,231
  nanostructured organic–inorganic hybrid
    composites synthesized using
    simultaneous processes, 89,93–95
6-Propionyl-2-(dimethylamino)naphthalene
  factors affecting sol–gel
    microenvironment, 319–322*f*
  properties, 318
Protein–inorganic interactions in
  ferrimagnetic nanocomposite synthesis,
  20–28
Pyrolytic graphite, highly ordered,
  oxidation, 358,359*f*

R

Radziszewski hydrolysis, synthesis of
  water-soluble polyacrylic-modified
  silica sol–gel materials, 131–132
Reaction conditions, polymer-rich system,
  237–245
Refractive index of suspension, 182
Reinforcement techniques, silicones, 237
Rhodamine 6G, properties, 318
Roughness, description, 67
Rubber–clay hybrid
  development, 61
  properties, 63–64
  reinforcement mechanism, 64,65*f*
  synthesis, 63

S

Scanning electron microscopy
  hexylene- and phenylene-bridged
    polysiloxane network materials,
    275,276*f*
  oxidized fibers, surface modification
    of carbon fibers for advanced
    composite materials, 356,357*f*
Scattered intensity
  calculation, 172
  for correlated systems, 102–107
  for noncorrelated domains, 102–103

Scattering, small-angle, 97–110
Scratch resistance, inorganic–organic hybrid coatings for metal and glass surfaces, 345–346
Semiorganic shell, description, 67
Si–C bond, stability in aqueous environment, 67
Silica
colloidal crystals, solidification, 181–190
depolymerization–polymerization, 249–250
functionalized, nucleation of composite polymer colloid, 112–124
occurrence, 248
production from tetraethoxysilane, 1
solidification of colloidal crystals, 181–190
structure, 264
synthesis, 292
use in organic polymer composites, 181
Silica-based organic–inorganic hybrid composites, 210–224
Silica materials, organic–inorganic hybrid, 210–224
Silica–polyamide hybrid materials, high clarity, 292–296
Silica–polybenzoxazole hybrid materials, 297–305
Silica–polystyrene hybrid materials, sulfonation, 132
Silica–siloxane composites
sol–gel derived, 237–245
using small-angle scattering, multiple size scale structures, 98–110
Silica surface, emulsion polymerization of ethyl acrylate, 115–119
Silicates, occurrence, 248
Silicon-derived materials, applications, 248,249f
Silicon metal, preparation, 248
Silicon oxide nanoparticles, in situ grown, 66–82
Siliconate materials, naturally occurring, examples, 218
Silicones, reinforcement techniques, 237

Siloxane-based hybrid organic–inorganic materials, structures and synthesis, 264–265
Siloxane–silica composites
sol–gel derived, 237–245
using small-angle scattering, multiple size scale structures, 98–110
Silsesquioxanes, structure, 265
Simultaneous processes, nanostructured organic–inorganic hybrid composite synthesis, 86–98
Single filament breaking strength of oxidizing fibers, surface modification of carbon fibers for advanced composite materials, 356,358
Small-angle scattering, multiple size scale structures in silica–siloxane composites, 97–110
Small-angle X-ray spectroscopy, polymer-filled glass, 172,174,175f
Softening behavior, polymer-filled glass, 170,171f
Sol–gel(s), factors affecting microenvironment, 319–322f
Sol–gel composites, nonshrinking, 89–95
Sol–gel-derived inorganic–organic hybrid materials, 318
Sol–gel-derived silica–siloxane composites, 237–246
catalyst activity, 240–243
development, 237
experimental procedure, 238
loading, 240,242–245
mechanical properties vs. loading, 243,245
molecular weight, 238–240
phase separation, 243,245
Sol–gel hybrid materials, vinyl polymer modified, 125–139
Sol–gel phase transition, considerations for metal–olefin coordination in diene polymers, 195–196
Sol–gel process
advantages, 1
chemically bonded organic–inorganic hybrids, 231,233–235

Sol–gel process—*Continued*
  crystalline ceramic oxides
    and oxide glasses, 226–227
  entrapped organic hybrids, 229
  for synthesis of ceramics and
    organic-modified hybrid
    materials, 1,125
  for synthesis of composites, 86
  impregnated organic–inorganic hybrid
    composites, 227
  organic–inorganic hybrid composites,
    226–236
  reactions, 87
  synthesis, 226–235
Sol–gel process for organic–inorganic
    hybrid composite preparation, 1–2
Sol–gel processing
  description, 317–318
  structural design of high-performance
    polymers, 280–290
Solid films from ternary solutions
  elastic moduli, 203–206
  evidence of strong chemically bonded
    gels, 200–201
  polarized optical microscopy,
    201–203,205*f*
Solid-state ¹³C-NMR spectroscopy,
    hexylene- and phenylene-bridged
    polysiloxane network materials,
    272,273*f*
Solid-state ²⁹Si-NMR spectroscopy,
    hexylene- and phenylene-bridged
    polysiloxane network materials,
    272,274*f*,275
Solidification of colloidal crystals of
    silica, 181–190
  Bragg diffraction, 184–189
  colloidal 3-(trimethoxysilyl)propyl
    methacrylate–silica particles, 184
  crystals in films, 189–190
  experimental procedure, 183
  morphology of crystallites, 184
  theory, 182–183
Solubility parameter, determination, 154
Soluble polymers, intercalation with
    MPS₃ layered compounds, 32–33

Solvent–diene polymer–bis(acetonitrile)
    dichloropalladium(II) system,
    isothermal ternary phase diagrams,
    196–200
Spacings of most common structures of
    colloidal crystals, definition, 182
Spiropolysiliconate and -polygermylate
    ionomers, hypervalent, 249–260
Stress–strain behavior, polymer-filled
    glass, 168–170
Structural design of high-performance
    polymers for sol–gel processing,
    280–290
  characterization, 282,285
  coupling agents, 282,285,287
  gel preparation, 285,288–290
  monomer synthesis, 281
  polymer synthesis, 281–286
  solubility in alcohol, 283
  thermal stability, 285–287*f*
Structural levels, multiple, accounting,
    98–102
Structure
  chemically bonded organic–inorganic
    hybrids, 231,233–235
  entrapped organic hybrids, 231,232*f*
  impregnated organic–inorganic hybrid
    composites, 227,229
  multiple size scale, 98–110
  polymer-filled glass, 170–173
Sulfonation, polystyrene–silica hybrid
    materials, 132
Surface(s)
  in situ grown silicon oxide nanoparticles,
    organic modification, 66–82
  metal and glass, inorganic–organic
    hybrid coatings, 331–346
Surface area analysis, hexylene- and
    phenylene-bridged polysiloxane
    network materials, 275
Surface–chemical property relationship,
    inorganic–organic hybrid coatings for
    metal and glass surfaces, 338–346
Surface–mechanical property relationship,
    inorganic-organic hybrid coatings for
    metal and glass surfaces, 338–346

Surface modification of carbon fibers for advanced composite materials, 348–360
  benzoate labeling and analysis, 352–354
  chemical safety of $KClO_3$–H_2SO_4, 358,360
  ESCA of oxidized fibers, 355–356
  experimental procedure, 350–352
  methods, 349
  oxidation of carbon fibers, 352
  oxidation of highly ordered pyrolytic graphite, 358,359f
  SEM of oxidized fibers, 356,357f
  single filament breaking strength of oxidized fibers, 356,358
  trimellitic anhydride labeling and analysis, 354–355
Swelling coefficient, definition, 149
Synthesis
  hexylene- and phenylene-bridged polysiloxane network materials, 264–270
  inorganic–polymer networks, 164
  magnetoferritin, 20
  metal foil–polyimide adhesive materials, 333–335
  molecular metals, 37–38
  MPS_3 intercalates, 33
  nanostructured organic–inorganic hybrid composites, 86–98
  polybenzoxazole–silica hybrid materials, 297–303
  poly(ethylene oxide)–layered solid nanocomposites, 43–53
  polymer–clay hybrids, 55–56,59,62–63
  polymer-containing nanocomposites, 44–45
  vinyl polymer modified sol–gel hybrid materials, 126–127
Synthetic composites, properties compared to those of bone, 6–7

T

Techniques, organic–inorganic hybrid composites, 2

Template-based systems, examples, 2
Tensile properties, organic–inorganic hybrid composites with crystalline polymer matrix, 15,16f
Ternary phase diagrams for diene polymer–bis(acetonitrile) dichloropalladium(II)–solvent system, See Isothermal ternary phase diagrams for diene polymer–bis(acetonitrile) dichloropalladium(II)–solvent system
Tetraethoxysilane, production of silica, 1
Tetraethoxysilane–hydroxyl-terminated polydimethylsiloxane system, one- and two-step synthesis, 105–110
(±)-5,5′,6,6′-Tetrahydroxy-1,1′-spiro-bis(indane), synthesis, 253
Thermal analysis, hexylene- and phenylene-bridged polysiloxane network materials, 275
Thermo-irreversible gelation in diene polymers, 193–206
Thiophene units inside organic–inorganic silica matrix, polymerization, 216–221
Third-order materials, nonlinear optical materials using organic–inorganic composites, 326–328
Third-order susceptibility, 328
Titania–polypropylene composites, preparation and properties, 6–17
Topotatic methods of synthesis, poly(ethylene oxide)–layered solid nanocomposites, 44–45
Toughness, polymer-filled glass, 168,170
Transmission electron microscopy
  magnetoferritin characterization, 20–22f,27
  polymer-filled glass, 174,176f
Transparent polyamides, commercial use, 291
Trimellitic anhydride labeling and analysis, surface modification of carbon fibers for advanced composite materials, 354–355
3-(Trimethoxysilyl)propyl methacrylate–silica particles, colloidal, 184

Trogamid T, *See* Poly(trimethyl-hexamethylene-terephthalamide)

U

Unified equation for small-angle scattering
accounting for multiple structural levels, 98–102
correlated systems, 102–107
one-step vs. two-step synthesis, 105–110

V

Vinyl-polymer-modified sol–gel hybrid materials, 125–140
application, 131
characterization, 130–131

Vinyl-polymer-modified sol–gel hybrid materials—*Continued*
experimental procedure, 138–139
photochemical synthesis, 133–137
polymer component reaction, 131–133
precursor synthesis, 128–130
synthesis, 126–127

W

Water-soluble polyacrylic–modified silica sol–gel materials, synthesis using Radziszewski hydrolysis, 131–132
Waveguides, nonlinear optical materials using organic–inorganic composites, 325–326
Wetting behavior, metal foil–polyimide adhesive materials, 332–33

Production: Meg Marshall
Indexing: Deborah H. Steiner
Acquisition: Anne Wilson

Printed and bound by Maple Press, York, PA

Bestsellers from ACS Books

The ACS Style Guide: A Manual for Authors and Editors
Edited by Janet S. Dodd
264 pp; clothbound ISBN 0–8412–0917–0; paperback ISBN 0–8412–0943–X

Understanding Chemical Patents: A Guide for the Inventor
By John T. Maynard and Howard M. Peters
184 pp; clothbound ISBN 0–8412–1997–4; paperback ISBN 0–8412–1998–2

Chemical Activities (student and teacher editions)
By Christie L. Borgford and Lee R. Summerlin
330 pp; spiralbound ISBN 0–8412–1417–4; teacher ed. ISBN 0–8412–1416–6

Chemical Demonstrations: A Sourcebook for Teachers,
Volumes 1 and 2, Second Edition
Volume 1 by Lee R. Summerlin and James L. Ealy, Jr.;
Vol. 1, 198 pp; spiralbound ISBN 0–8412–1481–6;
Volume 2 by Lee R. Summerlin, Christie L. Borgford, and Julie B. Ealy
Vol. 2, 234 pp; spiralbound ISBN 0–8412–1535–9

Chemistry and Crime: From Sherlock Holmes to Today's Courtroom
Edited by Samuel M. Gerber
135 pp; clothbound ISBN 0–8412–0784–4; paperback ISBN 0–8412–0785–2

Writing the Laboratory Notebook
By Howard M. Kanare
145 pp; clothbound ISBN 0–8412–0906–5; paperback ISBN 0–8412–0933–2

Developing a Chemical Hygiene Plan
By Jay A. Young, Warren K. Kingsley, and George H. Wahl, Jr.
paperback ISBN 0–8412–1876–5

Introduction to Microwave Sample Preparation: Theory and Practice
Edited by H. M. Kingston and Lois B. Jassie
263 pp; clothbound ISBN 0–8412–1450–6

Principles of Environmental Sampling
Edited by Lawrence H. Keith
ACS Professional Reference Book; 458 pp;
clothbound ISBN 0–8412–1173–6; paperback ISBN 0–8412–1437–9

Biotechnology and Materials Science: Chemistry for the Future
Edited by Mary L. Good (Jacqueline K. Barton, Associate Editor)
135 pp; clothbound ISBN 0–8412–1472–7; paperback ISBN 0–8412–1473–5

For further information and a free catalog of ACS books, contact:
American Chemical Society
Distribution Office, Department 225
1155 16th Street, NW, Washington, DC 20036
Telephone 800–227–5558